Topics in Energy
Edited by L. Bauer, W. K. Foell, M. Grenon, G. Woite

G. Kessler
Nuclear
Fission Reactors

Potential Role and Risks
of Converters and Breeders

Springer-Verlag Wien NewYork

Prof. Dr. Günther Kessler
Institut für Neutronenphysik und Reaktortechnik,
Kernforschungszentrum Karlsruhe, Federal Republic of Germany

© 1983 by Springer-Verlag Wien

Softcover reprint of the hardcover 1st edition 1983

With 99 Figures

Library of Congress Cataloging in Publication Data. Kessler, G. (Günther), 1934 –. Nuclear fission reactors. (Topics in energy.) Includes index. 1. Nuclear reactors. 2. Nuclear fuels. 3. Breeder reactors. I. Title. II. Series: Topics in energy (Springer-Verlag). TK9202.K43. 1983. 621.48′3. 82-10661.

ISSN 0723-4570
ISBN-13:978-3-7091-7624-5   e-ISBN-13:978-3-7091-7622-1
DOI: 10.1007/978-3-7091-7622-1

Foreword

Nuclear power offers an abundant energy supply for the long term and at reasonable costs. Both are badly needed in this world of limited energy reserves and rising energy prices. On the other hand, there are questions widely discussed in the public on nuclear safety, on acceptable means of nuclear waste disposal, and concern on the proliferation of nuclear weapon capabilities. Public confusion is widespread since facts are often overshadowed by emotions. Recognizing the need for reliable, factual and comprehensive information on nuclear energy, this book on Nuclear Fission Reactors is published to present the scientific and technical facts of nuclear fission reactors, and to analyse their potential role and risks.

The author, Professor Dr. G. Kessler, has worked in nuclear research and project management since 1963. From 1975 to 1978, he acted as project leader for the German/Belgian/Dutch Fast Breeder research and development activities. Since then, he has been Director of the Institute of Neutron Physics and Reactor Technology in the Karlsruhe Nuclear Research Centre.

The book is part of the series "Topics in Energy" issued by Springer Publishers. The intention of this series of in-depth analyses is to present the facts, inherent problems, trends and prospects of energy demand, resources and technologies.

The vital importance of energy for human activities has become apparent to the public particularly through dramatic events in the area of oil supply. On a world-wide basis and at current standards and consumption rates, the easily exploitable oil and gas reserves will last only for a few more decades. The scientific and technical efforts to develop alternative long-range energy resources are therefore attracting increasing attention. We must learn to live with changes in energy supply and demand patterns and with the resulting difficulties, including technical problems, rising costs and prices, vast long-term investment requirements, and political and social implications.

Nuclear fission presents an important alternative for the world's energy supply, and particularly for countries with insufficient conventional energy resources. Base load electricity generation by large nuclear power stations generally has an economic margin over conventional base load generation, except in a few areas of the world with very favourable conditions for hydroelectricity or coal mining. Nuclear energy can also be applied for ship propulsion, heat generation and production of hydrogen or other energy-intensive products.

The utilization of nuclear energy entails a number of special aspects which

are not encountered in other areas of energy utilization. These are mainly due
to

- long lead times and large capital investment requirements for technology
  development and plant construction,
- special safety and licensing requirements,
- unique features of the nuclear fuel cycle and radioactive waste management
and, last but not least,
- political implications including the necessary international arrangements and
  the public attitude towards nuclear power.

Long-term commitments are required before harvesting the benefits of low
fuel costs and supply security of nuclear power. Foremost, a long-term national
energy policy must be widely accepted and established. Other important prereq-
uisites for a successful nuclear power program are the timely training of qualified
persons and the establishment of the necessary technical and organizational
infrastructure (e.g. an electric grid capable of integrating big power units, research and industrial infrastructure, and a licensing authority). Large amounts
of capital are then to be invested for the construction of nuclear power plants
and of nuclear fuel cycle facilities. All these investments can only be recouped
in the long term by the pursuit of a stable energy policy in an environment
of economic stability.

The strategic importance and potential misuse of nuclear materials, facilities,
and know-how imply that no nuclear activity can be performed without government sanction. Other reasons for government involvement are the vital importance of a widely accepted and stable national energy policy, and international
trade with nuclear materials and technology. National security concerns have
led to the signing of the Non-Proliferation Treaty (NPT) and to a world-wide
system of safeguarding nuclear materials.

The potential impact of nuclear radiation on the biosphere entails special
and unique risks which are not encountered in other areas of energy utilization
or industry, and which raise fear in the minds of many people. These risks
stem mainly from the large radioactive inventories of reactor cores, part of
which may be released in the course of accidents, sabotage or war. Other risks
are inherent to radioactive effluents routinely released from uranium mining,
nuclear reactors, reprocessing plants, and other facilities in which radioactive
materials are handled or stored.

In order to protect the public and employees of nuclear facilities from nuclear
risks, governments have enacted legislation related to nuclear activities. They
have entrusted authorities with the licensing and inspection of nuclear plants
and activities, and have set standards of qualification for persons who are responsible for the safe operation of these plants.

The operating conditions and safety requirements of nuclear plants necessitate
very high quality levels, experience and cooperation in many technical and
scientific disciplines. The comprehensive knowledge which is available today
for the design of reactor cores and for the assessment of nuclear risks would
not exist without the dedicated theoretical and experimental work which thousands of physicists and engineers have performed for several decades now. Extended data bases, advanced mathematical methods and highly sophisticated

computer programs are required to design reactor cores and to assess potential accident consequences.

The safe, reliable and economic operation of nuclear reactors over their whole lifetime of 30 years necessitates the intelligence and experience of physicists, of mechanical, electrical, civil, and chemical engineers and the qualified work of many technicians and skilled workers, particularly welders, boilermakers, pipefitters, and the electrical, electronic, and civil construction crafts. The implementation of these quality requirements has led to an impressive record in the safe and economic operation of nuclear plants. This is a reliable basis for long-term nuclear energy supply.

A number of reactor and fuel cycle options is available to further extend the nuclear energy potential. The energy equivalent of the world's presently known uranium reserves of about 5 million tons is about 80×10^9 tons of coal equivalent (TEC) when used in light water reactors in the once-through mode, or about 120×10^9 TEC when uranium and plutonium are recycled. For comparison, the world's natural gas reserves correspond to about 90×10^9 TEC. The energy potential of uranium could be increased to about 8000×10^9 TEC, i.e. by a factor of 60–100, by the introduction of fast breeder reactors. Thus nuclear fission offers the potential for a nearly unlimited energy supply; it is of the same order of magnitude as the potential of nuclear fusion in the deuterium-tritium cycle.

This book describes the fundamental scientific, technical and strategic features of converter and breeder reactors and of their corresponding nuclear fuel cycle options in a consistent and comprehensive manner. It analyses their potential for long-term energy supply, their risks and environmental impacts. I hope that the book will be well accepted, and will prove to be a valuable source of information, in particular for engineers, economists, energy planners and scientists in research institutes, industrial and governmental planning departments.

G. Woite

Preface

Unlike existing textbooks of nuclear reactor engineering and nuclear chemical engineering, this publication in the "Topics in Energy" series contains a complete description and evaluation of the technology of power generation by means of nuclear fission reactions. It covers the whole nuclear fuel cycle, from the extraction of natural uranium from ore mines, uranium conversion and enrichment, up to the fabrication of fuel elements for the cores of various types of nuclear power plants. This is followed by the description and discussion of the design concepts of present lines of commercial nuclear power plants and future advanced converter and breeder reactors. Next, an analysis is made of the different nuclear fuel cycle options currently under discussion and of the final storage of radioactive waste.

Another major item is the brief outline of the safety concepts of various types of nuclear power plants and their associated nuclear fuel cycle facilities as well as their radioactive releases, under normal operating conditions and in major accidents, with the resultant environmental impacts. Finally, the politically motivated aspects of proliferation and the international safeguards system of IAEA for the whole fuel cycle are treated.

Numerically, light water reactors outweigh all other types of reactors, generating electricity in most parts of the world at considerably lower cost than coal or oil fired plants. Most probably, they will continue to hold the largest share of the market at least for the next two decades, when the contribution of nuclear power to the generation of electricity in most of the industrialized countries will rise considerably from its present level of 10–12%. The enrichment of natural uranium for the fuel of these nuclear power plants will initially be achieved mainly in gaseous diffusion plants, with gas ultracentrifuges coming in at a later date. It will be one or two decades from now when such enrichment techniques as nozzle separation or even laser enrichment, which are still in their development stages right now, will secure for themselves a certain share of the market (Chapter 3).

Assessments of the world uranium resources, as have been attempted within the International Uranium Resource Evaluation Project (IUREP, 1980) and by the international organizations of OECD and IAEA (1982), still meet with great uncertainties. These are further aggravated by the uncertainties associated with the quantities of natural uranium actually available on the world market which, ultimately, depend on the extraction capacities in various parts of the world and on the prices attainable for natural uranium. This could well lead to a growing scarcity of natural uranium in the next few decades. This threat can be counteracted by the replacement, in time, of today's light water reactors, which have the highest uranium consumption, with advanced converter reactors, and, above all, breeder reactors. This will drastically reduce the consumption of natural uranium. Especially fast breeder reactors operated in a symbiosis with light water reactors can curb the uranium requirement enough to assure the world energy generation for several millenia – a result also hoped of nuclear fusion reactor technology. If it is possible to introduce fast breeder reactors commercially in time after the turn of the millenium, even today's assured world uranium reserves would be sufficient to meet requirements. However, the commercialization of advanced reactor lines will imply considerable further development efforts and costs. Consequently, besides the nuclear power plants and advanced converter reactors already available commercially (Chapter 4), special emphasis was laid on the treatment of the liquid metal cooled fast breeder reactor line (LMFBR) with all its technical data (Chapter 5).

Advanced converter and breeder reactors require a closed nuclear fuel cycle and, in order to get started, plutonium or U-233 generated artificially. These man-made fissile materials must be produced by chemical reprocessing of the spent fuel elements of the present line of commercial nuclear power plants. This directly links the decision to be taken in many countries about the construction of reprocessing plants and installations for subsequent conditioning and final storage of the radioactive waste with the introduction and operation of advanced converter and breeder reactors. For this reason, Chapters 6 and 7 deal with the possible options in the uranium/plutonium and thorium/U-233 cycles, indicating the systems data of all converter and breeder reactors and of fuel reprocessing and refabrication as well as waste conditioning. This outline is also based on data elaborated within the International Fuel Cycle Evaluation (INFCE, 1980).

Chapter 8 is devoted to the environmental impacts and risks associated with the different types of nuclear power plants and with nuclear fuel cycle facilities. First of all, these are due to the releases of radioactive substances from various stages of the nuclear fuel cycle, such as uranium ore mines, enrichment plants, fuel fabrication plants, nuclear power plants, reprocessing plants and waste conditioning installations. Systematic comparisons are also made between today's light water reactors and future breeder reactors. The second part of Chapter 8 is concerned with the risks created by major accidents in nuclear power plants; this assessment is made on the basis of the corresponding risk studies of light water reactors in the USA (WASH-1400) and Europe (German Risk Study). Also more recent hazards evaluation studies relating to reprocessing plants, fast breeder reactors and advanced converter reactors have been taken into account. Finally, a brief account is given of the proliferation aspects of the civilian nuclear fuel cycle. A balanced outline has been attempted of the present underlying scientific and technical principles and the network of international agreements. Again, the results of the INFCE program, in which I was involved myself, and recent developments in the IAEA safeguards system have been incorporated.

The fundamentals of reactor physics necessary for comprehending Chapters 3 to 8 have been summarized in Chapter 2 for the benefit of those readers who have no background in reactor physics. This description has been based on current reactor design practice and refers only to items required to understand the following chapters.

In order to keep this outline brief and to the point, many aspects of detail had to be omitted. Also in the references, a selection had to be made which will help readers to familiarize with the problems in greater depth.

In covering the many interdisciplinary aspects discussed in this book I was able to make use of the broad base of knowledge and the facilities of the Karlsruhe Nuclear Research Center. A number of colleagues of the Institute for Neutron Physics and Reactor Engineering and the Applied Systems Analysis Department assisted me greatly in this work.

In the first place, I would like to mention D. Faude, who made valuable suggestions about the arrangement of chapters in the final drafting phase of the manuscript, wrote Section 1.3 on Economic Aspects, and revised Chapters 1, 3, 4, and 5. In the most painstaking way he also compiled and supervised the many drawings, tables, lists of references, and the numerous editorial changes in the script. E. Kiefhaber helped me greatly with many improvements in writing Chapter 2, and C. Broeders had some of the figures in that chapter plotted by means of a computer. G. Halbritter and E. Leßmann made available some unpublished results and provided valuable information for Chapter 8.1. M. Küchle revised and reshaped Sections 8.3.1 to 8.3.5 in the light of his many years of experience in safeguards technology.

In reviewing the final draft of the manuscript I was assisted in many important ways by a number of colleagues of the Karlsruhe Nuclear Research Center (KfK) and other organizations. In particular, I wish to thank P. Jansen (University of Vienna) for Chapters 1, 3, and 6., H. Henssen (Interatom, Bergisch-Gladbach) for Chapter 2, U. Ehrfeld (KfK) for Chapter 3, H. Märkl (Kraftwerk Union, Erlangen) and W.A. Williams (General Electric Co., San Jose) for Chapter 4.1, N. Kirsch (KFA, Jülich) for Chapter 4.2, G. Kugler (AECL, Toronto) for Chapter 4.3, F. Morgenstern and M. Köhler (Interatom, Bergisch-Gladbach) and W. Scholtyssek and E. Kiefhaber (KfK) for Chapter 5, D. Closs, W. Ochsenfeld, H.O. Haug, and G. Mühling (KfK) for Chapter 7, and R. Schröder, A. Bayer (KfK) and M. Hübel for Chapters 8.1 and 8.2.

This book could not have been published in the English language without the support and broad experience of R. Friese (KfK), who translated the original German draft with great care. G.A. Moses (University of Wisconsin) gave the English version its final polish. My secretary, Miss Ch. Kastner, was untiring in typing the many versions of the manuscript and in carefully arranging the tables.

The suggestion to write this book came from G. Woite who, as the editor, offered much advice and reviewed the complete manuscript. I am most grateful to them all.

Karlsruhe, January 1983   **G. Kessler**

Contents

List of Abbreviations

ADAM + EVA   energy transmission system: $CH_4 + H_2O + 49$ kcal/mole
         $\rightarrow 3\,H_2 + CO$;
         EVA (German: *E*inzelspaltrohr-*V*ersuchs*a*nlage) is the
         chemical reactor for the production of $H_2 + CO$ (energy
         in) – its counterpart, ADAM, being the burner of the
         gaseous fuel (energy out)

AECL   *A*tomic *E*nergy of *C*anada *L*td.
AGNS   *A*llied *G*eneral *N*uclear *S*ervices (USA)
AGR   *a*dvanced *g*as cooled, graphite moderated *r*eactor
ALARA   *a*s *l*ow *a*s *r*easonably *a*chievable
ANL   *A*rgonne *N*ational *L*aboratory (USA)
AUC   *a*mmonium *u*ranyl *c*arbonate
AUPuC   *a*mmonium *u*ranyl *p*lutonyl *c*arbonate
AVR   *A*rbeitsgemeinschaft *V*ersuchs*r*eaktor, experimental HTR
         plant (Federal Republic of Germany)
BES   Soviet fast critical facility
BMI   *B*undes*m*inisterium des *I*nnern, Ministry of Interior (Fed-
         eral Republic of Germany)
BN 350   prototype fast breeder reactor (USSR)
BN 600   advanced prototype fast breeder reactor (USSR)
BNFL   *B*ritish *N*uclear *F*uels *L*td.
BOR 60   Soviet experimental fast test reactor
BR   *b*reeding *r*atio
BR-1, BR-2, BR-5 Soviet experimental fast reactors
BWR   *b*oiling light *w*ater cooled and light water moderated *r*eac-
         tor
CANDU   *Can*ada *d*euterium *u*ranium reactor
CANDU-PHWR *Can*ada *d*euterium *u*ranium *p*ressurized *h*eavy *w*ater reac-
         tor
CDFR   *c*ommercial *d*emonstration *f*ast *r*eactor (UK)
CIVEX   analogous to PUREX: proliferation resistant reprocessing
         technique for "civilian" purposes
CLEMENTINE   the first Pu-fueled fast research reactor (USA)
CR   *c*onversion *r*atio
CRBR   *C*linch *R*iver *b*reeder *r*eactor, US fast breeder demonstra-
         tion reactor

c/s	containment/surveillance
DFR	*Dounreay fast reactor*, experimental fast test reactor (UK)
DWK	*Deutsche Gesellschaft für Wiederaufarbeitung von Kern-brennstoffen*, German reprocessing company
EBR-I, EBR-II	experimental *breeder reactors* (USA)
ECCS	*emergency core cooling system*
EFFBR	*Enrico Fermi fast breeder reactor* (USA)
ENDF	*evaluated nuclear data file*
EPA	see USEPA
EPRI	*Electric Power Research Institute* (USA)
EUROCHEMIC	*European* Company for the *Chemical* Processing of Irradiated Fuel (Belgium)
EURODIF	derived from *European diffusion*, international company for uranium isotope separation plants
EVA	see ADAM + EVA
FBR	*fast breeder reactor*
FCA	*fast critical assembly* (Japan)
FFTF	*fast flux test facility* (USA)
GALAXY	code system used in producing microscopic group constants for reactor calculations (UK)
GGR	*gas* cooled, *graphite* moderated *reactor*
GRS	*Gesellschaft für Reaktorsicherheit*, Institute for Reactor Safety (Federal Republic of Germany)
HLW	*high level waste*
HLWC	*high level waste concentration*
HEPA	*high efficiency particulate air* filter
HEU	*highly enriched uranium*
HM	*heavy metal*
HP	*high pressure*
HRB	*Hochtemperatur-Reaktorbau* GmbH, German manufacturer of HTR power plants
HTGR	*high temperature gas* cooled, graphite moderated *reactor* with prismatic fuel elements
HTR	*high temperature gas* cooled, graphite moderated *reactor* with spherical fuel elements (pebble bed reactor)
HTR-Th	HTR with thorium fuel
HWR	*heavy water reactor*
IAEA	*International Atomic Energy Agency*
ICRP	*International Commission on Radiological Protection*
IHX	*intermediate heat exchanger*
IIASA	*International Institute for Applied Systems Analysis* (Austria)
INFCE	*International Nuclear Fuel Cycle Evaluation*
IPS	*international plutonium storage*
IUREP	*International Uranium Resource Evaluation Project*
JOYO	fast experimental test reactor (Japan)
KFA	*Kernforschungs-Anlage* Jülich, Nuclear Research Center, Jülich (Federal Republic of Germany)

KfK	*Kernforschungszentrum Karlsruhe*, Nuclear Research Center, Karlsruhe (Federal Republic of Germany)
KNK-II	*Kompakte natriumgekühlte Kernreaktoranlage*, sodium cooled fast test reactor with thermal driver zone (Federal Republic of Germany)
LLW	*low level waste*
LEU	*low enriched uranium*
LGR	*light water cooled graphite moderated reactor*
LMFBR	*liquid metal cooled fast breeder reactor*
LOCA	*loss-of-coolant accident*
LOF	*loss-of-flow accident*
LWBR	*light water breeder reactor*
LWR	*light water cooled and light water moderated reactor*
LWR-Pu	LWR with plutonium fuel
MAGNOX	derived from magnesium alloy; gas cooled graphite moderated reactor using a Mg alloy as the cladding material
MASURCA	*maquette surrégénératrice Cadarache*, fast critical assembly, Cadarache (France)
MLW	*medium level waste*
MC2-2SDX	multigroup cross section constants, code system used to produce microscopic group constants for reactor calculations (USA)
MEU	*medium enriched uranium*
MIGROS	code system used in producing microscopic group constants for reactor calculations (Federal Republic of Germany)
MONJU	fast breeder prototype reactor (Japan)
MOX	PuO_2/UO_2 *mixed oxide* fuel
MSBR	*molten salt breeder reactor*
MUF	*material unaccounted for*
NEA	*Nuclear Energy Agency* of OECD
NFS	*Nuclear Fuel Services* Inc. (USA)
NPT	*Non-Proliferation Treaty*
NRC	see USNRC
OECD	*Organization for Economic Cooperation and Development*
OT	*once-through* fuel cycle
PFR	*prototype fast reactor* (UK)
PHENIX	fast breeder prototype reactor (France)
LP	*low pressure*
PCRV	*prestressed concrete reactor vessel*
PWR	*pressurized light water cooled and light water moderated reactor*
PWR-Pu	PWR with plutonium fuel
PUREX	*plutonium and uranium recovery by extraction*
rad	*radiation absorbed dose*
RAPSODIE	*rapide sodium*, experimental fast test reactor (France)

rem	*r*oentgen *e*quivalent *m*an
RNFCC	*r*egional *n*uclear *f*uel *c*ycle *c*enter
RPV	*r*eactor *p*ressure *v*essel
SAGSI	*S*tanding *A*dvisory *G*roup on *S*afeguards *I*mplementation
SAP	*S*ervice *A*telier *P*ilote, French pilot plant for reprocessing spent fast reactor fuel
SEFOR	*S*outhwest *e*xperimental *f*ast *o*xide *r*eactor (USA)
SGHWR	*s*team *g*enerating *h*eavy *w*ater *r*eactor
SGR	*s*elf-*g*enerated *r*ecycling
SNEAK	*S*chnelle *N*ullenergie-*A*nordnung *K*arlsruhe, fast critical zero power assembly, Karlsruhe (Federal Republic of Germany)
SNR 300	*s*chneller *n*atriumgekühlter *R*eaktor, prototype fast breeder reactor (Federal Republic of Germany/Belgium/Netherlands)
SS	*s*tainless *s*teel
SUPERPHENIX	fast breeder reactor power plant following after PHENIX in the French breeder reactor program
SWU	*s*eparative *w*ork *u*nit
TBP	*t*ri-n-*b*utyl *p*hosphate
THOREX	*t*horium *o*xide *r*ecovery by *ex*traction
THTR	*t*horium fueled *h*igh *t*emperature gas cooled *r*eactor
TOR	*T*raitement d'*O*xydes *R*apides, French demonstration plant for reprocessing spent fast reactor fuel
UKAEA	*U*nited *K*ingdom *A*tomic *E*nergy *A*uthority
URENCO	*U*ranium *E*nrichment *C*ompany Ltd. (UK)
USEPA	*U*nited *S*tates *E*nvironmental *P*rotection *A*gency
USNRC	*U*nited *S*tates *N*uclear *R*egulatory *C*ommission
VHTR	*v*ery *h*igh *t*emperature gas cooled *r*eactor
WAES	*W*orkshop on *A*lternative *E*nergy *S*trategies (USA)
WAK	*W*iederaufarbeitungs-*A*nlage *K*arlsruhe, pilot reprocessing plant, Karlsruhe, Federal Republic of Germany
WEC	*W*orld *E*nergy *C*onference
WL	*w*orking *l*evel
WLM	*w*orking *l*evel *m*onth
WOCA	*w*orld *o*utside *c*entrally planned economies *a*rea
ZEBRA	*z*ero *e*nergy *b*reeder *r*eactor *a*ssembly (UK)
ZPPR	*z*ero *p*lutonium *p*ower *r*eactor, fast critical assembly (USA)
ZPR	*z*ero *p*ower *r*eactor, fast critical assembly (USA)

1 Introduction

1.1 The Development of Nuclear Energy in the World

Since the first nuclear power plants were commissioned for electricity generation, namely Shippingport, Yankee and Dresden in the USA (1957–1961),
Calder Hall and Chapelcross in the UK (1956–1958), the exploitation of nuclear
fission energy for power generation has taken a powerful upward trend,
especially in the sixties and early seventies. By late 1981, some 260 nuclear
power plants with an aggregate power of 153,000 MW(e) were in operation
worldwide in 22 countries (Table 1-1). This power is generated chiefly in three
types of nuclear power plants: *l*ight *w*ater *r*eactors (LWR's), of which there
are the two variants of *p*ressurized *w*ater *r*eactors (PWR's) and *b*oiling *w*ater
*r*eactors (BWR's), and a variant of the *l*ight water cooled *g*raphite moderated
*r*eactor (LGR), which is being used especially in the USSR; the *h*eavy *w*ater
*r*eactors (HWR's), especially of the CANDU type (*Ca*nadian D_2O *u*ranium);
and the gas cooled reactors fueled with natural uranium (MAGNOX) or enriched uranium (AGR, *a*dvanced *g*as cooled *r*eactor). The light water reactor
line has made by far the greatest contributions to nuclear power generation,
its plants attaining unit sizes of 1300 MW(e).

Another 240 nuclear power plants with an aggregate 218,000 MW(e) have
been under construction and firmly committed worldwide by late 1981. This
means that in the late eighties approximately 500 nuclear power plants with
an aggregate 370,000 MW(e) will be in operation worldwide in 32 countries.

The future requirement of nuclear generating capacity strongly depends on
the level of the future cumulative world energy requirement which, in turn,
is determined very much by the growth of the world population and by economic
developments in industrialized and developing countries.

The market share that nuclear power will be able to gain will depend on
the economically available reserves of nuclear fuels, the reserves of the established sources of primary energy, i.e., coal, oil and natural gas, and on future
new technologies for the exploitation of renewable energies, especially solar
power. Finally, the rate at which nuclear power will be introduced will be
determined also by the solution of the acceptance problems in the public and
by the international non-proliferation policy.

This multitude of partly conflicting factors, some of which include major
uncertainties and/or regional and national differences, makes it extremely difficult to forecast the expected future worldwide nuclear generating capacity.

Table 1-1. *Nuclear power plants in the world (plants larger than 30 MW(e) as of December 31, 1981) (Nuclear News)*

	In operation		Under construction, on order		Total	
	Number	MW(e)	Number	MW(e)	Number	MW(e)
A) Breakdown by countries:						
Argentina	1	335	2	1,292	3	1,627
Belgium	3	1,650	4	3,800	7	5,450
Brazil			3	3,116	3	3,116
Bulgaria	3	1,320	1	440	4	1,760
Canada	10	5,476	14	9,856	24	15,332
Czechoslovakia	3	990	8	3,520	11	4,510
Egypt			1	622	1	622
Finland	3	1,500	1	660	4	2,160
France	29	20,118	30	32,480	59	52,598
Germany (Dem. Rep.)	5	1,830	2	880	7	2,710
Germany (Fed. Rep.)	11	8,576	14	15,220	25	23,796
Hungary			4	1,760	4	1,760
India	4	804	6	1,320	10	2,124
Italy	4	1,387	3	2,004	7	3,391
Japan	23	15,047	12	10,025	35	25,072
Korea	1	564	8	6,834	9	7,398
Mexico			2	1,308	2	1,308
Netherlands	2	495			2	495
Pakistan	1	125			1	125
Philippines			1	620	1	620
Poland			2	880	2	880
Rumania			3	1,840	3	1,840
South Africa			2	1,844	2	1,844
Spain	4	2,003	12	11,424	16	13,427
Sweden	9	6,400	3	3,010	12	9,410
Switzerland	4	1,940	2	1,867	6	3,807
Taiwan	3	2,159	3	2,765	6	4,924
Turkey			1	440	1	440
United Kingdom	32	8,048	10	6,340	42	14,388
USA	76	56,790	74	82,207	150	138,997
USSR	30	15,515	11	9,280	41	24,795
Yugoslavia			1	615	1	615
World total	261	153,072	240	218,269	501	371,341
B) Breakdown by reactor types:						
Light water reactors						
– PWR's	119	84,557	149	142,534	268	227,091
– BWR's	61	38,678	46	47,767	107	86,445
– LGR's	18	10,760	5	5,000	23	15,760

Table 1-1 (continued)

	In operation		Under construction, on order		Total	
	Number	MW(e)	Number	MW(e)	Number	MW(e)
Heavy water reactors (all kinds)	18	6,614	26	14,537	44	21,151
Gas cooled reactors	40	10,700	10	6,340	50	17,040
High temperature gas cooled reactors (HTGR's, HTR's)	1	330	1	296	2	626
Fast breeder reactors (FBR's)	4	1,433	3	1,795	7	3,228
Total	261	153,072	240	218,269	501	371,341

An additional 28 power plants with a total of 30,735 (MW(e) are not included, since the expected date of commercial operation has been deferred indefinitely.

Table 1-2. *Development of installed nuclear capacity in the world (in GW(e)) (IAEA/INFCE)*

	1980	1985	2000
North America	58.0	110 −125	300– 450
Western Europe	41.2	90 −100	270– 400
Japan, Australia, New Zealand	14.5	25 – 30	100– 150
Eastern Europe, USSR	18.3	50 – 80	250– 450
Asia	2.5	8.5– 9.6	60– 75
Latin America	0.3	3.1– 5.3	40– 100
Africa + Middle East	–	2.4– 3.0	10– 25
World total	134.8	289 −352.9	1030–1650
– Industrialized countries	132.0	276.8–336.8	925–1460
– Developing countries	2.8	12.2– 16.1	105– 190

Table 1-2 represents an IAEA estimate based on data elaborated within the INFCE (*I*nternational *N*uclear *F*uel *C*ycle *E*valuation) program up until the year 2000 for various regions of the world. These estimates show major differences in some respects, due to the great uncertainties inherent in such projections. Besides a powerful upward trend in the highly industrialized countries of the Western world, a similar growth of nuclear generating capacity is expected to occur in the centrally planned economies of Eastern Europe and the USSR. However, nuclear power is not likely to achieve comparable shares in the other countries by the year 2000.

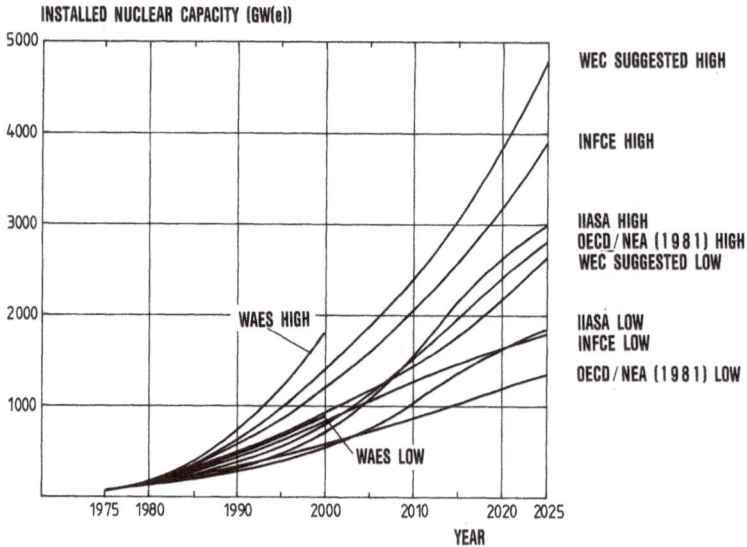

Fig. 1-1. Projections of installed nuclear capacity in WOCA (world outside centrally planned economies area) (INFCE, IIASA, OECD/NEA)

Looking ahead into the more distant future naturally implies even greater uncertainties. However, for the market penetration of a technology as new and complex as nuclear power, a span of at least fifty years must be considered: a nuclear power plant has a life of twenty or thirty years, and the future development of nuclear power, including the industrial fabrication capacities required, can be discussed and analyzed meaningfully only over a period spanning several power plant generations.

Fig. 1-1 shows some forecasts of the development of nuclear generating capacity up until 2025 for the world, excluding centrally planned economies. In addition to the

INFCE   – International Nuclear Fuel Cycle
               Evaluation (1980)

estimates, these are

WEC   – World Energy Conference, Report of the Conservation Commission (1977),
WAES   – Workshop on Alternative Energy Strategies (1977),
IIASA   – International Institute for Applied Systems Analysis, Energy in a Finite World (1980),
OECD/NEA   – Nuclear Energy Agency of the Organization for Economic Cooperation and Development (1981).

Estimates vary between 1350 GW(e) – OECD/NEA low – and 4,800 GW(e) – WEC suggested high – in the year 2025. Even the lowest estimate, OECD/NEA low, implies the completion and commissioning of an average of thirty nuclear power plants of 1000 MW(e) per annum in the Western world.

Measured by the aggregate world energy consumption, the fraction of nuclear power is still modest today, amounting to less than 2%. However, this nuclear capacity already generates 7% of the electricity worldwide. In the industrialized countries using nuclear power, this percentage is around 10–12%, in some cases even higher.

It is to be expected that the percentage fraction of nuclear power generation on a worldwide average will rise to 25 or 35% and, in the industrialized countries, to 30 or 40% by the year 2000. On a longer term basis, a 50 or 60% share is possible. Basically, nuclear fission energy is a source of power which, in view of a continuously rising world energy consumption, can make a sizable long term contribution and thus reduce the use of fossil sources of energy, namely oil, natural gas and coal.

1.2 Technical Applications of Nuclear Fission Energy

1.2.1 Nuclear Power for Electricity and Process Heat Generation

The energy released in the nuclear fuel of the reactor core by the fission of uranium or plutonium nuclei mostly consists of kinetic energy of the fission products with the result, first of all, that the nuclear fuel is heated. Consequently, the primary energy in nuclear fission is thermal energy (heat), which can be extracted from the reactor core by means of a coolant and used either directly as process heat or converted into electricity by a thermodynamic water/steam process.

Almost all nuclear reactors built and operated so far have been used as power plants for electricity generation; for commercial electricity generation, nuclear power plants today are mostly built in unit sizes of 1000 to 1300 MW(e) and operated in the base load regime. Nuclear power plants equipped with LWR's attain saturated steam conditions slightly below 300 °C and approx. 70 bar (thermal efficiency, 30–33%); nuclear power plants incorporating advanced types of reactors use superheated steam at slightly more than 500 °C and 160 bar (thermal efficiency, approx. 40%). The use of electricity for lighting, power and direct ohmic heating in industry, transport and private households has become widespread. The percentage fraction of electricity in the total consumption of final energy is likely to increase also in the future. In addition to direct ohmic heating, heating by means of electric heat pumps may well achieve growing importance in the future.

Besides the use of nuclear power for electricity generation, wider applications directly utilizing the heat are quite possible in the future, especially under the incentive of finding substitutes for oil and natural gas as primary sources of energy. In principle, the waste heat of nuclear power plants, as of coal fired power plants, can be exploited to supply district heat to cities or industrial regions. Optimum utilization of nuclear heat will be achieved in dual purpose nuclear power plants, in which the steam generated will first be partly expanded

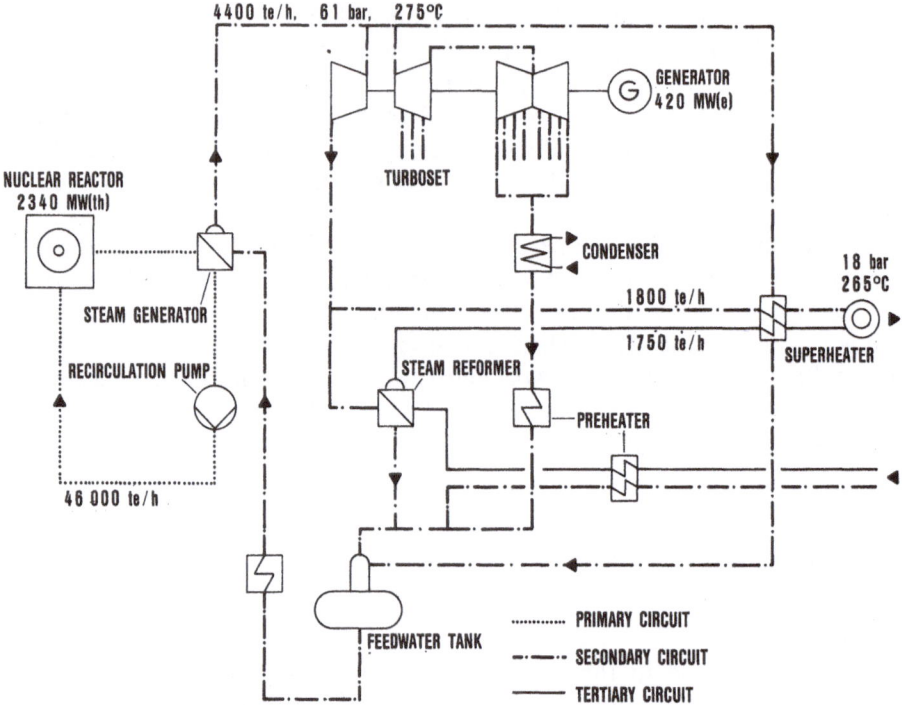

Fig. 1-2. Basic flowsheet of a dual purpose LWR plant for electricity and process heat generation (Kraftwerk Union)

in turbines for electricity generation and then extracted either partly or entirely from the final turbine stage for purposes of heat supply (back pressure turbine process). The combined generation of power and heat in dual purpose nuclear power plants not only offers advantages in terms of energy (overall thermal efficiency, 75–85%), but is especially attractive also from an economic point of view. Fig. 1–2 shows the basic flowsheet of a dual purpose nuclear power plant.

Since nuclear power plants are built in large units for economic reasons, and since heat cannot be transported over long distances economically (the costs of the necessary distribution system decisively influence the costs of district heat), the use of nuclear power in combined dual purpose plants offers economic advantages only in areas of high and concentrated heat requirements. A number of plans and some first applications in Europe and the USSR already demonstrate the supply of heat to cities from small compact LWR's or pool type reactors.

A large part of industrial process heat is generated in the range of temperatures between 200 and 400 °C, especially in the chemical industries. For this application, nuclear power from dual purpose power plants would constitute a solution. A special area of application of nuclear process heat is the generation of fresh water by sea water distillation. The world's fresh water requirement increases very much like the energy requirement. In many developing countries, the supply of fresh water by this "synthetic" technique will very rapidly become a vital necessity. Since, for technical reasons, a steam quality not exceeding 150 °C is sufficient for the desalination process, a combined dual purpose system again would be most reasonable economically. The high capital outlay required for the power plant plus the distillation plant today still hampers the use of nuclear power for sea water desalination. The Russian BN 350 fast breeder demonstration power plant at Shevshenko on the Caspian Sea is the only dual purpose plant of this type presently existing.

1.2.2 Nuclear Ship Propulsion

Nuclear reactors were used for ship propulsion initially to drive warships (submarines, aircraft carriers). Today's PWR's, which rank at the top in the list of reactor concepts, are a product of this marine reactor development in the fifties. In commercial shipping, nuclear propulsion has not been developed beyond a few demonstration projects.

The vessel to be mentioned first in this respect is the Russian icebreaker, N.S. "Lenin", which serves to keep the Western Arctic route open for the Soviet marine in winter. The N.S. "Lenin" was operated with three reactors in the period 1959 to 1966 and has been run on two improved reactors since 1970.

Over a period of eight years of operation, between 1962 and 1970, the American N.S. "Savannah" accumulated the necessary operating experience and was then decommissioned for cost reasons.

The N.S. "Otto Hahn", the German nuclear power research vessel, was operated between 1968 and early 1979 and, like the N.S. "Savannah", has produced excellent operating results. The Japanese N.S. "MUTSU" has not yet completed its trial phase of operation.

The market penetration of commercial vessels with nuclear propulsion systems is now mainly an economic question. Studies have shown that nuclear power could become economically attractive in very large and very fast vessels (such as container vessels with powers of 100,000 shp (shaft horsepowers)). In addition, the future of commercial nuclear shipping depends very much on the establishment of international agreements about port entry permits for the most important commercial ports in the world.

1.2.3 Nuclear High Temperature Process Heat

High temperature gas cooled reactors (HTGR's) can greatly expand the use of nuclear process heat. They attain coolant outlet temperatures of 700–1,000 °C. In that range of temperature, especially processes of direct nuclear coal gasification are of interest in which the process heat required for conversion is supplied as nuclear heat from an HTGR in a temperature range up to 950 °C. Fig. 1-3 shows one example of a coal gasification process, namely steam coal gasification: high temperature heat is transported to a steam gasifier through a second helium system. Hot steam is fed into a coal bed, gasifying the coal into H_2 and CO by the addition of reactor heat; subsequent methanation of

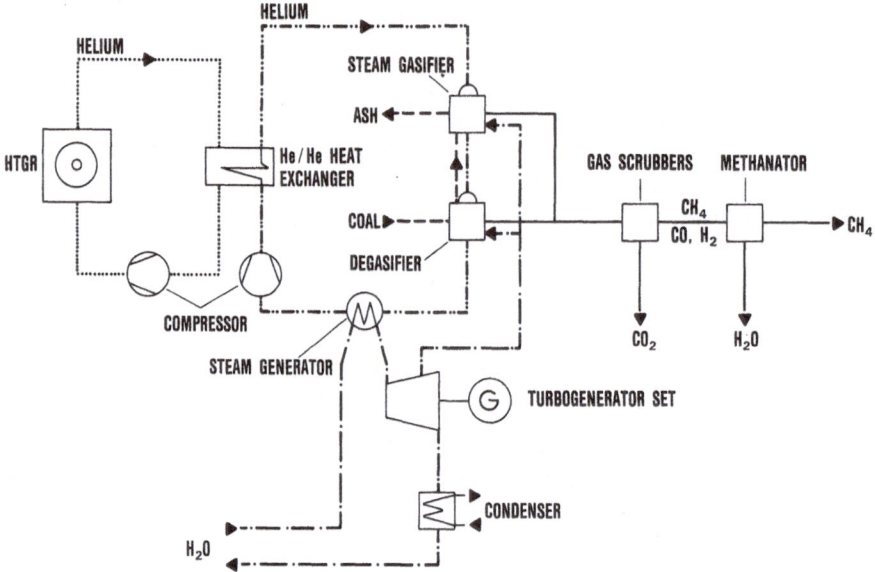

Fig. 1-3. Basic flowsheet of the steam coal gasification process by means of HTGR process heat (KFA Jülich)

the synthetic gas allows synthetic natural gas to be produced. The reactor heat in the medium and low temperature ranges generates process steam in an intermediate system, which is used for the gasification process, and live steam for the steam power plant.

Another area of application of process heat from HTGR's is the process known in Germany as "ADAM+EVA". The problem of transferring heat from a large central plant to the consumers over long distances is to be solved in this concept by transporting a chemically reactive system. The thermal energy of an HTGR is converted into the chemical energy of an endothermal chemical reaction in a tube reformer heated with hot helium (EVA in German, standing for *Einzelspaltrohr-Versuchs-Anlage*, which translates into single reforming tube test facility):

$$\text{EVA: } CH_4 + H_2O + \text{energy} \quad \rightarrow \quad 3H_2 + CO;$$

the mixture of gases ($3\,H_2 + CO$) is transported to the consumer in pipelines, the reverse reaction at the point of consumption releasing the contained heat (ADAM):

$$\text{ADAM: } 3H_2 + CO \quad \rightarrow \quad CH_4 + H_2O + \text{energy}.$$

The energy released can be used to generate industrial process heat, hot water for space heating, or for electricity generation close to load centers.

1.2.4 Nuclear Power for Hydrogen Generation

Hydrogen as a future secondary fuel can make major contributions to the supply of energy in all areas of consumption of final energy. Nuclear power plants can produce hydrogen both by electrolysis and by thermochemical water splitting processes at high temperatures. Hydrogen can be carried to load centers over long distances in pipelines of the type sucessfully applied today in the chemical industry. According to experience gathered in France, hydrogen can be stored, e.g., in underground cavities.

For water electrolysis, large electrolytic plants of several 100 MW(e) power will have to be developed in the future with low capital costs and high efficiencies (80–90%) for hydrogen generation. One main incentive in electrolysis processes may lie in the utilization of off-peak electricity, which means that the surplus electricity generated in nuclear power plants outside peak load times can be used to produce hydrogen. The problems inherent in thermochemical water splitting processes today still lie in the choice and demonstration of economically viable processes. They can work successfully only in a range of temperatures offered at present by high temperature gas cooled reactors.

Hydrogen can be applied mainly in direct combustion, to generate industrial process heat, as an addition to synthesis gas ($CO + H_2$), e.g., in methane production, or it can be used as a constituent in the synthesis of methanol or gasoline. Besides gasoline or methanol, hydrogen can make quite considerable future contributions as a direct motor fuel, either for direct combustion or in fuel cells with electric motors. This would therefore enable nuclear power to contribute to the substitution of oil in the transport sector.

1.3 Economic Aspects of Nuclear Energy

The future of nuclear power decisively depends on its economic prospects. Economics in this case not only implies the operation of nuclear power plants, but also the facilities going with them to supply nuclear fuel and dispose of nuclear waste.

The economic advantage of nuclear power lies in its relatively low fuel costs, which means that, e.g., changes in the uranium price will only have moderate effects on the overall electricity generating costs of a nuclear power plant. This applies in particular to fast breeder reactors with their extremely low fuel consumption. This characteristic has a stabilizing influence on the energy market. On the other hand, nuclear power plants have capital costs clearly higher than those of fossil fired power plants. However, nuclear power plants have a much more pronounced cost degression for larger units. Fig. 1-4 (right side) shows a comparison of the specific costs of nuclear and coal fired power plants.

1.3.1 Electricity Generating Costs

Economic assessments of nuclear power today can most reliably be made for LWR's, because this power plant line has been built and operated in large numbers in the United States and Europe. The figures quoted below for the electricity generating costs of a nuclear power plant are based on a PWR of 1255 MW(e) net power, which will be commissioned in the Federal Republic of Germany in 1989 after six years of construction.

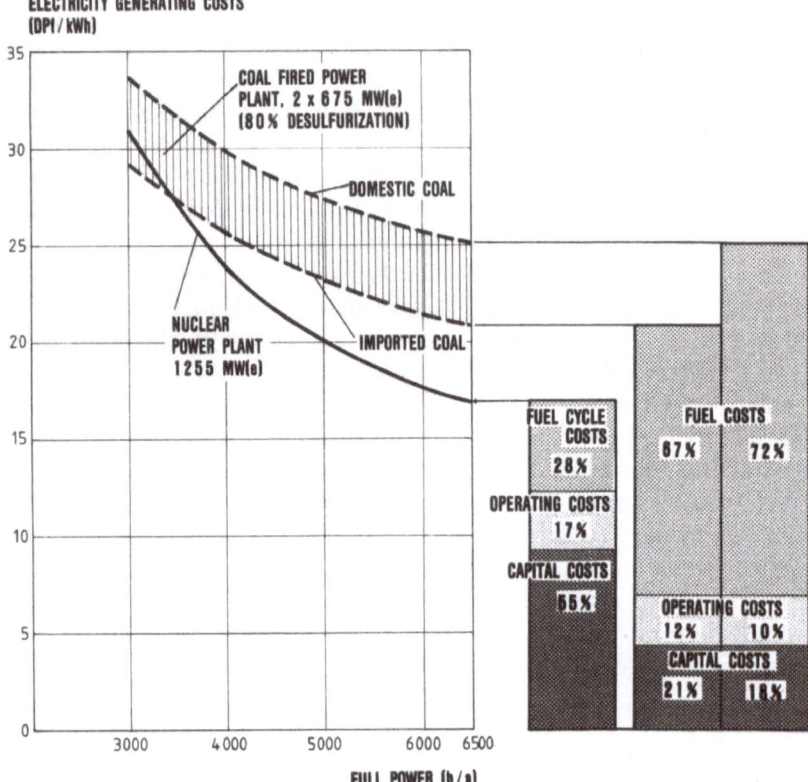

ELECTRICITY GENERATING COSTS
(DPf / kWh)

Fig. 1-4. Structure of the electricity generating costs of nuclear power plants and coal fired power plants (D. Schmitt)

Electricity generating costs have been subject to violent changes in the past and will probably remain so in the future. Moreover, it must be taken into account that comparisons of energy costs may include economic conditions with vast regional differences (USA, Europe, Japan). The figures quoted below therefore should not be regarded as absolute quantities, but rather as relative orders of magnitude.

For the capital costs of a plant, including owner's contributions, DM 2550/ kW(e) is assumed (cost base 1981); added to this must be interest and taxes over the period of construction. Decommissioning after twenty years of operation is taken into account to the tune of DM 350×10^6.

The operating costs include personnel costs for running the plant, costs of maintenance and repair, taxes and insurance, totaling approximately DM 60/ kW(e) per year.

The fuel cycle costs incorporate all activities associated with the production and processing of nuclear fuel, from the extraction of uranium ore up to final storage of the nuclear waste.

Calculations of the fuel cycle costs are made on the basis of the following assumptions (cost base 1981, the appropriate escalation factors to be applied for the future):

Natural uranium   DM   60/lb U_3O_8 *
Conversion and enrichment   DM   200/kg SWU *
Fuel element fabrication   DM   500/kg of fuel
Reprocessing and waste treatment   DM 2200/kg of fuel

Table 1-3. *Breakdown of electricity generating costs of a LWR power plant of 1255 MW(e) (load factor 0.74) (D. Schmitt)*

		DPf/kWh
Capital costs		9.1
Decommissioning costs		0.3
Operating costs		2.6
Fuel cycle costs		4.8
– Natural uranium	1.2	
– Enrichment	0.7	
– Fabrication	0.6	
– Reprocessing and waste treatment	2.3	
Total		16.8

At a rate of plant utilization of 6500 h/a, the capital costs, operating costs and fuel cycle costs result in average electricity generating costs calculated on the basis of investments over the life of the plant of DPf 16.8/kWh, which break down as indicated in Table 1-3.

Fig. 1-4 shows the strong dependence on annual utilization of the electricity generating costs. For comparison, the electricity generating costs of a coal fired plant of comparable size using either imported or domestic coal are indicated. (In the Federal Republic of Germany, imported coal at present is some 33% cheaper than coal mined in the country.) Fig. 1-4 highlights the economic advantage of LWR's in the range of utilization above 3500 h/a.

1.3.2 Load Factors of Nuclear Power Plants

Because of the high percentage of fixed costs in electricity generating costs (see Fig. 1-4), the load factor very strongly determines the level of electricity generating costs. Nuclear power plants will be able to exploit their economic advantages only if they reach high utilization levels.

The load factor is the ratio between the energy actually generated in a year and the energy generation theoretically possible at design power and full load (8760 h/a).

The load factor takes into account downtimes of the plant for refueling and repair and the fact that the plant, for other reasons, may not have reached its design power for certain periods of operation. Fig. 1-5 shows the load factor for various reactors in the period 1970–1980.

---

* $1 ≙ DM 2.00.

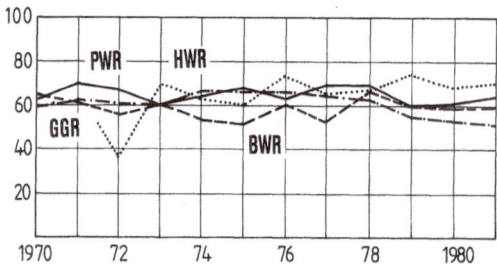

Fig. 1-5. Annual capacity utilization of nuclear power plants (A. Szeless)

The average load factor of LWR's in the world was about 60%, which corresponds to 5300 hours of full load operation. PWR's in general had load factors roughly 6% higher than boiling water reactors. The average load factor of LWR's in Western Europe, was 7% above the corresponding worldwide average. However, these differences are not due to technical reasons, but rather to insufficient statistical data in the early development phase of nuclear power. In some cases, PWR's in Western Europe have already reached load factors of 80%. The cost analysis in the previous section was based on a load factor of $0.74 \cong 6500$ h/a.

As more operating experience is accumulated and technical improvements are made, the average load factor of nuclear power plants in the world is likely to rise. For HWR's, average load factors of 70% are indicated; in some cases they may reach 78%. Similar values are quoted for AGR's.

Selected Literature

Baujat, J.: La propulsion nucléaire en France; Ulken, D., *et al.:* Nuclear Ship Propulsion; and Levine, Z.: The US Nuclear Merchant Ship Program. In: Nuclear Energy Maturity, Proc. European Nuclear Conference, Paris, 21–25 April 1975 (Zaleski, P., ed.), Vol. 9, pp. 392–403. Oxford: Pergamon Press. 1976.

Eklund, S.: Nuclear Power Development – The Challenge of the 1980s. IAEA Bulletin *23* (3), 8–18 (1981).

Galvez, L., *et al.:* Water Desalination in Mexico. In: Nuclear Energy Maturity, Proc. European Nuclear Conference, Paris, 21–25 April 1975 (Zaleski, P., ed.), Vol. 9, pp. 187–211. Oxford: Pergamon Press. 1976.

Häfele, W.: Energy in a Finite World. 1) Paths to a Sustainable Future; 2) A Global Systems Analysis. Cambridge, Mass.: Ballinger. 1981.

Häfele, W., *et al.:* Application of Nuclear Power Other Than for Electricity Generation. In: Nuclear Energy Maturity, Proc. European Nuclear Conference, Paris, 21–25 April 1975 (Zaleski, P., ed.), Vol. 9, pp. 362–374. Oxford: Pergamon Press. 1976.

International Nuclear Fuel Cycle Evaluation, Fuel and Heavy Water Availability. Report of INFCE Working Group 1. Vienna: International Atomic Energy Agency. 1980.

Krymm, R., Charpentier, J.P.: Nuclear Power Development: Present Role and Medium-Term Prospects. IAEA Bulletin *22* (2), 11–22 (1980).

Low-Temperature Nuclear Heat. Papers presented at the ANS/ENS Topical Meeting, Helsinki University of Technology, Otaniemi, Finland, 21–24 August 1977. Nuclear Technology *38*, 1–333 (1977).

Nagel, O.: Nuclear Low Temperature Process Heat for Chemical Production Sites. In: Nuclear Energy Maturity, Proc. European Nuclear Conference, Paris, 21–25 April 1975 (Zaleski, P., ed.), Vol. 9, pp. 431–434. Oxford: Pergamon Press. 1976.

Nuclear Energy and Its Fuel Cycle – Prospects to 2025. Paris: Nuclear Energy Agency/ OECD. 1982.

Quadea, R.N., Meyer, L.: High Temperature Nuclear Heat Source for Hydrogen Production. Int. Journal of Hydrogen Energy *4*, 101–110 (1979).

Schmitt, D., Junk, H.: Kostenvergleich der Stromerzeugung auf der Basis von Kernenergie und Steinkohle. Atomwirtschaft/Atomtechnik *27*, 270–274 (1982).

Schulten, R., Kugeler, K.: High Temperature Reactors and Application to Nuclear Process Heat; and Quadea, R.N., Meyer, L.: The HTGR for Nuclear Process Heat Application. In: Nuclear Energy Maturity, Proc. European Nuclear Conference, Paris, 21–25 April 1975 (Zaleski, P., ed.), Vol. 9, pp. 1–42. Oxford: Pergamon Press. 1976.

Szeless, A., Oszuszky, F.: Verfügbarkeit der Kernkraftwerke in der Welt im Jahre 1981. Atomwirtschaft/Atomtechnik *27*, 375–380 (1982)

Wilson, C.L.: Energy: Global Prospects 1985–2000. Report of the Workshop on Alternative Energy Strategies (WAES). New York: McGraw-Hill. 1977.

World Energy Demand. The Full Reports to the Conservation Commission of the World Energy Conference. Published for the World Energy Conference by Guildford, UK. New York: IPC Science and Technology Press. 1978.

World List of Nuclear Power Plants. Nuclear News *25* (2), 83–102 (1982).

2 Some Basic Physics of Converter and Breeder Reactors

2.1 Basic Nuclear Physics

The important nuclear reactions in the cores of fission reactors are primarily caused by neutrons interacting with atomic nuclei of the fuel, the coolant, the structural materials and absorber materials. Four main neutron interactions have to be considered:
- elastic scattering,
- inelastic scattering,
- neutron capture,
- nuclear fission.

2.1.1 Elastic Scattering

Neutrons can be scattered elastically by atomic nuclei. In such scattering processes, neutrons will change their flight path and velocity (kinetic energy). Neglecting the effects of chemical bonding and the influence of crystalline materials observed at fairly low neutron energies allows such nuclear reactions to be treated by the collision laws of classical mechanics. According to these laws, neutrons generated by nuclear fission (fast neutrons) lose energy with each elastic collision they suffer until their energy becomes comparable to that of the nuclei. The average loss of energy per collision varies inversely with the atomic weight A of the nucleus involved. For atomic nuclei with $A = 100$, for instance, it is only approximately 2%, but in an elastic collision with a hydrogen nucleus with $A = 1$, it is 50%. Multiple elastic collisions finally slow down high velocity neutrons to kinetic energies, where they are in thermal equilibrium with the atomic nuclei. Their velocity spectrum then follows approximately a Maxwellian distribution with an average kinetic energy of 0.0253 eV* at a temperature of 20 °C in a weakly absorbing, predominantly scattering material.

The average number of collisions needed to slow down (moderate) fission neutrons to thermal energies of about 0.0253 eV is about 16 for moderation in light water, 28 in heavy water, and 91 in graphite.

---

* 1 eV $= 1.602 \times 10^{-19}$ J is the kinetic energy acquired by an electron passing through a potential gradient of 1 V. 1 keV is equal to 10^3 eV and 1 MeV is equal to 10^6 eV. The energy of 0.0253 eV corresponds to a neutron velocity of 2200 m/s.

2.1.2 Inelastic Scattering

In inelastic scattering interactions, the incident neutron is briefly absorbed by the atomic nucleus. The resulting compound nucleus will be in a highly excited state, which is partially relieved by emitting a neutron of a lower kinetic energy. The remaining excited nucleus usually returns to its original ground state by emitting γ-radiation. Inelastic scattering is a threshold reaction, i.e., it can occur only if the energy of the incident neutron exceeds a certain minimum kinetic energy. This energy threshold is around a few MeV for light atomic nuclei and decreases to a few keV for intermediate and heavy atomic nuclei.

2.1.3 Neutron Capture

Another possible neutron interaction with matter is the so-called capture process where the nucleus involved and the capture neutron form a compound nucleus, which usually is in an excited state so that it may

(a) change into its stable state after the emission of γ-radiation ((n, γ)-reaction),

(b) become unstable and emit an electron (β^--decay), a proton ((n,p)-reaction) or an alpha particle ((n,α)-reaction).

Occasionally, two or more neutrons may be emitted when a nucleus is struck by a high energy neutron ((n, 2n)- or (n,3n)-reactions).

As an example of neutron capture followed by β^--decay, the conversion or breeding process in the U-238/Pu-239 cycle and in the Th-232/U-233 cycle is shown below:

$$^{238}_{92}U \xrightarrow{(n,\gamma)} {}^{239}_{92}U \xrightarrow[23.5\,\text{min}]{\beta^-} {}^{239}_{93}Np \xrightarrow[2.35\,\text{d}]{\beta^-} {}^{239}_{94}Pu$$

$$^{232}_{90}Th \xrightarrow{(n,\gamma)} {}^{233}_{90}Th \xrightarrow[22.1\,\text{min}]{\beta^-} {}^{233}_{91}Pa \xrightarrow[27.0\,\text{d}]{\beta^-} {}^{233}_{92}U$$

2.1.4 Nuclear Fission

If a neutron is absorbed by a heavy nucleus, e.g., U-235, the resulting compound nucleus may become highly unstable and split into two (sometimes even three) fragments. The two fragments (fission products) generated in nuclear fission are unlikely to be identical, but are produced in accordance with a certain probability distribution. Fig. 2-1 shows this probability distribution as a double-humped yield curve. In addition to the fission products, 2–3 neutrons are produced immediately after the fission event. The kinetic energy of these prompt fission neutrons also follows a certain distribution curve. This fission neutron spectrum (Fig. 2-2) and the correlated dependence on mass number of the fission product yields depend on the kinetic energy of the neutron (fast or thermal) initiating fission and on the heavy atomic nucleus undergoing fission.

In certain heavy nuclei, nuclear fission can only be initiated by incident neutrons above a minimum energy (threshold):

Atomic nucleus	Th-232	U-233	U-234	U-235	U-238	Pu-239
Minimum neutron energy required (MeV)	> 1.3	0	> 0.4	0	> 1.2	0

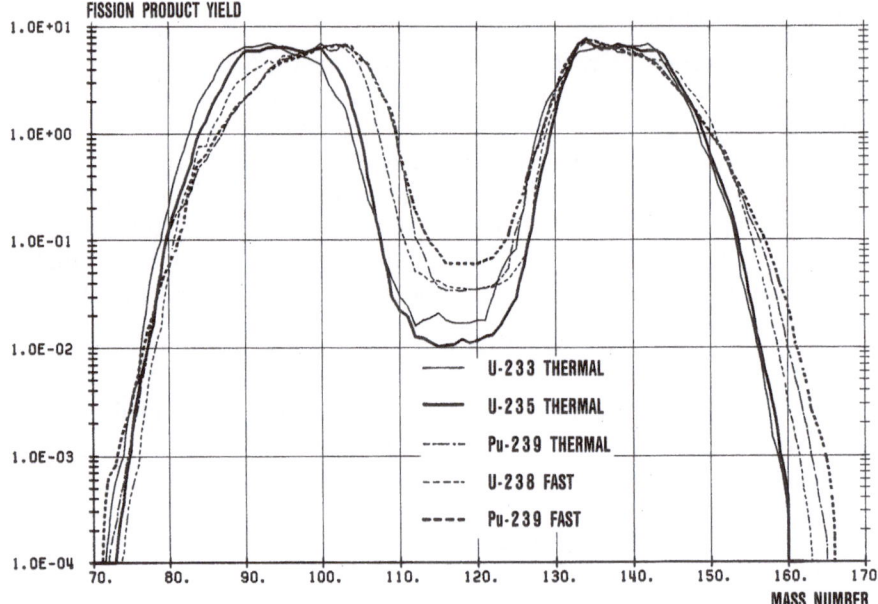

Fig. 2-1. Fission product yield (%) for fission reaction by thermal (T) and fast (F; E > 1.2 MeV) neutrons (KfK)

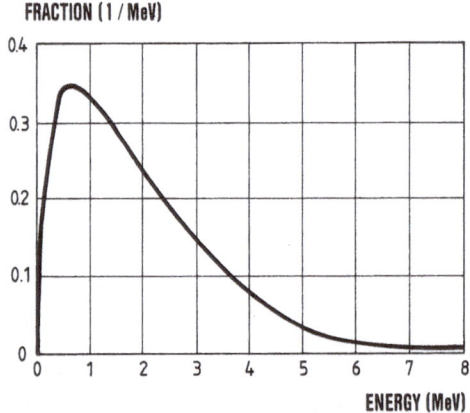

Fig. 2-2. Fission neutron energy distribution for fission of U-235

Fission products may be gaseous, volatile or solid. Differences in fission product mass yields for different fuel isotopes, e.g., U-235 or Pu-239, have to be considered in the safety analysis of fission reactors and can be used for detection purposes.

2.1.5 Energy Release in Nuclear Fission

The energy in nuclear fission, Q_{tot}, is released as the kinetic energy, E_f, of the fission products, as the kinetic energy, E_n, of the prompt fission neutrons,

Table 2-1. *Total energy release per fission and component energies (in MeV)*
(A. Michaudon)

Target nucleus	Incident neutron energy	E_f	E_n	E_β	E_γ	E_ν	Q_{tot}	Q_{th}
Th-232	3.35	161.79	4.70	8.09	14.01	10.87	196.11	185.24
U-233	thermal	168.92	4.90	5.08	12.53	6.82	198.25	191.43
	0.5	169.37	4.9	5.05	12.52	6.79	198.13	191.34
U-235	thermal	169.75	4.79	6.41	13.19	8.62	202.76	194.14
	0.5	169.85	4.8	6.38	13.17	8.58	202.28	193.7
U-238	3.10	170.29	5.51	8.21	14.29	11.04	206.24	195.2
Pu-239	thermal	176.07	5.90	5.27	12.91	7.09	207.24	200.15
	0.5	176.09	5.9	5.24	12.88	7.05	206.66	199.61
Pu-240	2.39	175.98	6.18	5.74	12.09	7.72	206.68	198.94
Pu-241	thermal	175.36	5.99	6.54	14.22	8.80	210.91	202.11
	0.5	175.62	6.0	6.49	14.19	8.73	210.53	201.8
Pu-242	2.32	176.79	4.59	6.62	12.98	8.90	206.15	197.25

as β-radiation, E_β, and γ-radiation, E_γ. In addition, a small amount of energy, E_ν, is emitted as neutrino radiation which does not produce any heat in the reactor. The thermal energy, Q_{th}, released per fission of different fissile nuclei and the different component energies are listed in Table 2-1. On the average, about

$$194 \text{ MeV/fission or } 3.11 \times 10^{-11} \text{ J/fission}$$

are released as thermal energy, Q_{th}, for U-235. Since 1 g of U-235 metal contains about 2.56×10^{21} atoms, complete fission of 1 g of U-235 results in

$$7.96 \times 10^{10} \text{ J or } 2.21 \times 10^4 \text{ kWh}$$

of thermal energy, which corresponds to

$$0.92 \text{ MWd(th) or } \approx 1 \text{ MWd(th)}.$$

For other fissile materials, e.g., U-233 or Pu-239, the result is roughly the same. Usually, the energy produced by the fuel in the reactor core is indicated in MWd(th) per te of fuel. Taking the above relation into account, this figure then roughly corresponds to the mass (in grams) of heavy atoms split in one tonne of fuel.

2.1.6 Decay Constant and Halflife

Radioactive decay changes the number of isotopes, N(t), existing per cm³ as a function of time, (t). This change can be described by the exponential law of

$$N(t) = N_o \cdot \exp(-\lambda \cdot t) \qquad \text{(Eq. 2-1)}$$

where λ is the decay constant and N_o the number of atomic nuclei per cm^3 at the time $t = 0$. Instead of the decay constant, λ, one can also use the halflife, $T_{1/2} = (\ln 2)/\lambda$, which is the time by which half of the nuclei existing at $t = 0$ have decayed. The decay rate, $\lambda \cdot N(t)$, is called the activity of a specimen of radioactive material. This activity is measured in units of Curie or Becquerel. One Becquerel, denoted Bq, is defined as one disintegration per second. One Curie, denoted Ci, is defined as 3.7×10^{10} disintegrations per second, which is approximately the activity of 1 g of radium. Low activities are also measured in $mCi = 10^{-3}$ Ci or $\mu Ci = 10^{-6}$ Ci.

2.1.7 Prompt and Delayed Neutrons

More than 99% of the neutrons generated by fission appear within some 10^{-14} s of the splitting of the atomic nuclei (prompt neutrons). The fission products (fragments) generated, however, will be in a highly excited state and some of them may emit a so-called delayed neutron with delay times on the order of seconds. Thus, a small fraction of less than 1% of the total fission neutron yield is produced by disintegrating fission fragments (precursors of delayed neutrons), a process following the exponential law mentioned above. These delayed neutrons are generally combined in six groups according to the disintegration characteristics of their precursors (parent nuclei). The halflives, $T_{1/2}$, of these six groups of delayed neutron precursors differ between 56 and about 0.2 seconds (Table 2-2). Their average kinetic energies are between 400 and 500 keV, which is below those of prompt fission neutrons (see Fig. 2.2). The delayed neutrons allow a reactor core to be controlled fairly easily by relatively slow movements of the absorber rods (see Section 2.8.3).

The average kinetic energies of delayed neutrons and the disintegration characteristics of the precursors are roughly in the same ranges of energy and time for the most important fuel isotopes, i.e., Th-232, U-233, U-235, U-238, Pu-239,

Table 2-2. *Delayed neutron data for 6 groups of delayed neutron precursors (fission by fast neutrons)*

Group i	Halflife $T_{1/2}(s)$			Decay constant λ_i (1/s)		
	U-235	U-238	Pu-239	U-235	U-238	Pu-239
1	55.72	52.38	54.28	0.0124	0.0132	0.0128
2	22.72	21.58	23.04	0.0305	0.0321	0.0301
3	6.22	5.00	5.60	0.111	0.139	0.124
4	2.30	1.93	2.13	0.301	0.325	0.325
5	0.610	0.49	0.618	1.14	1.41	1.12
6	0.230	0.172	0.257	3.01	4.02	2.69

Table 2-3. *Total fractions of delayed neutrons for different fuel isotopes (in %)*

Fissionable isotopes	Th-232	U-233	U-235	U-238	Pu-239	Pu-240
Fraction of delayed neutrons						
– thermal fission	0	0.266	0.641	0	0.204	0
– fast fission	2.03	0.267	0.65	1.48	0.212	0.266

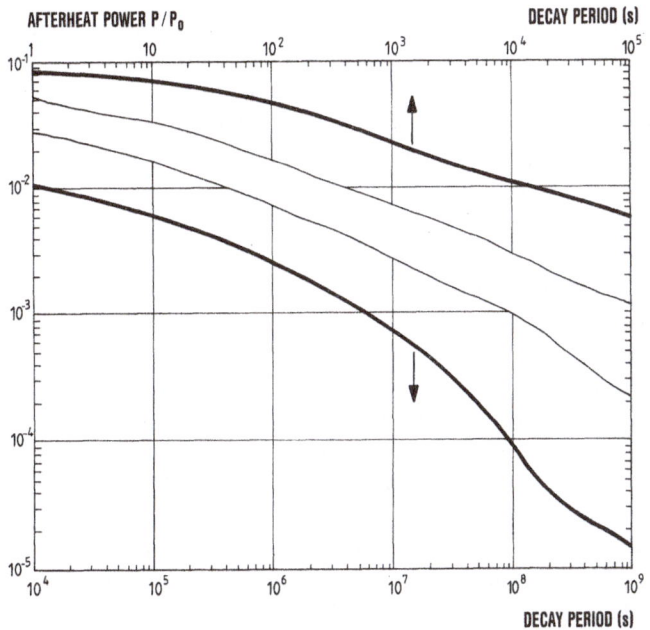

Fig. 2-3. Post-shutdown afterheat of a PWR core as a function of time (initial enrichment 3.2% U-235, burnup 32,000 MWd(th)/te) (KfK)

Pu-240, Pu-241, Pu-242. In addition, they are nearly independent of the energy of the neutron inducing fission. However, the fraction of delayed neutrons occurring per fission differs, as indicated in Table 2-3, and depends on the energy of the neutron inducing fission.

2.1.8 Afterheat of the Reactor Core

The gradually decaying fission products generate heat in the reactor core, even if the neutron fission chain reaction has been interrupted (after shutdown of the reactor core). This afterheat, or decay heat, is composed of the contributions by the decay chains of the fission products and of contributions by U-239, Np-239, and the higher actinides, which are unstable. It is a function of the power history of the reactor core before shutdown and is thus strongly influenced by the burnup of the fuel. Fig. 2-3 by way of example shows the relationship between the power of the fuel elements in the reactor core of a PWR after shutdown, $P(t)$, and the power during operation, P_0. This afterheat, $P(t)$, drops very sharply as a function of time. Immediately after shutdown it is slightly below 10%, after 10 s it is still 7%, after one hour 1.5%, after one month about 0.15%, and after one year it is 0.03% of the reactor power during operation, P_0.

2.2 Neutron Flux and Reaction Rates

The neutrons produced in nuclear fission have a certain speed (kinetic energy) and direction of flight. During their lifetime they may be scattered elastically or inelastically or absorbed by atomic nuclei. In some cases, they may induce nuclear fissions so that successive generations of fission neutrons are produced and a fission chain reaction is established. The number of neutron-nuclear reactions per cm^3 at the point $\vec{r}$ of the reactor core is calculated according to the following considerations:

Let $n(\vec{r},v,\vec{\Omega})$ be the number of neutrons at point $\vec{r}$ with the speed, v, and the direction of flight, $\vec{\Omega}$. Within a volume element, dV, they can react with $N \cdot dV$ atomic nuclei, N being the number of atomic nuclei per cm^3 of reactor volume. The number of reactions per second and cm^3 between neutrons and atomic nuclei is proportional to $v \cdot n(\vec{r},v,\vec{\Omega})$ and $N \cdot dV$, the proportionality factor, $\sigma(v)$, being a measure of the probability of the nuclear reaction. $\sigma(v)$ is called the microscopic cross section of the nucleus for the respective reaction and is indicated in units of 10^{-24} $cm^2 \cong 1$ barn. It is a function of the speed or kinetic energy of the neutron, the type of reaction, and differs for every type of atomic nucleus. For scattering processes, it also depends on the angle between the direction of the incident and scattered neutron but, except for very special cases, not on $\vec{\Omega}$.

Integration over all flight directions furnishes the nuclear reactions of all neutrons with the speed v. The reaction rate per cm^3 of reactor volume at point $\vec{r}$ then turns out to be:

Reaction rate:

$$R(\vec{r}, v) = \sigma(v) \cdot N(\vec{r}) \cdot v \cdot n(\vec{r}, v) = \Sigma(\vec{r}, v) \cdot \Phi(\vec{r}, v) \qquad \text{(Eq. 2-2)}$$

The quantity

$$\Sigma(\vec{r}, v) \, [cm^{-1}] = N(\vec{r}) \, [cm^{-3}] \cdot \sigma(v) \, [cm^2] \qquad \text{(Eq. 2-3)}$$

is called the macroscopic cross section. The quantity

$$\Phi(\vec{r}, v) = v \cdot n(\vec{r}, v) \, [n/cm^2s] \qquad \text{(Eq. 2-4)}$$

is usually called the neutron flux or, sometimes, more appropriately labeled the neutron flux density.

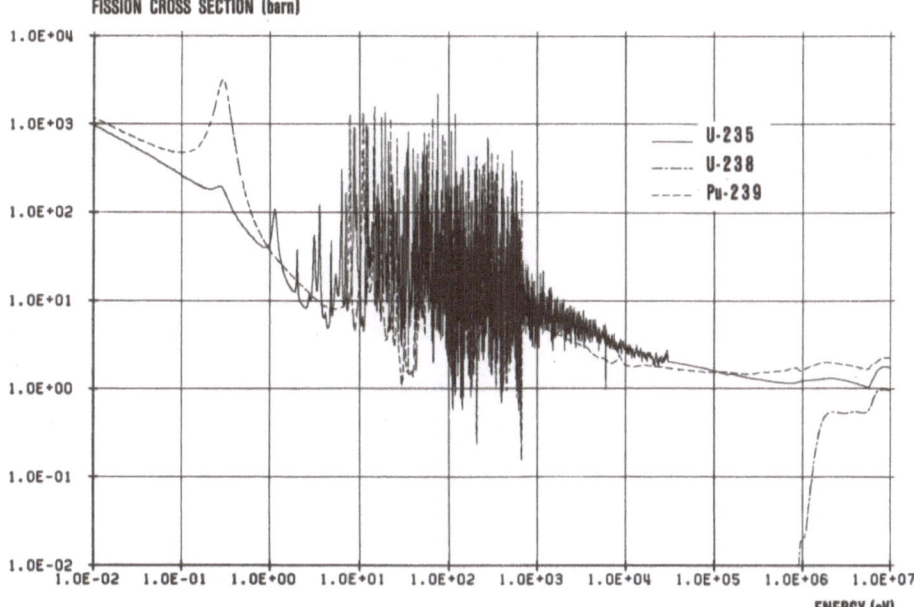

Fig. 2-4. Fission cross sections of U-235, U-238 and Pu-239 as a function of kinetic energy of the incident neutron (KfK)

Generally, the microscopic cross section, $\sigma(v)$, decreases with increasing neutron speed. Exceptions to this rule are the resonance reactions and threshold reactions. As an example, Fig. 2-4 shows the fission cross sections for U-235, U-238 and Pu-239.

2.3 Spatial Distribution of the Neutron Flux in the Reactor Core

The number of neutrons of a certain speed or energy and direction of flight is determined as a function of space and time from the solution of an integro-differential equation with boundary conditions. This Boltzmann neutron trans-port equation is essentially a balance equation counting the number of neutrons gained and lost via different reaction processes, such as scattering, fission, capture and spatial neutron migration (for time dependent problems, also the delayed neutrons and the change in time of the number of neutrons need to be considered). For many practical applications the migration of the neutrons may be described by the neutron diffusion equation, which is an approximation to the Boltzmann neutron transport equation. The neutron diffusion equation is derived as a coupled system of G partial differential equations for G neutron groups. Usually it can not be applied near local neutron sources, external bound-aries, internal interfaces between material regions with different nuclear proper-ties and within strongly neutron absorbing materials. In these cases, the Boltz-mann neutron transport equation must be applied. Fig. 2-5 shows the neutron

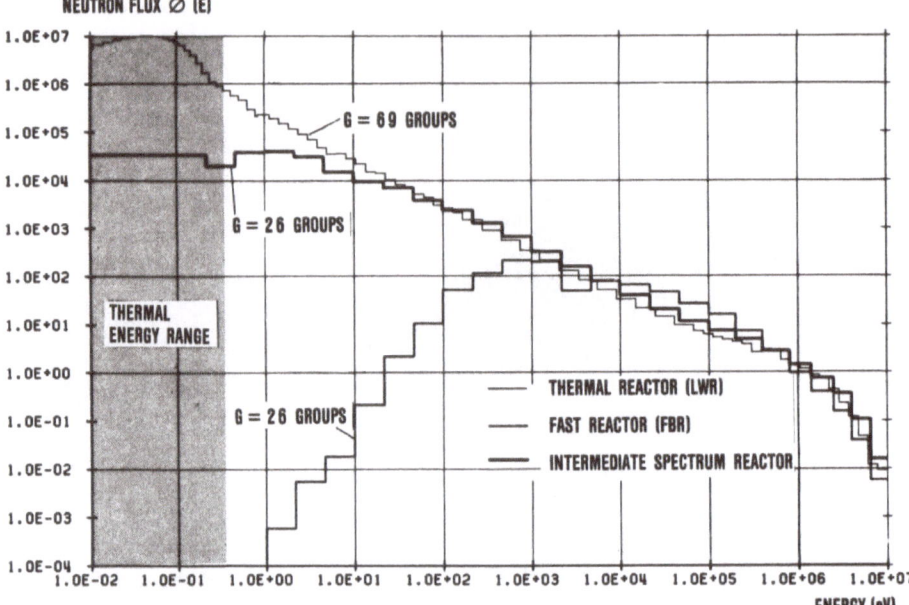

Fig. 2-5. Neutron flux, Φ (E), as a function of neutron energy for a LWR core, an FBR core and an intermediate spectrum reactor core (KfK)

energy spectra in a thermal reactor core (LWR), a fast reactor core (FBR), and an intermediate spectrum reactor core. In this figure, the neutron flux spectra are normalized and plotted as a function of the neutron kinetic energy, E.

Solutions for the spatial distribution of the neutron flux, $\Phi_g(\vec{r})$, of a specific energy group, g, are usually determined numerically by means of computer programs run on digital computers. Fig. 2-6 shows the spatial distribution of the thermal neutron flux, $\Phi_{th}(\vec{r})$, in a PWR core with the control rods partly inserted. The thermal neutron flux is tilted somewhat in the vicinity of the neutron absorbing control rods and decreases rapidly at the outer core boundaries.

The ratio between the number of neutrons absorbed in the reactor core and escaping from the reactor and the number of neutrons newly generated characterizes the so-called criticality parameter or effective multiplication factor, k_{eff}. For $k_{eff} = 1$, the reactor core is just critical and can be operated in a steady state condition. In this case, the statistical average of the number of neutrons appearing in successive neutron generations does not change with time.

At $k_{eff} < 1$, the reactor core remains subcritical. With $k_{eff} > 1$, more neutrons are produced than are consumed, i.e., the number of neutrons increases steadily with time. However, in the latter case it must be taken into account that the rate of neutron multiplication under normal operating conditions, and thus the increase in the number of neutrons, is determined by the properties of the delayed neutrons. This enables the reactor core to be controlled (see Section 2.8.3 for more details).

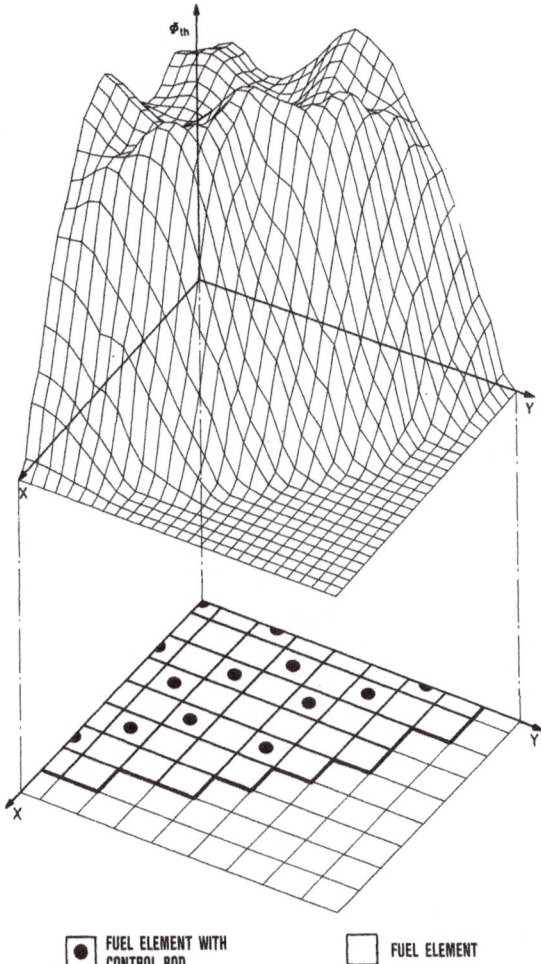

FUEL ELEMENT WITH
CONTROL ROD   FUEL ELEMENT

Fig. 2-6. Spatial distribution of the thermal neutron flux, $\Phi_{th}(\vec{r})$, in a PWR core with partially inserted control rods (W. Oldekop)

The steady state condition, $k_{eff} = 1$, must be established by the following design parameters of the reactor core, which are significant for the neutron balance in the reactor core:
– geometric dimensions of the fuel rods and of the entire reactor core,
– volume fractions and types of coolant and/or moderator, structural and absorber materials in the reactor core and their geometrical arrangement,
– choice of enrichment, i.e., the isotopic composition of U-235, U-238, Pu-239, Pu-240, etc., or Th-232, U-233, U-234, etc. in the fuel rods.
Usually the choice of the coolant and/or moderator, the structural material and the type of fuel with its fissile isotopes (U-235, U-233, Pu-239) is made first. The type of coolant or moderator and fuel, by its inherent materials properties determines the maximum possible power density per cm^3 of core volume. For a given fuel rod diameter this leads to a certain power per cm length

of fuel rod. Assuming a certain power output of the reactor core (e.g. 3000 MW(th)), one then calculates the required core volume (radius and height) or the necessary number of fuel rods of a given length. Thermohydraulic conditions determine the coolant mass flow through the reactor core at specific coolant inlet and outlet temperatures (see Chapters 4 and 5). Given these design parameters, the next step is to calculate the necessary enrichment in fissile isotopes. This is done for the given reactor geometry by solving numerically the steady state multigroup diffusion equation for its smallest eigenvalue, $1/k_{eff}$, and the neutron flux distribution in energy group g and space, $\Phi_g(\vec{r})$. The fissile fuel enrichment is determined such that k_{eff} is slightly above 1, to allow for control throughout the whole period of operation of the reactor core. Enrichment and overall core size furnish the total quantity of fissile fuel needed within the core.

In those reactor cores where fission events are predominantly caused by thermal neutrons (thermal reactors), the substructure of the neutron flux in fuel rods and the ambient coolant must be taken into account by subdividing the core into subcells (Wigner-Seitz cells) (Fig. 2-7). In that case, the neutron flux is determined by superposition of the microdistribution in the subcell and the macrodistribution over the entire core. The microdistribution in the subcell is obtained by solving the multigroup diffusion equations for characteristic subcells consisting of a fuel rod and the respective moderator. In more precise calculations, the Boltzmann neutron transport equation is applied. For the

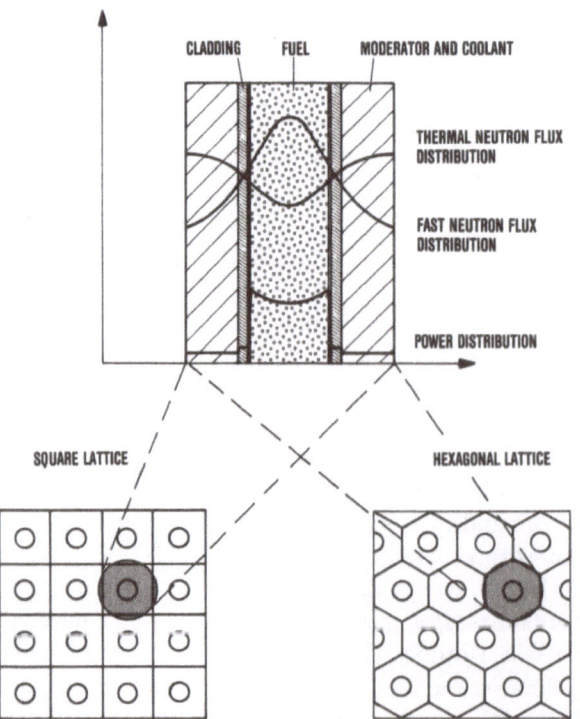

Fig. 2-7. Sub-cells in thermal ractor cores (Wigner-Seitz cells)

boundary conditions of the subcells the simplifying assumption is made that a periodic lattice system of infinite extension is involved. In a two-energy group model, fast fission neutrons are generated in the fuel rod, from where they move into the surrounding moderator. Although they may penetrate the fuel region several times, they are slowed down to thermal energies, thus avoiding to a large extent undesired capture processes of epithermal neutrons in the fuel. Such capture processes in the resonance region otherwise would reduce the multiplication factor, k_{eff}. From the moderator, thermal neutrons diffuse back into the fuel rod where they are absorbed by fuel nuclei, thus producing fission reactions. The above discussion explains the distributions shown in Fig. 2-7, where the fast neutron flux is high in the fuel region and the thermal neutron flux is high in the moderator region. In a well thermalized reactor, the life of a neutron from birth (as a fast neutron) to death (by absorption or leakage out of the reactor) can also be described by the socalled four-factor formula for k_{eff}. Knowing the microdistribution of the neutron flux, this allows average material constants of the subcell to be determined by some kind of homogenization procedure. These average parameters are subsequently used in the calculation of the macroflux distribution.

To determine the macrobehavior of the neutron flux in the reactor core, averaged materials constants are established by detailed calculations of the fine structures of materials and the neutron fluxes in the subcells. Afterwards, the multigroup diffusion equations are solved with the macrostructure of the reactor core and its boundary conditions taken into account.

The local power distribution, $P(\vec{r})$, in the reactor core is determined from the fission reaction rate as

$$P(\vec{r}) = 3.11 \times 10^{-11} \sum_{g=1}^{G} \Sigma_f(\vec{r})_g \, \Phi_g(\vec{r}) \, [\text{W/cm}^3] \qquad \text{(Eq. 2-5)}$$

where $\Sigma_f(\vec{r})_g$ is the macroscopic fission cross section and $\Phi_g(\vec{r})$ is the neutron flux in energy group g at location $\vec{r}$.

Thermal reactor cores, such as PWR's, typically have a U-235 enrichment of about 3%. As the microscopic fission cross sections in the thermal energy range (0.025 eV) are relatively high (about 580 barns) (see Fig. 2-4), the corresponding average thermal neutron flux, Φ_{th}, is in the range of 3×10^{13} n/cm²s, to achieve power densities of about 100 kW(th)/l of core volume. However, fast reactor cores with average neutron energies of about 100 keV require a fissile enrichment of about 15–25%. As the microscopic fission cross sections in the 100 keV range are rather low (about 1.8 barns (see Fig. 2-4)), a much higher average neutron flux of about 3×10^{15} n/cm²s is required to produce an average power density of about 300 kW(th)/l.

For practical applications in nuclear reactor core design, predictions of the local energy spectrum of the neutrons, the spatial distributions of neutron flux and power, the value for k_{eff} and other parameters essentially rely on two and three-dimensional numerical methods involving computer programs run on fast digital computers. In special cases, the neutron transport equation is solved by numerical methods in one and two-dimensional geometries or three-dimensional Monte Carlo methods are applied. The number of neutron energy groups to be used depends upon the problem to be treated. Frequently, instead of the neutron energy, E, a corresponding variable, lethargy, $u = \ln (E/E_o)$, is used where E_o is the upper limit of the energy scale. This logarithmic energy scale is suggested by the fact that the average logarithmic energy loss per collision of a neutron with a nucleus is an energy independent constant. In reactors with fast neutrons, basically 20–30 neutron energy groups are treated, which are often condensed into 6–12 groups. Determining the fine structure of the neutron spectrum may take hundreds or even thousands of neutron groups. In thermal reactors, such as LWR's, the problem may well be reduced to two or four energy groups. In reactors with fast neutrons, the microstructure of the neutron flux in the subcells is usually less pronounced than in thermal reactors; in most cases it can therefore be taken into account by a heterogeneity correction of the so-called group constants.

As indicated in Fig. 2-7, the thermal power is low in the cladding of the fuel rod and in the moderator and coolant. It is mainly generated there by

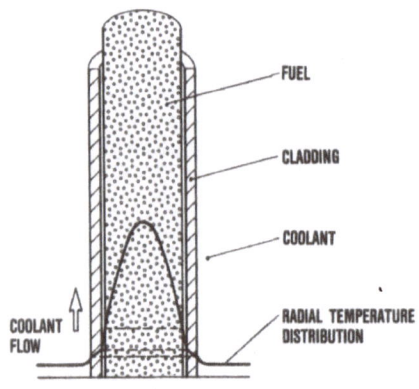

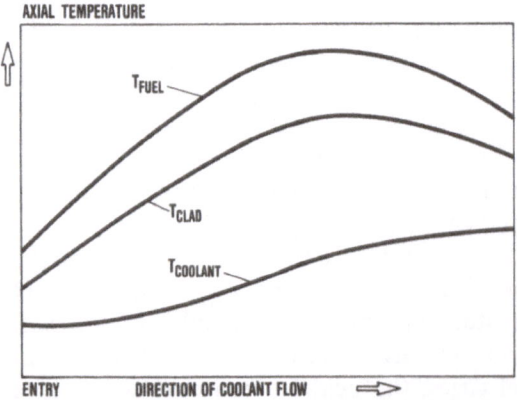

Fig. 2-8. Radial and axial temperature distributions in a fuel rod and the surrounding coolant channel

elastic and inelastic collisions of neutrons and by absorption of γ-radiation. The radial temperature distribution within the fuel rod and the axial temperature increment in the cooling channel surrounding the fuel rod is indicated by Fig. 2-8. During normal operation, the fuel must not reach its melting temperature and the maximum cladding temperature must not exceed certain limits specific to materials (see Chapter 4 and 5).

2.4 Fuel Burnup, Fission Product and Actinide Buildup

The fission of atomic nuclei and generation of fission products, neutron capture and radioactive α- and β-decay change the concentration of various isotopes in the reactor core during reactor operation. The change in concentration, $N^i(\vec{r},t)$, of the isotope i as a function of time at location $\vec{r}$ per cm^3 of reactor volume can be written as a balance equation between the production rate, R_p^i $(\vec{r},\ t)$, and the loss rate rate, R_l^i $(\vec{r},\ t)$, taking into account all neutron reaction processes as well as radioactive decay processes of the nuclei. This leads to an ordinary differential equation for each isotope i. Integration over a given time period furnishes the concentration of the isotope i at location $\vec{r}$ and time t.

The possible chains of production and decay of fuel isotopes or higher actinides in the uranium/plutonium cycle and the thorium/uranium cycle are presented in Fig. 2-9. Neutron capture of a fuel isotope increases the mass number of the isotope by 1 (vertical step). The radioactive β^--decay, with the mass number unchanged, means an increase by 1 in the atomic number (protons) (horizontal step). The radioactive α-decay implies a reduction by 4 in the mass number and by 2 in the atomic number.

In nuclear reactors with U-235/U-238 fuel, isotopes of the elements uranium, neptunium, plutonium, americium and curium are built up over the period of operation. Reactors containing fertile thorium in addition to U-235/U-238 fuel (HTGR, CANDU) build up possible isotopes of thorium and protactinium and their decay chains, while reactors operated only in the Th-232/U-233 fuel cycle build up almost no plutonium, americium and curium isotopes, but correspondingly larger amounts of uranium, thorium and protactinium isotopes.

In thermal reactors, the concentrations of Xe-135 and Sm-149 fission products must be observed carefully because of their high absorption cross sections. Their variation as a function of space and time during reactor operation and shutdown affect considerably the criticality, k_{eff}, of the reactor core and the spatial neutron distribution.

Fig. 2-10 shows the decrease of fissile U-235 nuclei (burnup), the buildup of U-236 and of the different plutonium isotopes, as well as the buildup of Am-243, Cm-244, Sr-90 and Cs-137 in one tonne of LWR fuel over a period of about three years of operation. It is seen that the isotopic distribution of plutonium (Pu-239:Pu-240:Pu-241:Pu-242) also changes as a function of burnup and time. While the enrichment of U-235 has dropped to roughly 0.8% or 8 kp in 1 te of fuel by the end of the period of operation, the plutonium

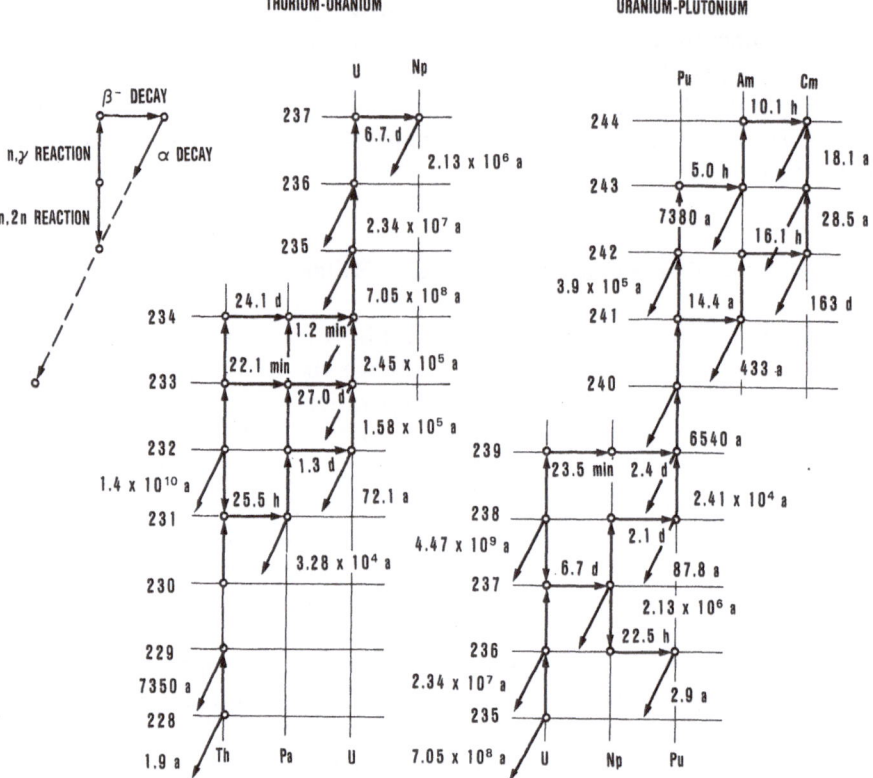

Fig. 2-9. Actinide chains in the uranium-plutonium and uranium-thorium nuclear fuel cycles (KfK)

content (all plutonium isotopes) has risen to almost 0.9% over the same period. The plutonium buildup is due to the continuous in situ conversion of U-238 and Pu-240 by neutron capture. In this way, the fissile isotopes, Pu-239 and Pu-241, produced during reactor operation make major contributions to the fission reaction rate and, hence, to heat production. A corresponding process of conversion into fissile U-233 is possible as a result of the capture of neutrons in Th-232.

To determine the change in time of the isotopic composition of the fuel, including actinides and fission products, ordinary differential equations must be solved for all relevant isotopes in the actinide and fission product chains. This is done by taking into account the initial concentrations for each isotope for a sufficiently large number of space points and material zones in the reactor core. After preparation of the corresponding average one-group cross sections programs are run on digital computers to follow these concentrations in time. In this way, the isotopic concentrations in the reactor core and their repercussions upon k_{eff} are determined for various time steps of a power cycle. The burnup of fissile isotopes and the buildup of neutron absorbing fission products and actinides mostly cause k_{eff} to decrease over a given operating cycle of a reactor core.

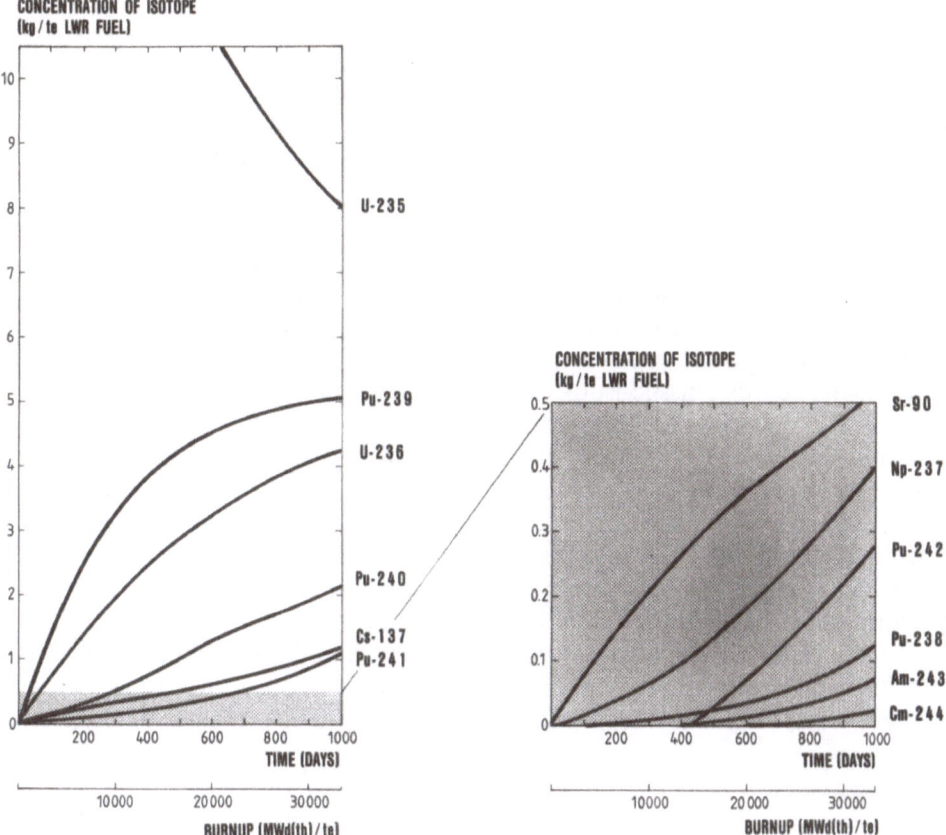

Fig. 2-10. Buildup of various actinides and fission products in LWR fuel during three years of reactor operation (KfK)

2.5 Conversion Ratio and Breeding Ratio

The ratio between the production rate of new fissile material (U-233, Pu-239, Pu-241) continuously generated from fertile Th-232, U-238 and Pu-240 and the continuous annihilation rate of fissile U-233, Pu-239 and Pu-241 atomic nuclei is called the *conversion ratio* (CR). It is called *breeding ratio* (BR), if it attains values $\geqq 1$.

Most thermal reactors (LWR, CANDU, HTGR) have conversion ratios between 0.5–0.9 and thus are consumers of fissile material. Because of their relatively low conversion ratios they are called converter reactors. If CR always equals 1, the fissile core inventory does not change during reactor operation. Breeding ratios close to 1 can be attained by near breeder reactors. Breeding ratios of approximately 1.15 to 1.30 are only reached in breeder reactors with a fast neutron spectrum (FBR) and using uranium-plutonium fuel.

The conversion ratio or breeding ratio of a reactor core can be described by the relation

$$\left.\begin{array}{c} CR \\ BR \end{array}\right\} = \bar{\eta} - 1 - \bar{a} - \bar{l} + \bar{f} \qquad \text{(Eq. 2-6)}$$

where $\bar{\eta}$ is the neutron yield, i.e., the total number of fission neutrons generated per neutron absorbed averaged over all fissile isotopes, the neutron spectrum and the whole reactor. The quantities $\bar{a}$, $\bar{l}$ and $\bar{f}$ are equivalent corresponding averages; $\bar{a}$, the so-called parasitic absorption, describes the neutron loss fraction arising from absorption in coolant, structural and control materials; $\bar{l}$ is the neutron leakage from the reactor core (or, in FBR's, from the surrounding breeding blanket); $\bar{f}$ is the fractional contribution of U-238 or Pu-240 fissions (fast fission effect) (see Section 2.1.4).

The dominating quantity in the above relationship is the neutron yield, $\bar{\eta}$. Fig. 2-11 shows the individual neutron yields for the fissile fuel isotopes, U-233, U-235, Pu-239, and Pu-241. It is seen that Pu-239 assumes the highest possible η-values in the range of neutron energies > 100 keV and that U-233 has higher values than the other fissile isotopes in the thermal energy range. With U-235 as the fissile material, CR reaches relatively low values. In thermal reactors using previously bred U-233 fissile material, it is expected that CR can attain a maximum possible value of approximately 0.9 to 1.03 at $\eta = 2.28$. The highest values of CR are obtained with plutonium as a fissile material in reactors having neutron spectra in the range of several hundred keV. This is achieved by employing in the cores of those reactors relatively small fractions of coolants and structural materials with relatively high atomic mass numbers and low capture cross sections so that neutron slowing-down (moderation) is diminished and parasitic neutron absorption remains low. These reactors can clearly attain BR $>$ 1, which is why they are called breeder reactors. As will be explained in Chapter

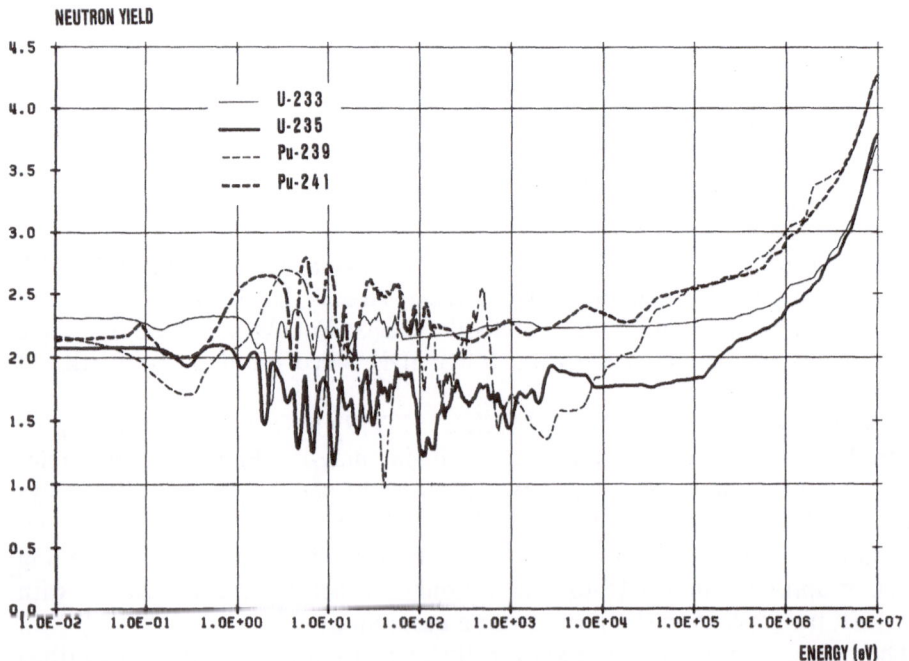

Fig. 2-11. Neutron yield as a function of neutron energy for various fissile fuel isotopes (KfK)

5, the net production of new fissile material (Pu) occurs essentially in axial and radial blankets surrounding the core. Because of their neutron spectrum (see Fig. 2-5) with predominantly fast neutrons they are usually called *fast breeder reactors* (FBR's). FBR's with plutonium/uranium mixed oxide fuel, steel as a structural material and sodium as a coolant reach BR values of 1.15 to 1.30 at an average $\eta \approx 2.4$.

The fast fission effect, $\bar{f}$, i.e., the fraction of fissions of fertile atoms by fast neutrons, is only around 0.01–0.03 in thermal reactors. In FBR cores, the fast fission effect may well reach levels of 0.1–0.15.

2.6 Conversion Ratio and Fuel Utilization

The higher the conversion ratio, the more fertile Th-232 or U-238 nuclei will be converted into fissile U-233 or Pu-239 nuclei. Some of these converted nuclei will be utilized in situ in the reactor. However, this conversion of fertile material is fully exploited in the fuel cycle only if the newly formed, man-made fissile U-233 or Pu-239 is recovered by chemical reprocessing after unloading the fuel elements from the reactor core and is then used for fabricating new fuel elements (recycle mode). This involves small losses in chemical processing and fabrication of the fuel.

In converter reactors (LWR, CANDU, HTGR), the fuel elements must be unloaded, after an operating period of about three years due to burnup of the fissile fuel and buildup of fission products. Fission products capture neutrons and decrease criticality, k_{eff}. In FBR's, unloading of the fuel after some three years is mainly necessary because of radiation damage to structural materials and buildup of gaseous fission products in the fuel rods. The latter effect leads to a steady increase in pressure that the clad of the fuel pin has to withstand.

Fuel utilization is the fraction of the original nuclear fuel that can be ultimately converted into thermal energy by nuclear fission after having passed through the whole fuel cycle once or several times, taking into account possible losses during reprocessing and refabrication of the fuel.

Fuel utilization depends on
– the tails assay and the enrichment of natural uranium,
– the neutronic properties of the reactor core, which determine the conversion or breeding ratios. These factors, above all, are the design parameters of the reactor core (fractions and types of coolant, structural material, absorber), the choice of fissile material (U-233, U-235, Pu-239) and fertile material (U-238, Th-232), and the neutron spectrum present in the reactor core (thermal neutrons, fast neutrons).

If the reactor operates in the recycle mode, fuel utilization also depends on
– the burnup of the fuel when unloaded from the reactor core, which determines the frequency of reprocessing and refabrication cycles (the lower the fuel burnup, the less energy will be generated while the fuel remains in the reactor core and the more frequently the fuel will pass through the fuel cycle when recycled),
– the fuel losses occurring during reprocessing and refabrication of the fuel.

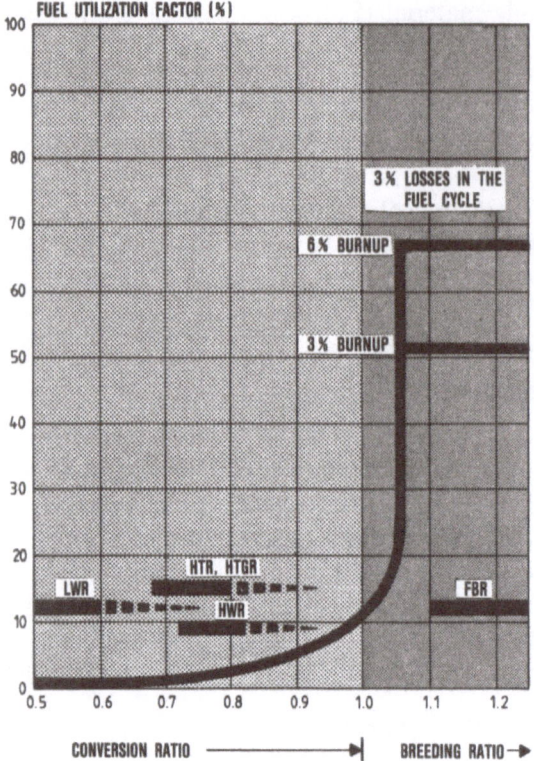

Fig. 2-12. Utilization of uranium or thorium as a function of the conversion or breeding ratios (KfK)

Fuel utilization can be determined by analyzing these functional relationships and listing in a balance sheet the amount of fuel which can be utilized (fissioned) in each operating cycle of the reactor core. This leads to Fig. 2-12, which represents fuel utilization as a function of the conversion ratio or breeding ratio for various types of reactors. Reactors not operated in the recycle mode, such as present LWR's, only have conversion ratios of about 0.6, attaining fuel utilizations as low as 0.6%. CANDU's and AGR's reach conversion ratios around 0.68 and, correspondingly, slightly better fuel utilizations.

Reactors operating in a recycle mode and having appropriately optimized core neutron physics properties may reach conversion ratios of 0.9 or close to 1.0. In that case, fuel utilization may rise to several percent (e.g., for advanced CANDU's, HTGR's).

Fuel utilization shows a steep increase in the range of low breeding ratios (BR = 1 to 1.05), attaining a constant level at a certain limit. At this limit of the breeding ratio, the reactor system is just able to make up for the fuel cycle losses by breeding sufficient fissile material. Below this limit of the breeding ratio, fuel utilization is severely affected by relatively slight changes in the design and operating parameters of the reactor core and the fuel cycle (reprocessing and refabrication).

FBR's employing the uranium/plutonium cycle operate beyond these limits. In this region, fuel utilization is a function only of the average fuel burnup (averaged over the reactor core and the blanket) and of the fuel losses in the fuel cycle. FBR's with breeding ratios BR ≥ 1.1 achieve a fuel utilization of roughly 60% and above. This is approximately a factor of 100 above the level reached by present LWR's operating in the once-through cycle.

This high utilization of the fuel in FBR's is the ultimate reason for the low fuel consumption of breeder reactors. FBR's started on a core containing plutonium/U-238 fuel will need only small amounts of U-238 or depleted uranium for further operation.

Improvements in the conversion ratio and in fuel consumption still attainable by optimizing the reactor physics of converter reactor cores will be outlined in Chapter 6.

2.7 Radioactive Inventories in Fission Reactors

As outlined in Section 2.1 and 2.4, the processes of nuclear fission and the capture of neutrons in the fuel, coolant and structural material generate both chains of radioactive fission products and higher actinides. Radioactivity is measured in Becquerel or Curie. The units of Bq or Ci only indicate the number of disintegrations, not the type of radioactive decay (β^--, γ- or α-decays). Energy and the range of penetration through matter can be very different for the various decay products.

The quantities of fission products, actinides and the radioactivities of structural materials and coolants are a function of the time for which the fuel elements are kept in the reactor core and generate heat by nuclear fission. They also depend on the respective core loading conditions (fresh fuel elements or fuel elements with higher burnups) and on the kind and purity of materials used as the fuel, coolant or structural material. Table 2-4 lists the most important fission products and actinides responsible for the radioactivity accumulated in the core of a PWR. These radioactivity data are calculated for 1 te of fuel charged to the reactor core at its target burnup of 33,000 MWd(th)/te. Only one third of the fuel elements in a PWR core reach this maximum burnup at the end of each cycle, because the reactor core of a PWR is loaded in such a way that, in the course of a year, one third of the fuel elements increase their burnups from 0 to 11,000 MWd(th)/te, another third from 11,000 to 22,000 MWd(th)/te, and the last third from 22,000 to 33,000 MWd(th)/te. At the end of the burnup cycle, the third of the fuel reaching its maximum burnup is unloaded and replaced by fresh fuel elements. Accordingly, the maximum radioactivity has accumulated at the end of the burnup cycle prior to unloading. PWR's contain roughly 80 te of fuel per GW(e) and, at the end of the burnup cycle, they contain some 10^{10} Ci or 3.7×10^{20} Bq of radioactivity, mainly in the fission products. The bulk of the fission products disintegrate relatively quickly, while the actinides are extremely longlived. The time dependence of the activities of the different fission products and actinides plays a major role in the nuclear fuel cycle. Fission products and actinides are separated chemically

Table 2-4. *Activities of radionuclides in spent PWR fuel. (Initial enrichment 3.2% U-235, burnup 33,000 MWd(th)/te) (KfK)*

Radio-nuclide	Half-life	Specific radioactivity (Ci/te$_{HM}$ charged to reactor)				
		At discharge	After 1 year	After 3 years	After 5 years	After 7 years
Fission products						
H-3	12.4 a	4.98E+02	4.70E+02	4.21E+02	3.76E+02	3.36E+02
Kr-85	10.7 a	9.40E+03	8.82E+03	7.75E+03	6.81E+03	5.98E+03
Sr-90	29.1 a	7.47E+04	7.30E+04	6.96E+04	6.63E+04	6.32E+04
Y-90	64.0 h	7.86E+04	7.30E+04	6.96E+04	6.63E+04	6.33E+04
Zr-95	64.0 d	1.47E+06	2.81E+04	1.03E+01	3.77E−03	1.38E−06
Nb-95	35.2 d	1.46E+06	6.32E+04	2.28E+01	8.36E−03	3.06E−06
Ru-106	368.2 d	5.00E+05	2.51E+05	6.36E+04	1.61E+04	4.06E+03
Rh-106	29.9 s	5.46E+05	2.51E+05	6.36E+04	1.61E+04	4.06E+03
Cs-134	2.1 a	1.57E+05	1.12E+05	5.71E+04	2.92E+04	1.49E+04
Cs-137	30.0 a	1.07E+05	1.04E+05	9.94E+04	9.49E+04	9.06E+04
Ba-137m	2.6 min	1.01E+05	9.85E+04	9.40E+04	8.98E+04	8.57E+04
Ce-144	284.3 d	1.10E+06	4.50E+05	7.59E+04	1.28E+04	2.15E+03
Pr-144	17.3 min	1.11E+06	4.50E+05	7.59E+04	1.28E+04	2.15E+03
Pm-147	2.6 a	1.53E+05	1.23E+05	7.26E+04	4.28E+04	2.52E+04
Eu-154	8.6 a	1.23E+04	1.13E+04	9.64E+03	8.20E+03	6.98E+03
Activation products						
Mn-54	312.5 d	7.13E+03	3.17E+03	6.28E+02	1.24E+02	2.46E+01
Fe-55	2.6 a	6.21E+04	4.76E+04	2.79E+04	1.64E+04	9.61E+03
Co-60	5.3 a	2.05E+04	1.80E+04	1.38E+04	1.06E+04	8.17E+03
Ni-63	92.1 a	2.48E+03	2.46E+03	2.42E+03	2.39E+03	2.35E+03
Sr-90	29.1 a	2.25E−03	2.20E−03	2.09E−03	2.00E−03	1.90E−03
Y-90	64.0 h	2.58E+03	2.20E−03	2.09E−03	2.00E−03	1.90E−03
Zr-95	64.0 d	6.23E+04	1.19E+03	4.37E−01	1.60E−04	5.86E−08
Nb-95	35.2 d	6.10E+04	2.67E+03	9.62E−01	3.52E−04	1.29E−07
Ta-182	115.0 d	1.90E+05	2.10E+04	2.57E+02	3.15E+00	3.86E−02
Actinides						
U-234	2.45E+05 a	9.06E−01	9.13E−01	9.26E−01	9.40E−01	9.54E−01
U-235	7.05E+08 a	1.78E−02	1.78E−02	1.79E−02	1.79E−02	1.79E−02
U-236	2.34E+07 a	2.62E−01	2.62E−01	2.62E−01	2.62E−01	2.62E−01
U-238	4.47E+09 a	3.17E−01	3.17E−01	3.17E−01	3.17E−01	3.17E−01
Np-237	2.13E+06 a	3.00E−01	3.07E−01	3.07E−01	3.08E−01	3.09E−01
Np-239	2.4 d	1.95E+07	1.67E+01	1.67E+01	1.67E+01	1.67E+01
Pu-238	87.8 a	2.26E+03	2.43E+03	2.43E+03	2.39E+03	2.36E+03
Pu-239	2.41E+04 a	3.35E+02	3.40E+02	3.40E+02	3.40E+02	3.40E+02
Pu-240	6.54E+03 a	5.08E+02	5.08E+02	5.08E+02	5.09E+02	5.09E+02
Pu-241	14.4 a	1.25E+05	1.19E+05	1.08E+05	9.85E+04	8.95E+04
Pu-242	3.87E+05 a	1.86E+00	1.86E+00	1.86E+00	1.86E+00	1.86E+00
Am-241	4.33E+02 a	1.27E+02	3.23E+02	6.86E+02	1.02E+03	1.31E+03
Am-243	7.38E+03 a	1.67E+01	1.67E+01	1.67E+01	1.67E+01	1.67E+01
Cm-242	163.2 d	4.11E+04	8.78E+03	3.98E+02	2.09E+01	3.89E+00
Cm-244	18.1 a	1.94E+03	1.87E+03	1.73E+03	1.60E+03	1.49E+03

E + 02 ≙ 10^{-2} etc.

during reprocessing and then stored in waste repositories after having been conditioned. Some of the fission products, such as tritium and certainly krypton and xenon, appear in gaseous form, while iodine and cesium are highly volatile. All other fission products and actinides are non-volatile (for further details, see Chapter 7).

2.8 Inherent Safety Characteristics of Converter and Breeder Reactor Cores

2.8.1 Reactivity and Non-Steady State Conditions

It has been explained in Section 2.3 that $k_{eff} = 1$ corresponds to the steady state condition of the reactor core, in which case the production of fission neutrons is in a state of equilibrium with the number of neutrons absorbed and the number of neutrons escaping from the reactor core. For $k_{eff} \neq 1$, either the production or the loss term becomes dominant, i.e., the number of neutrons varies as a function of time. The neutron transport equation or, by way of approximation, the multigroup diffusion equation for calculation of the time dependent neutron flux must then be solved for the non-steady state case. However, in most cases it is a sufficiently good approximation to solve the so-called point kinetics equations in connection with equations describing the temperature field and its impacts on k_{eff}.

Axial movements of the absorber rods in the core change the loss term for neutrons and influence k_{eff}. The relative change as a function of time of $k_{eff}(t)$ is called reactivity:

Reactivity: $$\rho(t) = \frac{k_{eff}(t) - 1}{k_{eff}(0)} = \frac{\Delta k_{eff}(t)}{k_{eff}(0)} \qquad \text{(Eq. 2-7)}$$

The point kinetics equation describing the reactor power as a function of time can be derived from the time dependent multigroup diffusion equation under the assumption that the space distributions of the neutron flux or power and the concentrations of delayed neutron precursors always equal those in steady state conditions. The point kinetics equations read

$$\frac{d P(t)}{dt} = \left[\frac{\rho(t) - \beta_{eff}}{l_{eff}} \right] \cdot P(t) + \lambda \cdot C(t) \qquad \text{(Eq. 2-8)}$$

$$\frac{d C(t)}{dt} = -\lambda \cdot C(t) + \frac{\beta_{eff}}{l_{eff}} \cdot P(t) \qquad \text{(Eq. 2-9)}$$

with the initial conditions of
$P(t=0) = P_0$ (steady state reactor power)
$C(t=0) = C_0$ (steady state concentration of the parent nuclei (precursors, see Section 2.1.7) for all delayed neutrons combined in one group).

In these equations, $P(t)$ is the reactor power, $C(t)$ describes the averaged concentration of parent nuclei (precursors) of the delayed neutrons, λ is the average decay constant of all parent nuclei of delayed neutrons, β_{eff} the effective fraction of all delayed neutrons (integrated over all fissile isotopes and averaged over the reactor), l_{eff} the lifetime of the prompt neutrons in the reactor, i.e., the average time required for a prompt fission neutron to induce a new fission process. Values of β_{eff} and λ are calculated from those indicated in Section 2.1.7. Values of l_{eff} are in the range of 10^{-3} to 10^{-5} s for thermal reactors and around 4×10^{-7} s for FBR's. The average decay constant for a one-group treatment is in the range of about $\lambda = 0.08$ s^{-1} for all fissile fuel isotopes.

The reactivity, $\rho(t)$, is composed of the superimposed or initiating reactivity, $\rho_i(t)$, which can be caused, e.g., by movements of absorber rods or fuel, and the feedback reactivity, $\rho_f(t)$, which takes into account all repercussions of temperature changes in the reactor core,

$$\rho(t) = \rho_i(t) + \rho_f(t) \qquad \text{(Eq. 2-10)}$$

Movements of absorbers produce local changes in the macroscopic cross sections and the neutron flux in certain material zones in the reactor core and, accordingly, also in $k_{eff}(t)$ and $\rho_i(t)$ (superimposed or initiating perturbations). The resultant change as a function of time of the neutron field and the power level alters the temperatures in the reactor core. Temperature changes provoke changes in material densities (expansion and displacement) and microscopic cross sections by the Doppler broadening of resonances (see Section 2.8.2.1). Also the neutron flux spectrum can be shifted by changing the moderation of the neutrons. Moreover, the dimensions of the reactor core and its components are changed by thermal expansion. All these feedback reactivities, $\rho_f(t)$, resulting from changes in power and temperature together with external perturbation reactivities constitute a feedback circuit (Fig. 2-13).

For numerical treatment, the feedback reactivity, ρ_f, is split up into individual contributions by different temperature effects:

$$d\rho_f = \sum_{i=1}^{I} \frac{\partial k_{eff}}{\partial T_i} dT_i \qquad \text{(Eq. 2-11)}$$

where dT_i are the average changes in the temperatures of the fuel, moderator, coolant, structural or absorber materials. In LWR's, the coolant is also the moderator. In other types of reactors (HTGR), the moderator (graphite) is distinct from the coolant (gas).

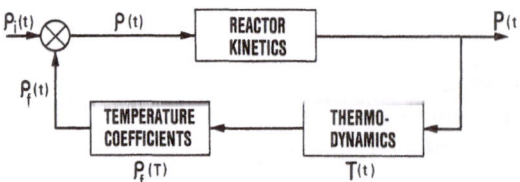

Fig. 2-13. Schematic of the reactor dynamics feedback loop

2.8.2 Temperature Reactivity Coefficients

2.8.2.1 Fuel Doppler Temperature Coefficient

The fuel Doppler temperature coefficient is due to the fact that the neutron resonance cross sections depend on the temperature of the fuel and the relative velocities, respectively, of neutrons and atomic nuclei.

The resonance cross sections, $\sigma(E,T)$, for U-238, Th-232 and U-233, U-235, Pu-239, etc. show very pronounced peaks at certain neutron kinetic energies (see Fig. 2-4). An increase in fuel temperature, T_f, broadens this shape of the resonance curve which, in turn, results in a change in fine structure of the neutron flux spectrum in these ranges of resonance energy. The reaction rates are changed as a consequence. Above all, the resonance absorption for U-238 increases as a result of rising fuel temperatures, while the effect of a temperature change in the resonance cross sections of the fissile materials, U-235 and Pu-239, is so small that it can generally be neglected if the fuel enrichment is not extremely high. The increases in fission and capture reactions in U-235 and Pu-239 partly compensate each other. For these reasons, temperature increases in the fuel result in a negative temperature feedback effect (Doppler effect) brought about by the increase in neutron absorption in U-238. For Th-232, the effects are similar. The Doppler effect is somewhat less pronounced at very high fuel temperatures because adjacent resonances will overlap more and more. The resonance structure then is no longer as pronounced as at low temperatures, which leads to a reduction of the negative Doppler effect.

Due to the specific energy distribution of the neutron spectrum (see Fig. 2-5), the Doppler effect in thermal reactors follows

$$\frac{1}{k_{eff}} \frac{\partial k_D}{\partial T_f} \sim \frac{-1}{\sqrt{T_f}} \qquad \text{(Eq. 2-12a)}$$

whereas in FBR's it follows the relation

$$\frac{1}{k_{eff}} \frac{\partial k_D}{\partial T_f} \sim -\frac{1}{T_f^x} \quad \text{with} \quad \tfrac{1}{2} \leqq x \leqq 3/2. \qquad \text{(Eq. 2-12b)}$$

The Doppler coefficient is always negative in power reactor cores because, given the relatively low enrichment in U-235 and Pu-239, respectively, the resonance absorption of U-238 will always dominate. It is an instantaneous negative feedback coefficient of reactivity, which immediately counteracts increases in power and temperature.

Besides the fuel Doppler coefficient, the fuel expansion coefficient also leads to a negative feedback coefficient of reactivity. Especially in fast breeder cores it may well attain an importance equal to the Doppler coefficient.

2.8.2.2 Coefficients of Moderator or Coolant Temperatures

The main contributions to the coefficients of moderator or coolant temperatures stem from changes in the densities of the moderator or coolant and from resultant shifts in the neutron spectrum. Temperature rises decrease the density of the coolant and accordingly reduce the moderation of neutrons. The neutron spectrum is shifted towards higher energies. As a result of the lower moderator density and the correspondingly higher transparency to neutrons of the core it is also possible that far more neutrons escape from the reactor core and neutron absorption will be reduced.

For the present line of PWR's, the sum total of the individual contributions to changes in various energy ranges finally leads to a negative coefficient of the moderator temperature which, however, also depends on the concentration of boric acid dissolved in the coolant and the burnup condition of the reactor core. In graphite moderated HTGR's containing U-233, the moderator temperature coefficient is usually positive. Also in sodium cooled FBR's with core sizes in excess of about 200 MW(e), the coolant temperature coefficient is positive because the neutron spectrum is shifted towards higher energies as a consequence of the reduced moderation. The resultant increased contribution by the fast fission effect of U-238 as well as the higher η-values (see Fig. 2-11) add to the reactivity. These positive reactivity contributions cannot compensate all negative contributions coming from an increase in the leakage rate of neutrons escaping from the core (which is the dominating effect in small sodium cooled FBR's with power levels of less than approx. 200 MW(e)).

2.8.2.3 Structural Material Temperature Coefficient

Especially in FBR's, the structural material temperature coefficient also plays an important role. Increasing temperatures cause the core structure to expand radially and axially and, in this way, result both in indirect changes in material densities and in changes of size of the reactor core and, as a consequence, of neutron leakage. The structural material temperature coefficient must be determined by detailed analyses of all expansion and bowing effects for given core and fuel element structures also taking into account the core restraint (clamping) system. For FBR's, the structural material coefficient is also negative. This is accomplished by the specific design of the core support plate and the core restraint system.

For analysis of the control behavior of a reactor core and its behavior under accident conditions, non-steady state neutron flux, power, temperature and all feedback reactivities must be considered in detail. Negative feedback reactivities or temperature coefficients always counteract increases in power and temperature. Positive coefficients of moderator or coolant can be tolerated as long as all the other temperature coefficients, above all the sufficiently fast prompt Doppler coefficient, are negative and larger in magnitude than the positive coefficients of moderator or coolant. Table 2-5 shows typical temperature coefficients of reactivity for various types of reactors.

PWR's or BWR's have highly negative coolant or moderator temperature coefficients, whereas HTGR's at the end of their operating cycle and also large sodium cooled FBR's throughout their whole operating cycle have positive coolant temperature coefficients. Thermal reactors, such as PWR's, BWR's and HTGR's, have more negative Doppler coefficients than FBR's.

Table 2-5. *Typical temperature coefficients of reactivity for various reactor lines*

Temperature coefficient $\Delta k/K$	PWR	BWR	HTGR[a]	LMFBR
Moderator or coolant (fresh fuel)	-9×10^{-5}	-10×10^{-5}	$+0.5$ to 1.7×10^{-5}	$+5 \times 10^{-6}$
Doppler coefficient (500–2800 °C)	-1.7 to -2.7×10^{-5}	-2.5 to -1.3×10^{-5}	-4 to -2×10^{-5}	-1.1×10^{-5} to -2.8×10^{-6}

[a] end of operating cycle

2.8.3 Reactor Control and Safety Analysis

2.8.3.1 Reactivity Changes During Startup and Full Power Operation

As the reactor core is slowly being started up from zero power to full power the temperatures of coolant and core structure rise by several 100 °C. At the same time, the fuel temperature increases by more than 1000 °C. This causes a negative reactivity effect, which must be overcome by moving absorber (control) rods out of the reactor core. In LWR's, this reactivity span is in the range of several percent. In sodium cooled FBR's, it is smaller mainly because of the lower value of the negative Doppler coefficient.

The buildup of fission products and actinides as well as the burnup of fissionable isotopes leads to a reactivity loss of up to 12% in LWR's and about 3% in sodium cooled FBR's. Sufficient excess reactivity, i.e., $k_{eff} > 1$, therefore must be provided in a core with fresh (non-irradiated) fuel and zero power at the beginning of an operating cycle. At this time, the excess reactivity is counterbalanced by the insertion into the core of such absorber materials as boron, cadmium, gadolinium, or indium, which provide a sufficient reactivity span for reactor control. Due to the burnup effects as well as fission product buildup mentioned in Section 2.4, the negative reactivity must be reduced during the operating cycle. This is accomplished by several methods, e.g., withdrawing absorber rods, reducing the concentration of soluble poisons, such as boric acid, and by the diminishing absorption effect of burnable poisons, such as cadmium or gadolinium contained in fixed rods.

The fraction of excess reactivity for fissile isotope burnup and fission product buildup designed into the fresh core determines the length of operation of a core (operating cycle). This length of the operating cycle is usually chosen in the light of an optimization of core physics properties and fuel cycle economics. The reactor core is shut down by moving into the core absorber rods with sufficient negative reactivity. In this case, the reactivity span from full power (high temperature) to zero power (low temperature) has to be overcome. In addition, the reactor core must be held subcritical, which means that it has to attain and maintain a k_{eff} well below 1.

2.8.3.2 Qualitative Description of a Reactor Core
Under Transient Power Conditions

Reactivity changes of the reactor core lead to power changes. They can be described by the point kinetics equations (2-8, 2-9), which must be solved in conjunction with the respective equations for temperature and feedback reactivity changes (2-10, 2-11) by means of programs run on digital computers.

A simplified qualitative description of the behavior of a reactor core under transient power conditions can be derived from the solution of the point kinetics equations (2-8 and 2-9) for the simple case of a step function of the reactivity, ρ, neglecting the temperature and reactivity feedback effects.

In this case of a step function for the initiating reactivity, we obtain $\rho_i \cong \rho$ (Eq. 2-10 for the total reactivity) and the reactor power, P(t), can be described by the relation:

$$\frac{P(t)}{P_0} = \frac{b-c}{b-a} \exp\left(a \cdot t\right) + \frac{c-a}{c-b} \exp\left(b \cdot t\right) \qquad \text{(Eq. 2-13)}$$

where

$$a = \frac{\rho \cdot \lambda}{\lambda \cdot \beta_{eff} + \beta_{eff} - \rho}; \quad b = \frac{\rho - \beta_{eff}}{l_{eff}}; \quad c = \frac{\rho}{l_{eff}}$$

P_0 is the steady state power at $t=0$; λ, β_{eff}, l_{eff} are described in Section 2.8.1 above.

Several important reactivity and transient power ranges can be distinguished in discussing the above solution (Eq. 2-13) of the point kinetics equations:

(a) $\rho < 0$

If the reactivity step, ρ, is negative, both exponents, a and b, of Equation (2-13) are negative. The reactor power, P(t), will decrease exponentially. For a sufficiently large negative reactivity step the power decreases rapidly (prompt jump) and remains at the level of the afterheat (reactor shut down).

(b) $0 < \rho < \beta_{eff}$

For a positive reactivity step where $0 < \rho < \beta_{eff}$, the reactor core is called delayed critical. The power increase as a function of time is mainly determined by the delayed neutrons. In this case, the second term of Equation (2-13) decays rapidly, because b is negative and l_{eff} is relatively small. After a small rapid increase in the reactor power (prompt jump), its time behavior is dominated by the first term of Equation (2-13). The reactor power then increases exponentially with the time period $T = 1/a$. As long as ρ does not exceed the value of 1/2 β_{eff}, the time period $T = 1/a$ is larger than $1/\lambda = 1/0.08$, which is in the range of more than 12 seconds. This allows easy and reliable control of the reactor core by slowly moving absorbers. As the decay constants for delayed neutron precursors, λ_i, are very similar for all fissile isotopes as well as for thermal and fast neutron spectra (see Table 2-2), the design data of control systems of thermal reactors, e.g., LWR's and FBR's, are very similar.

(c) For $\rho = \beta_{eff}$, the reactor core is called prompt critical; for $\rho > \beta_{eff}$ it is called supercritical. The reactor power, P(t), in that case is dominated entirely by the prompt neutrons. In Equation (2-13) only the second term is responsible for a fast increase in reactor power. The time period $T = 1/b$ is determined

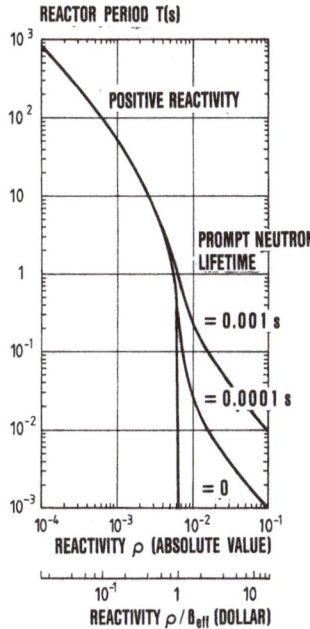

Fig. 2-14. Reactor period as a function of positive reactivity steps for a U-235 fueled reactor (J.R. Lamarsh)

by

$$T = 1/b = \frac{l_{eff}}{\rho - \beta_{eff}}$$ (Eq. 2-14)

As l_{eff} is small, the time period becomes short and the power would increase very rapidly as long as the effective reactivity input is not reduced by counteracting inherent negative feedback effects or negative reactivity effects of the shutdown system.

Fig. 2-14 shows the time period, T, as a function of positive reactivity steps, ρ, and for different prompt neutron lifetimes, l_{eff}, as a parameter. Instead of absolute values, the reactivity is usually also indicated in units of β_{eff}. The ratio of ρ/β_{eff} is defined as a reactivity unit known as a dollar.

As explained above, Eq. 2-13 allows only a simplified description of the transient power range. In power reactors, counteracting feedback reactivity effects are always involved. This is especially valid for case (c), where the core may become prompt critical or superprompt critical. Therefore, in quantitative analyses of the dynamic behavior of reactor cores, the point reactor kinetic equations are solved by means of sophisticated computer codes also describing the temperature and feedback mechanisms. If, in the course of such analysis, the core becomes prompt critical and the power increases very rapidly, also the fuel temperature increases and a negative Doppler reactivity accumulates. This Doppler reactivity decreases the total reactivity below prompt critical and the reactor power, after having attained a certain peak level, rapidly drops again. If the reactor core were not shut down shortly after this power peak and if the initiating reactivity increased further, an oscillating behavior would result. If sufficient energy had been accumulated during such a postulated excursion accident, the core temperatures would become so high that the fuel and the structural material would melt and material relocations would result. In the latter case, the core would be destroyed and the coolant vaporized and expelled from the core. The reactor core would then shut itself down by material relocations and would turn subcritical.

Such prompt critical conditions of a reactor core could only be envisaged if, in the course of an accident sequence, sufficient positive reactivity could be generated and if the diverse and redundant control and shutdown systems of a reactor core failed completely on demand. Such questions, among a number of others, are generally analyzed for each type of reactor in detailed safety analyses (see Chapters 4 and 5).

Selected Literature

American National Standard for Decay Heat Power in Light Water Reactors. LaGrange Park, Ill.: American National Standard Institute/American Nuclear Society, ANSI/ANS-5.1 (1979).

Fission Product Nuclear Data (FPND) – 1977. Proc. Second Advisory Group Meeting on Fission Product Nuclear Data, Energy Centrum Netherlands, Petten, 5–9 September 1977. Vienna: International Atomic Energy Agency, IAEA-213 (1978).

Glasstone, S., Edlund, M.C.: The Elements of Nuclear Reactor Theory. New York: Van Nostrand. 1965.

Glasstone, S., Sesonske, A.: Nuclear Reactor Engineering. New York: Van Nostrand. 1981.

Greenspan, H., et al.: Computing Methods in Reactor Physics. New York: Gordon and Breach. 1968.

Hummel, H.H., Okrent, D.: Reactivity Coefficients in Large Fast Power Reactors. Hinsdale, Ill.: American Nuclear Society. 1970.

Keepin, G.R.: Physics of Nuclear Kinetics. Reading, Mass.: Addison-Wesley. 1965.

Lamarsh, J.R.: Introduction to Nuclear Reactor Theory. Reading, Mass.: Addison-Wesley. 1966.

Lamarsh, J.R.: Introduction to Nuclear Engineering. Reading, Mass.: Addison-Wesley. 1975.

Lewis, E.E.: Nuclear Power Reactor Safety. New York: John Wiley. 1977.

Meghreblien, R.v., Holmes, D.K.: Reactor Analysis. New York: McGraw-Hill. 1960.

Michaudon, A.: Nuclear Fission and Neutron Induced Fission Cross Sections. Oxford: Pergamon Press. 1981.

Nicholson, R.B., Fischer, E.A.: The Doppler Effect in Fast Reactors. Advances in Nuclear Science and Technology, Vol. 4, pp. 109–195. New York – London: Academic Press. 1968.

Oldekop, W.: Einführung in die Kernreaktor- und Kernkraftwerkstechnik, Teil I. München: Karl Thiemig. 1975.

Reactor Physics Constants. Argonne National Laboratory, ANL-5800 (1958).

Smidt, D.: Reaktortechnik. Karlsruhe: G. Braun. 1971.

Smidt, D.: Reaktorsicherheitstechnik. Berlin-Heidelberg-New York: Springer. 1979.

Weinberg, A.M., Wigner, E.P.: The Physical Theory of Neutron Chain Reactors. Chicago, Ill.: The University of Chicago Press. 1958.

Wiese, H.W., Fischer, U.: KORIGEN – Ein Programm zur Bestimmung des nuklearen Inventars von Reaktorbrennstoffen im Brennstoffkreislauf. Kernforschungszentrum Karlsruhe, KfK-3014 (1981).

3 Nuclear Fuel Supply

3.1 Introduction (The Nuclear Fuel Cycle)

The exploitation of nuclear power in power plants begins with the only fissile isotope occurring naturally, U-235. This isotope is contained in natural uranium in an abundance of 0.72%, the balance being 99.28% of U-238. Natural uranium can be found in uranium ores in varying concentrations ranging from fractions of a percent up to several percent. It must be extracted as uranium oxide by open pit mining or underground mining and subsequent ore dressing (Fig. 3-1). Since most reactor types require U-235 fuel with a low enrichment, uranium oxide is converted into gaseous uranium hexafluoride (UF_6) and raised to the desired enrichment level in isotope enrichment plants. This produces depleted

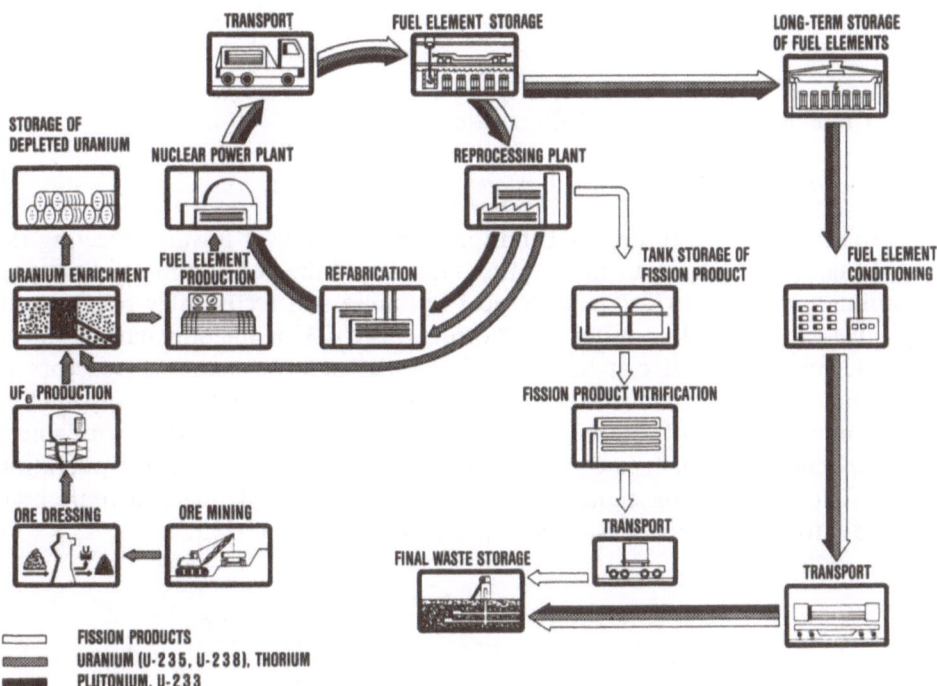

Fig. 3-1. Nuclear fuel cycle options (KFA Jülich)

uranium with a U-235 content of approx. 0.2%, which is first stored and can later be used as a fuel, e.g., in fast breeder reactors. Enriched UF_6 is reconverted into UO_2 in a chemical conversion process, fabricated into cylindrical pellets, which are stacked in zircaloy or steel tubes and then assembled into fuel elements. These fuel elements are loaded in the core of the reactor plant, where they generate nuclear power through fission processes. In this process, the enrichment in the U-235 isotope is reduced continuously, while radioactive fission products and transuranium isotopes are produced. After having generated energy in the core, the fuel elements are unloaded and, after short interim storage in the reactor plant, shipped to an intermediate storage facility for spent fuel.

After several years of storage of the spent fuel elements, two basic options are open for further treatment:

(1) The fuel elements can be directly put into a repository in deep geological strata after prolonged temporary storage and special conditioning (see Section 7.6.2).

(2) In a chemical reprocessing plant, the fission products and higher actinides can be separated from uranium and plutonium or U-233 in the spent fuel elements. After special waste treatment, the fission products and higher actinides can be shipped to a repository and stored in deep geological strata. The valuable uranium and plutonium can be recycled and reused as new fuel for energy production in the cores of nuclear reactors (see Sections 7.2 and 7.3).

In the following sections of this Chapter, the front end of the fuel cycle will be treated, while the different reactor lines will be described in Chapters 4 and 5. The remaining part of the fuel cycle (the back end) will be outlined in Chapters 6 and 7.

3.2 Uranium Resources and Requirements

3.2.1 Uranium Consumption in Various Reactor Systems

Projections of the future contribution of nuclear power towards meeting the world energy requirement invoke the question of the consumption of natural uranium, and the worldwide geological resources and recoverable reserves, respectively, of natural uranium and thorium. As U-235 is the only natural heavy atomic nucleus which can be split by thermal neutrons, nuclear reactors in the initial phase of a nuclear power economy must be operated on nuclear fuel containing the U-235 uranium isotope in its natural, or a higher, enrichment (isotope enrichment). It is only by nuclear reactions during reactor operation that, following neutron capture, U-238 and Th-232 (fertile materials) are converted into artificial fissile isotopes, such as Pu-239 and U-233:

$$^{238}_{92}U \xrightarrow{(n,\gamma)} {}^{239}_{92}U \xrightarrow[23.5\,min]{\beta^-} {}^{239}_{93}Np \xrightarrow[2.35\,d]{\beta^-} {}^{239}_{94}Pu$$

$$^{232}_{90}Th \xrightarrow{(n,\gamma)} {}^{233}_{90}Th \xrightarrow[22.1\,min]{\beta^-} {}^{233}_{91}Pa \xrightarrow[27.0\,d]{\beta^-} {}^{233}_{92}U$$

The artificial fissile isotopes, U-233 and plutonium can be chemically separated from the spent fuel (reprocessed) and returned (recycled) to the nuclear reactors either together with or instead of U-235.

In principle, any type of reactor can be operated on U-235 fuel. However, the development of nuclear power so far has shown that the initial phase was dominated by three different reactor lines (see Table 1-1). LWR's have gained a significant lead over HWR's in the nuclear power market, and gas-graphite reactors (GGR's, AGR's), originally the market leaders, now only play a minor role. Among the advanced reactors it is the Pu recycling converter (PWR-Pu) and the FBR which can reuse the plutonium generated in today's "convention-al" reactors that have received the greatest attention. In principle, however, plutonium can also be recycled in HWR's or AGR's. Within the development of HWR's and HTGR's thorium has been considered as a fertile material. The artificial fissile U-233 produced in this case can be burnt in advanced HWR's and high temperature thorium reactors (HTR-Th), which achieve good fuel economies.

Natural uranium consumption in various reactor types is covered in detail in Chapter 6. At this point, only a few figures will be singled out to explain the orders of magnitude involved.

The consumption of natural uranium in the present LWR design is still rela-tively high: 4220 te/GW(e) over thirty years of operation at an average load factor of 0.7. The consumption in the AGR is even slightly higher, amounting to 4490 te/GW(e). The HWR burns 3720 te/GW(e) of natural uranium in thirty years of operation. Fuel recycling reduces the natural uranium consumption of LWR's and HWR's to levels around 2700 and 1500 te/GW(e), respectively, over thirty years of operation.

Advanced reactors have even lower natural uranium consumption levels. They attain those levels with optimized reactor cores and by recycling the artificial fissile fuels (for details, see Chapter 6).

FBR's do not need natural uranium when started up with plutonium, for instance from LWR's. In that case, FBR's only need some 35 te of U-238 over an operating period of thirty years (see Chapter 5). This U-238 is available in the large quantities of depleted uranium accumulating as waste from the enrichment of fuel for LWR's. However, also for FBR's, fuel reprocessing and recycling is an absolute necessity.

On the basis of the aforementioned data the consumption of natural uranium can be determined for various nuclear energy scenarios and reactor types. As an example of such considerations, the following reactor strategies will be taken into account:

– LWR (no recycling of fuel)
– HWR (no recycling of fuel)
– LWR with recycling of self-generated plutonium
– HWR with recycling of self-generated plutonium or U-233
– LWR in symbiosis with FBR to be introduced from the year 2000 on.

For the energy scenario, the "INFCE low" curve shown in Fig. 1-1 will be assumed. In addition, two variants will be considered for some reactor types. One variant is based on present technology, while the other variant assumes

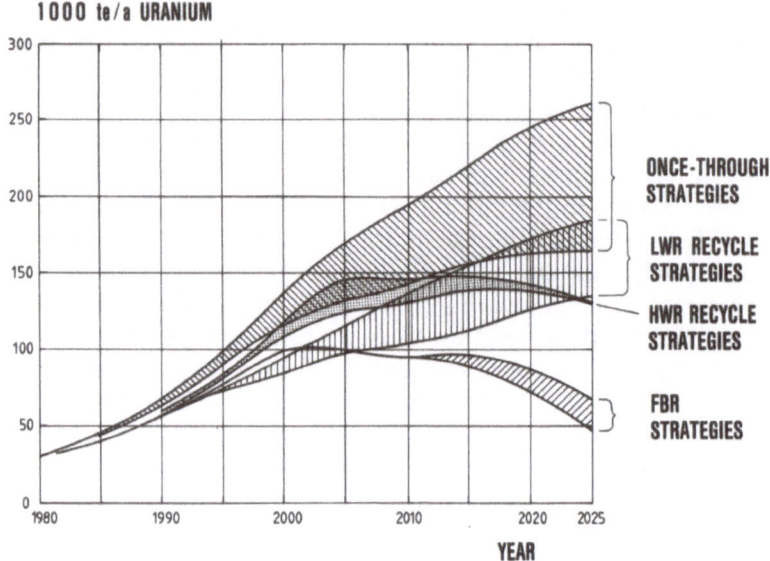

Fig. 3-2. Annual natural uranium requirements of representative reactor strategies (INFCE low for WOCA) (INFCE)

Table 3-1. *Cumulative natural uranium requirements of representative reactor strategies (Nuclear power growth according to INFCE low) (INFCE)*

Reactor strategy	Cumulative natural uranium requirements (10^6 te U)	
	2000	2025
Once-through LWR	1.5–1.6	5.4–6.8
Once-through HWR	1.5	5.2–5.8
LWR with recycle	1.3	4.0–5.0
HWR with recycle	1.4	4.7–4.9
Large scale FBR	1.3	3.5–3.7

a technically upgraded reactor design. Consequently, a band with an upper and a lower limit of natural uranium consumption will result.

Fig. 3-2 shows the annual uranium requirements of the different reactor types up to the year 2025 for the energy projection mentioned above. Table 3-1 adds the corresponding cumulative natural uranium requirements. It is seen that, without reprocessing of the fuel, the LWR has the highest annual uranium requirement. By the year 2025, this type of reactor will have accumulated a total requirement of 5.4–6.8 million tonnes for the "INFCE low" forecast. In 2025, "INFCE low" will require an annual uranium production capacity of 190,000–260,000 tonnes. HWR's without fuel reprocessing have a slightly lower natural uranium consumption.

Reprocessing and recycling of the fuel can help to curb natural uranium consumption. However, a major cutback in the consumption of natural uranium can be achieved only by the introduction of the FBR around the year 2000. In the best case, the breeding characteristics of FBR's could reduce to zero the annual natural uranium requirement some 35 years after their introduction. For the time afterwards, the depleted uranium left over from uranium enrichment will be sufficient to produce fission energy in breeder reactors for a long time (see Section 5.1).

3.2.2 Available Uranium and Thorium Reserves

3.2.2.1 Worldwide Available Uranium Reserves

In the previous Section, the quantities of natural uranium required for some representative strategies of reactor deployment have been discussed; now it is necessary to contrast the worldwide available uranium reserves with those findings.

Fig. 3-3 indicates the locations of important uranium deposits in the world. Uranium was discovered first in Czechoslovakia in the pitchblende mineral and was found also in Africa and northwestern Canada. Pitchblende is a dense, impure form of uranium oxide, which occurs mixed with ores of copper, tantalum or other minerals.

Natural uranium ores are also found in sandstone deposits. The most important deposits of this type occur in the USA, Gabon and Niger. The uranium content of the mineralized rock is commonly between 0.04 and 0.25% of U_3O_8. The size of the deposits ranges between a few hundred tonnes and over 60,000 tonnes of uranium. At Elliot Lake in Canada and the gold-uranium deposits of Witwatersrand in South Africa, uranium can also be found in quartz pebble conglomerates. Their uranium content varies between 0.03 and 0.13%. Deposits may be as large as 75,000 tonnes of uranium. Another type found at Lake Athabasca, Canada, and at Alligator Rivers of Australia is called proterozoic

Fig. 3-3. Worldwide uranium deposits

Table 3-2. *Uranium reserves in WOCA (OECD/IAEA)*

	Reasonably assured reserves (10^3 te U)		Estimated additional reserves (10^3 te U)	
	< $ 80/kg U	$ 80 to $ 130/kg U	< $ 80 kg/U	$ 80 to $ 130 kg/U
Africa				
Algeria	26	0	0	0
Namibia	119	16	30	23
Niger	160	0	53	0
South Africa	247	109	84	91
Others	39	9	2	18
Total	591	134	169	132
North America				
Canada	230	28	358	402
USA	362	243	681	416
Total	592	271	1039	818
South America				
Brazil	119	0	81	0
Others	28	5	7	19
Total	147	5	88	19
Asia	40	11	1	24
Australia	294	23	264	21
Europe				
France	59	16	28	18
Sweden	0	38	0	44
Others	24	48	15	38
Total	83	102	43	100
Grand total	1747	546	1604	1114

unconformity. These deposits range up to 150,000 tonnes of uranium with contents up to several percent of uranium. Other deposits are associated with igneous and metamorphic rocks, such as granite-pegmatites or carbonatites. Such deposits are known to be located in Namibia, Greenland, Alaska and Brazil. Calcrete deposits have been discovered in Namibia and Australia. They contain up to 40,000 tonnes of uranium at contents of 0.13%.

Uranium reserves are commonly classified by the cost of uranium recovery, which includes direct costs of mining and processing and cost of the production unit. It is commonly accepted to use two cost categories: up to $ 80/kg U and up to $ 130/kg U*. However, it should be noted that these cost categories do not reflect the prices at which uranium will be available to the user. Uranium reserves are also classified by the extent of geologic knowledge and the confidence in the estimates. The two categories commonly applied to this classification are "reasonably assured" and "estimated additional" reserves.

Table 3-2 shows the uranium reserves currently known to exist in the Western world.

They include 1.75 million tonnes of reasonably assured reserves in the cost category up to $ 80/kg U and 0.55 million tonnes in the cost category of $ 80–130/kg U. In addition, there are estimated additional reserves of 1.60 million tonnes in the cost category of up to $ 80/kg and 1.10 million tonnes in the cost category of $ 80–130/kg U. This adds up to 2.30 million tonnes of reasonably assured reserves and 2.72 million tonnes of estimated additional reserves. The largest uranium reserves exist in the USA, in Canada, Africa and Australia. Europe and Japan have only minor uranium ore reserves. No official estimates are available about the USSR, China and other countries with centrally planned economies.

In addition to the uranium reserves known today, speculative reserves also play a role in resource estimation. Unlike the known reserves, speculative reserves either have not yet been localized or no ideas have yet been developed about ways and means of making them available. Discussions about speculative reserves are based on the assumption that the reserves presently known are not likely to make up all deposits actually existing in nature. To put the concept of speculative reserves into proper focus, geological studies have been conducted and estimates made by IAEA/NEA.

These estimates, which are summarized in the IUREP study (*International Uranium Resource Evaluation Project*), are represented in Table 3-3. According to those data, between 6.6 and 14.8 million tonnes of speculative reserves exist in WOCA (*World outside CPE/centrally planned economies/areas*) which could be extracted in the cost category up to $ 130/kg U.

Eastern Europe, the USSR and China have an additional estimated potential of speculative reserves of 3.3 to 7.3 million tonnes. The high uncertainty involved in these estimates is evident from the large difference between the upper and the lower bounds. Therefore, the amounts of speculative reserves indicated in Table 3-3 are meant to convey only a general impression of the order of magnitude of "speculative reserves" existing in the world. It is also important to emphasize that even if these "speculative reserves" really exist, there is no guarantee that they will be discovered, or if discovered, can be made available.

Low-grade reserves with U contents down to 50 ppm U in cost categories higher than $ 130/kg U are also discussed in the literature. These reserves include uranium in sandstones, granites, carbonatites and phosphate rock. The uranium content in phosphate rock ranges between 50 and 130 ppm U.

If uranium were extracted from such low-grade ores with 70 ppm U content (e.g., Chattanooga shales in the USA) and utilized in LWR's, the total amount

---

* Cost basis in 1981 US dollars.

Table 3-3. *Speculative uranium reserves (INFCE)*

Continent	Number of countries	Speculative reserves (10^6 te U)
Africa	51	1.3– 4.0
North America	3	2.1– 3.6
South and Central America	41	0.7– 1.9
Asia and Far East[a]	41	0.2– 1.0
Australia and Oceania	18	2.0– 3.0
Western Europe	22	0.3– 1.3
Total	176	6.6–14.8
Eastern Europe, USSR, People's Republic of China	9	3.3– 7.3[b]

[a] Excluding People's Republic of China and the eastern part of the USSR.
[b] The potential shown here is the 'Estimated Total Potential' and includes an element for 'Reasonably Assured Reserves' and 'Estimated Additional Reserves.'

of energy which could be extracted would be only about seven times that needed to mine, process, convert, enrich and fabricate the uranium fuel. For 20 ppm low-grade ores (granites), this factor would be around 2. This makes mining such ores doubtful merely from the "energy harvesting" point of view. In addition, the extraction of uranium from such low grade ores may involve environmental impacts of such magnitude that the scale of necessary mining operations becomes unrealistic.

For reasons of completeness, it must be mentioned that there is uranium also in seawater. The average concentration is 3.4 µg/l. The extraction technology is still highly uncertain and the estimated cost range is above \$ 600/lb U_3O_8 (\$ 1550/kg U) for terrestrially sited recovery systems and above \$ 300/lb U_3O_8 (\$ 775/kg U) for seaborne recovery systems. In the light of the above considerations, the extraction of uranium from seawater is not envisaged at the present time.

3.2.2.2 Uranium Production

On the basis of the uranium reserves described above, it is mainly the existing production capacity of uranium mines, which controls the availability of uranium on the world market. In 1979/80, the main uranium producers of the Western world were the United States of America, Africa, and Canada (Fig. 3-4). In Europe, France ranked at the top with an annual production of roughly 3500 tonnes of natural uranium.

Future increases in uranium production will depend on a variety of parameters. Capital must be available for exploration as well as uranium mine and mill constructions. Markets and prices must be attractive for investors. In the USA, Canada, Australia, and Africa, expansions of uranium mines and mills

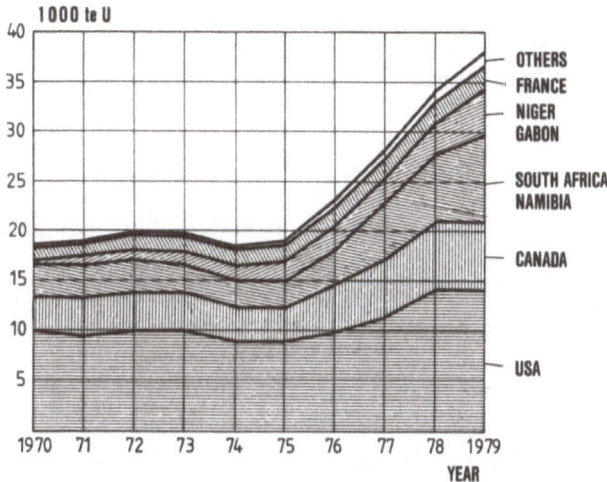

Fig. 3-4. Worldwide uranium production 1970–1979 (NUKEM)

Table 3-4. *Future uranium production capability in WOCA (Based on "reasonably assured" reserves) (INFCE)*

	Uranium production capability (te U/a)			
	1985	1990	1995	2000
USA	31,400	40,800	46,700	51,600
Canada	14,400	15,500	15,400	12,500
South Africa	10,600	10,400	10,000	10,000
Niger	6,000	8,500	10,200	5,500
Namibia	5,000	5,000	4,600	4,600
Australia	12,000	20,000	17,000	10,000
France	4,000	4,400	3,100	1,600
Remaining producers (Europe, Asia, South America)	6,800	6,300	10,000	11,000
Total	90,200	110,900	117,000	106,800

as well as the opening of new uranium mines are planned or underway. A projection of possible future uranium production capabilities was made within INFCE and is displayed in Table 3-4 for the main uranium producing countries.

Besides the USA and Africa it is above all Canada and, increasingly, Australia which will contribute to production increases. By 1985, uranium will be produced mainly from reserves in the cost category up to $ 80/kg U. After 1990, however, larger quantities of uranium in the cost category up to $ 130/kg U will also have to be incorporated in the production and will come onto the uranium market. Beyond 1990, already roughly one third and, after 1995, roughly half of the uranium production will have to come from estimated additional reserves (see Table 3-2). Under optimum economic and political conditions, the maximum estimated uranium production in the period 1990 to 2000 may reach an annual 120,000 tonnes of natural uranium. Increasing the uranium production above this level would be possible only if major contributions came from the "speculative reserves."

3.2.2.3 Thorium Reserves

Thorium contains only fertile material, no fissile isotopes. This assigns thorium reserves a quality different from that of natural uranium reserves. Thorium can be compared with U-238. However, the limits to the reserves of natural uranium when used in the present line of converter reactors are imposed by U-235.

Thorium can be recovered as a byproduct from minerals mined for the extraction of titanium, tin and zirconium. Monazite is the main thorium-bearing mineral ($ThSiO_4$). Most of the reasonably assured reserves are located in India, Brazil and the USA, with large estimated additional reserves in Canada, Egypt, Australia and the USA.

Monazite sands in India, Brazil, Australia and Egypt contain 4.6–7% thorium. In the United States and Canada, thorium reserves are also found in vein deposits. Table 3-5 indicates the reasonably assured and the estimated additional reserves in the world for two cost categories. Opportunities exist to expand known reserves and discover new areas. However, due to limited markets for thorium there has been little exploration to discover new reserves.

The world production of thorium in 1978–1980 was estimated at some 150 tonnes of ThO_2 per annum. However, the world monazite production was about 16,000 tonnes per annum, which would have allowed an annual production of 800 tonnes of ThO_2.

Table 3-5. *Thorium reserves (OECD/IAEA)*

	Reasonably assured reserves (10^3 te)		Estimated additional reserves (10^3 te)	
	< $ 75/kg Th	> $ 75/kg Th	< $ 75/kg Th	> $ 75/kg Th
Brazil	68	0	1200	0
Canada	0	0	293	0
Denmark (Greenland)	0	54	0	32
Egypt	15	0	280	0
Finland	0	0	60	0
India	319	0	0	0
Norway	132	0	132	0
Turkey	0	330	0	440
USA	108	14	261	17
Others	33	0	30	0
Total	675	398	2256	489

In addition to monazite sand, thorium also occurs in uranium ores processed for uranium production. The thorium recoverable from such uranium mines in Canada has reached a production level of several thousand tonnes per annum. Further increases should be attainable from uranium deposits specially mined for their thorium value.

3.2.3 Uranium Requirement vs. Uranium Reserves

In Chapter 1, the future nuclear power requirement has been found to depend on a number of parameters adding to the range of uncertainty inherent in requirement forecasts over a span of fifty years. This chapter has shown similar ranges of uncertainty to exist in assessments of the natural uranium reserves. Comparisons of uranium requirements (see Fig. 3-2 or Table 3-1) and uranium reserves (see Tables 3-2 and 3-3) indicate that strong requirements for nuclear power, e.g., nuclear energy forecasts according to "INFCE high" (see Fig. 1-1), and the use of reactors with relatively high uranium consumptions (LWR or HWR without recycling) would mean that the natural uranium reserves known today would have been spent already between 2000 and 2025. The speculative reserves would have to be available by then. This would require extraordinary efforts of exploration and production to be made by the uranium mining industry. But also for requirement forecasts according to "INFCE low," the uranium production capacities (see Table 3-4) and the capacities of enrichment plants (see Section 3.4) would have to be stepped up considerably after the year 2000, if LWR's and HWR's without recycling were to be used. The situation improves slightly, if the fuel is recycled in these reactors. Advanced converter reactors improve conditions even further. However, a fundamental improvement will be brought about only by the use of breeder reactors. In that case, the annual natural uranium requirement will be limited to some 100,000 tonnes of natural uranium in "INFCE low." After having passed through this maximum, the annual natural uranium requirement will eventually converge towards zero. With a commercial introduction of FBR's around 2000–2010, the cumulative requirement will be limited to some 3.0–4.5 million tonnes of natural uranium. The uranium reserves presently known in the "reasonably assured" and "estimated additional" categories up to a cost category of US $ 130/kg U would be sufficient for the "INFCE low" requirement forecasts. Accordingly, the transition from the present once-through LWR to advanced converters and breeder reactors would have to be made around the year 2000. This applies in particular to countries without uranium reserves and enrichment plants of their own which, therefore, need to purchase natural or enriched uranium on the world market. In this way, high converters and breeder reactors can greatly relieve the tension in the uranium market and clearly add to the continuity of supply in a number of countries.

3.3 Concentration, Purification and Conversion of Uranium

Uranium ores usually contain only a few tenths of a percent of uranium and must be separated from byproducts to reduce the weight for subsequent

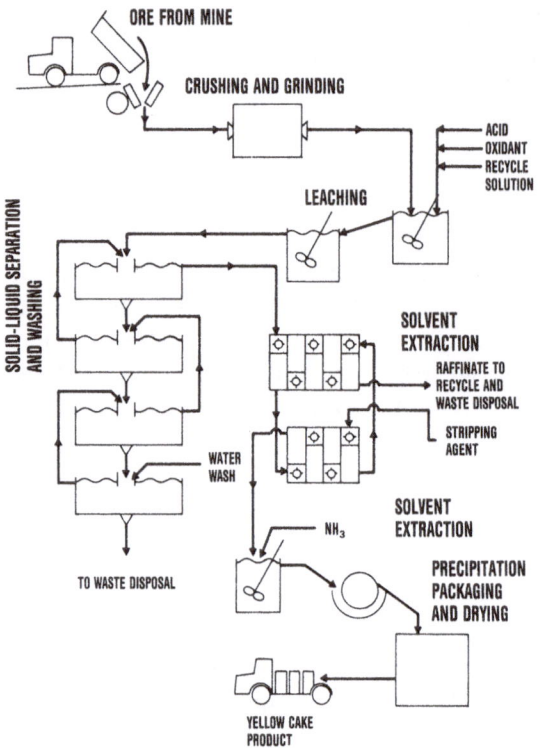

Fig. 3-5. Flowsheet of uranium ore processing (D.C. Seidel)

shipment. This concentration stage is predominately performed by leaching processes. However, physical concentration methods are also applied including crushing and sizing, gravity, magnetic, electrostatic and flotation types of separation. Roasting may be employed to improve the solubility of the uranium. The type of leaching agent used depends on the uranium-bearing mineral and can be either sulfuric, nitric, hydrochloric acids or alkaline carbonates. If acid solutions are employed, solvent extraction or ion exchange are the preferred methods.

Fig. 3-5 shows a uranium ore processing flowsheet for acid leaching. The ore arriving from the mine is crushed and then ground to the consistency of fine sand. If wet grinding is used, the resulting slurry is fed to a leaching circuit where acid is added. With many ores, an oxidant must be added to convert the uranium to the hexavalent state, which is readily soluble. After leaching, the solids and liquids are separated. The solids are washed to recover the adhering leach solution. Uranium is extracted from the leach solution by solvent extraction or ion exchange. In the solvent extraction process, the active agent is usually an organic amine salt diluted in kerosene that can selectively extract the uranium ions into an organic complex insoluble in water. This organic phase is separated from the aqueous phase by continuous settling. Afterwards, the uranium is again stripped from the organic complex by contacting it with an inorganic salt solution. The uranium is then precipitated from the strip solution and the resultant concentrate (yellow cake) is dried and packaged for shipment to a refinement plant.

For its application as a reactor fuel, these natural uranium concentrates must then be purified. For purification, uranium concentrates are dissolved in nitric acid. The resulting uranyl nitrate is then extracted in a solution of tributyl phosphate in kerosene. This purification step removes to a level of a few ppm such neutron absorbing elements as boron, cadmium and the rare earths. In addition, the contents of many other elements must be below rather low levels. The product of purification is usually one of the oxides of uranium, UO_2, UO_3 or U_3O_8.

These uranium oxides are converted into gaseous uranium hexafluoride, UF_6, for the necessary enrichment steps in isotope separation plants. Uranium hexafluoride, UF_6, is the most volatile compound of uranium, with a boiling point (sublimation point) of 56.5 °C at 1 bar.

3.4 Uranium Enrichment

3.4.1 Introduction

For neutron physics reasons, the use of natural uranium as a reactor fuel requires the presence of weakly neutron absorbing materials as moderators, coolants and structural materials. Such materials, for instance, are heavy water, graphite or gaseous media. Conversely, if steel is used as a structural material, or normal light water as a moderator and coolant, it is not possible to burn natural uranium as a nuclear fuel.

For practical purposes, this means that heavy water reactors and gas-graphite reactors (e.g., those of the MAGNOX type) are the only reactor lines to be run on natural uranium. All other types of reactors, especially light water reactors, need the uranium fuel to be present with higher enrichments of uranium-235.

LWR's and AGR's require *low enriched uranium* (LEU) with an enrichment in the range of 2–3% of uranium-235. Separative work requirements are on the order of 100 tonnes of *separative work units* (SWU) per GW(e)·a of installed nuclear power. Advanced reactor types, e.g. HTGR's, need fuel with 8–20% U-235 enrichment (*medium enriched uranium*: MEU). Advanced converters with high conversion ratios operating in the uranium-thorium fuel cycle even need uranium fuel with a uranium-235 enrichment of 93% (*highly enriched uranium*: HEU). More detailed data are given in Chapter 6.

3.4.2 Designs of Enrichment Plants

Isotope separation plants contain as an elementary unit the "separating element," in which the feed material is fractionated into a "head fraction" enriched in the desired isotope and a "tails fraction" depleted in the same isotope (Fig. 3-6a).

One or more separating elements connected in parallel are called a "stage" (Fig. 3-6b). In all elements of one stage, the feed, head, and tails have the

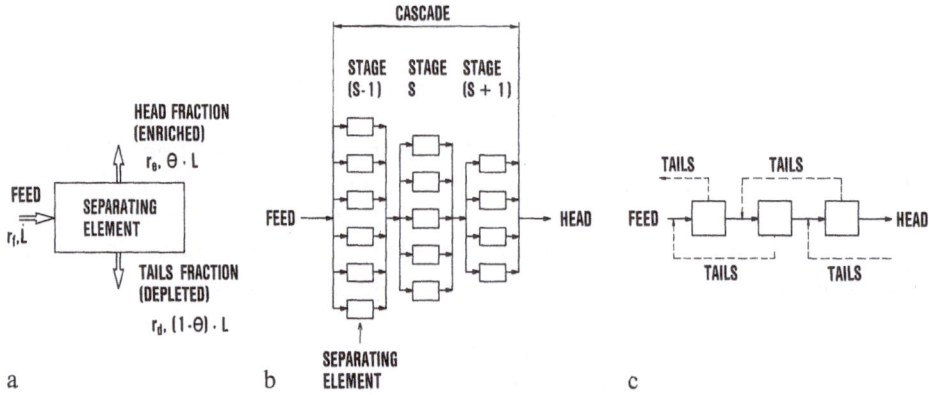

Fig. 3-6. Separating element, stage and cascade arrangement in enrichment plants (S. Villani)

same isotopic composition. The desired isotope concentration (enrichment) can be achieved by connecting many stages in series. This stage arrangement is known as a "cascade." In isotope separation plants, a countercurrent flow cascade is generally adopted in which the tails fraction of each stage is subjected to further fractionation in the next lower stage (Fig. 3-6c).

The effectiveness of the enrichment process is expressed in terms of the separation factor. If r designates the ratio of the numbers of U-235 atoms to U-238 atoms (abundance ratio) in the uranium (UF_6) gas, and if r_e defines this ratio at the head (enriched) side, r_d at the tails (depleted) side and r_f at the feed side of the separation unit or stage, the separation factor is given by

$$f = r_e/r_d$$

and the enrichment factor by

$$\alpha = r_e/r_f.$$

The feed stream, L, is divided into the head stream, $\theta \cdot L$, and the tails stream, $(1-\theta)L$. The ratio of head to feed streams is known as the cut, θ. In the case of natural uranium as feed, the abundance ratio starts at $r_f = 0.0072$ (0.72% U-235) and a certain number of stages of cascade arrangement are needed to achieve the desired enrichment. The tails assay, r_d, of present enrichment plants is 0.2–0.3%. Lower tails assays down to the range of 0.1% are under discussion for future enrichment activities and would lead to lower natural uranium requirements (see Section 3.4.7).

The separative work is a measure of the amount of work necessary in the enrichment plant to produce a certain amount of enriched uranium. It has the dimension of mass and is indicated in kg SWU or tonne SWU. Similarly, the separative capacity of an enrichment plant is generally given in tonnes SWU/a. The energy consumption for a separative work unit is given in kWh/kg SWU.

Uranium enrichment is now performed almost exclusively by the gaseous diffusion process originally developed in the United States of America within the framework of the Manhattan Engineer District project. The gas ultracen-

Table 3-6. *World enrichment capacity (in 1000 te SWU/a (INFCE)*

	1980	1985	1990/95
Diffusion			
USA	10.5	25.6	25.6
France	6	10.8	10.8–20.8
USSR	3.9	3.4	2.4
Centrifuge			
Germany, UK, Netherlands	0.5	2.0	10.0–17.5
USA	–	–	4 –8.8
Japan	0.02	0.35	2.5–5.5
Aerodynamic			
Brazil-Germany, South Africa	–	0.5	0.5
Total Western World	17	39.25	53.4–78.7
Grand total	20.9	42.65	55.8–81.1

(Figures for the years 1990/95 include commited and planned capacities)

trifuge process is in the stage of commercialization in Europe, the USA and Japan. Other methods, such as the separation nozzle method or the Helikon method (both aerodynamic processes), have reached the technical demonstration stage. Laser isotope separation, chemical isotope separation and plasma isotope separation are still in the phase of scientific studies.

Table 3-6 shows the existing and planned enrichment capacities in the world. It indicates that the overwhelming percentage of enrichment capacity (98%) in 1980 was in gaseous diffusion plants. The contribution made by gas ultracentrifuge plants will not become noticeable before 1985–1990.

3.4.3 Uranium Enrichment by Gaseous Diffusion

The principle of enrichment by gaseous diffusion (Fig. 3-7) is the phenomenon of molecular diffusion through the micropores in a membrane (barrier). In a closed cell in thermal equilibrium all molecules of a gas mixture have the same average kinetic energy. Hence, the lighter molecules of $^{235}UF_6$ travel faster on the average and strike the cell wall more frequently than the heavier $^{238}UF_6$ molecules. Micropores in the cell wall will therefore preferably allow the passage (diffusion) of the lighter $^{235}UF_6$ molecules. The remaining non-diffused mixture is then depleted in the lighter isotope.

The elementary unit of the gaseous diffusion process is a diffusion cell (diffusor) divided into two compartments by a porous barrier. A compressor maintains a steady pressure at the cell inlet. A heat exchanger removes the heat of compression of the uranium hexafluoride gas. On the depleted gas side a

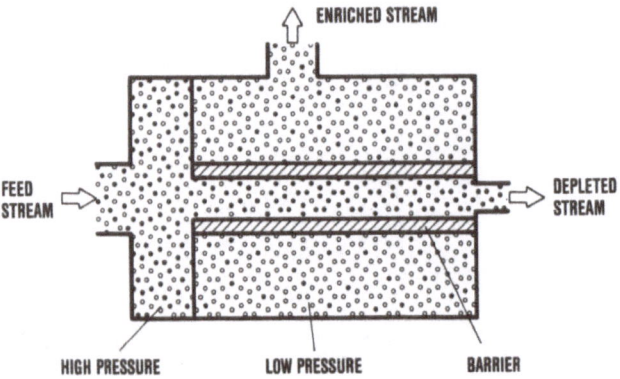

Fig. 3-7. Schematic of a gaseous diffusion cell (W. Ehrfeld)

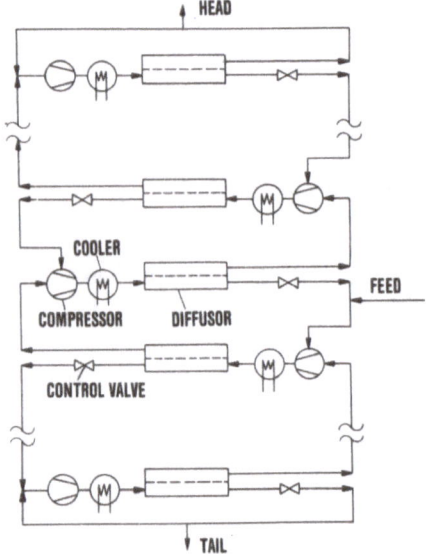

Fig. 3-8. Schematic of a gaseous diffusion cascade (W. Ehrfeld)

regulation valve controls the flow rate of the process gas (Fig. 3-8). The elementary enrichment factor of a gaseous diffusion cell is low (ideal enrichment factor $\alpha = 1.0043$). Gaseous diffusion plants therefore typically need 1000–1500 stages to produce low enriched uranium for LWR's. Several thousand stages are needed in series to produce highly enriched uranium.

Fig. 3-9 shows the different components of a large stage of a US gaseous diffusion plant. Multistage axial compressors are used. Protection of the process gas, UF_6, against leakage from the compressor or against the entrance of air into the compressor is achieved by special seals using nitrogen under overpressure. The barrier holding diffusor resembles the design of a heat exchanger. The cooler is integrated in the diffusor design.

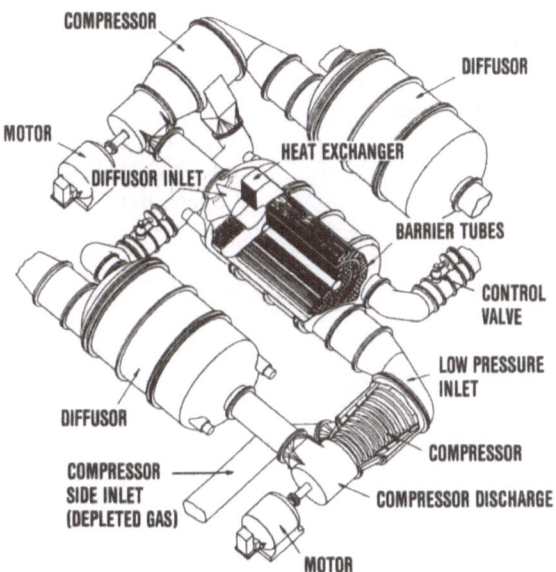

Fig. 3-9. Components of a large stage in a US enrichment plant (S. Villani)

The barrier design in diffusion cells must satisfy a number of conflicting requirements. It must be chemically resistant against the highly reactive uranium hexafluoride. It must be as thin as possible, have pores with very small radii of 100 Ångstrom or less and, at the same time, have a high porosity (more than 10^9 pores/cm^2). In addition, it must have sufficient mechanical strength at operating temperatures around 80 °C and a differential pressure on both sides of about 1 bar. Two kinds of barrier design are feasible:
– film type barriers, where pores are generated by leaching a certain constituent of a well-dispersed alloy or by electrolytic etching of thin aluminum foils,
– aggregate type barriers, where pores are generated by sintering powders (e.g., nickel or alumina powders).
The barriers of present US gaseous diffusion plants are sintered nickel tubes assembled in tube bundles housed in cylindrical diffusors (see Fig. 3-9).
The three gaseous diffusion plants presently in operation in the USA at Oak Ridge, Paducah and Portsmouth represent the greatest part of separative work capacity available on the market for years. France completed construction of its commercial gaseous diffusion plant at Tricastin in 1982 (see Table 3-6). The USSR offers separative work capacity of 3900 te SWU/a on the world market. Estimates in the literature indicate a total available enrichment capacity in the USSR of 7,000–10,000 tonnes SWU/a.
The specific power consumption of today's gaseous diffusion plants is on the order of 2400 kWh/kg SWU; this is relatively high, compared to the centrifuge process (cf. next Section). Gaseous diffusion, on the other hand, requires relatively low specific capital costs of $ 450–600/kg SWU/a (1979 dollars) compared to some $ 600–800/kg SWU/a (1979 dollars) for gas ultracentrifuge plants.

Table 3-7. *Comparison of technical data of commercial size gaseous diffusion and gas ultracentrifuge plants (INFCE)*

Process Plant name		Gaseous diffusion EURODIF/Tri-castin	Gas ultra-centrifuge URENCO/Almelo-Capenhurst
Capacity	1000 te SWU/a	10.8	1
Minimum commercial plant size	1000 te SWU/a	6	0.5–1
Enriched product assay	%	3.25/2.6/1.9/1.5	≈ 3
Feed material assay	%	0.72	0.72
Feed rate	te U_{nat}/a	14,050	1272
Tails assay	%	0.2	0.2
Tails production rate	te U/a	11,260	1040
Number of stages in series		1400	12
Number of elementary separations (units)		1400	tens of thousands
Power requirements	MW(e)	3000	20
Site requirements	10^4 m^2	250	15

The minimum plant size for commercial operation of gaseous diffusion plants appears to be in the range of 5000–6000 te SWU/a, while it is only some 1000 te SWU/a for gas ultracentrifuge plants. Table 3-7 compares the main technical data of a commercial size gaseous diffusion plant with those of a gas ultracentrifuge plant.

3.4.4 Gas Ultracentrifuge Process

The gas ultracentrifuge process is based on the separation effect of UF_6 isotopes in a strong centrifugal field, suitably combined with the cascading effect of a countercurrent axial flow circulation within the centrifuge. This process has been developed in several countries since the mid-1950's and 1960's. It is now being applied on an industrial scale, representing the strongest competitor of the gaseous diffusion process.

Within the high speed centrifuge, the centrifugal force causes the heavier isotopes, $^{238}UF_6$, to move closer to the wall of the centrifuge than the somewhat lighter isotopes, $^{235}UF_6$. This produces partial separation of the isotopes in the radial direction. Depending upon the peripheral speed (400–700 m/s), very high pressure ratios between the centrifuge axis and centrifuge wall are attained (Table 3-8). As this generates vacuum conditions at the centrifuge axis, uranium

Table 3-8. *Elementary separation factor of uranium isotopes and UF$_6$ pressure ratio for various peripheral speeds for a given gas ultracentrifuge (T = 310 K) (S. Villani)*

Peripheral speed (m/s)	Elementary separation factor	Pressure ratio between axis and wall
400	1.0975	5.5×10^4
500	1.156	2.5×10^7
600	1.233	4.6×10^{10}
700	1.329	3.3×10^{14}

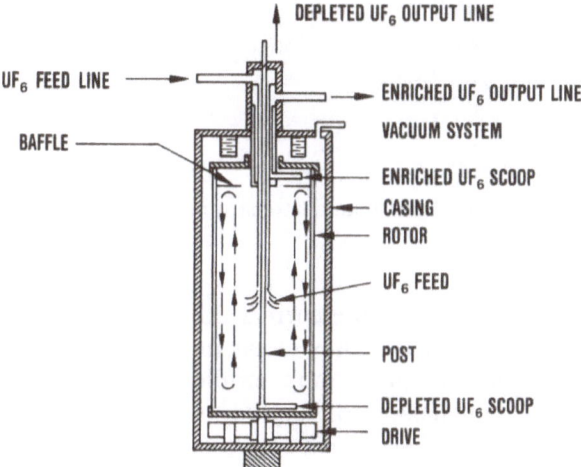

Fig. 3-10. Schematic representation of a countercurrent gas ultracentrifuge (S. Villani)

hexafluoride gas must be extracted close to the centrifuge wall. The elementary separation factor indicated in Table 3-8 not only depends upon the peripheral speed, but also upon the temperature. Ultracentrifuges of present enrichment plants are operated with an internal countercurrent flow. Superposition of the axial countercurrent flow on the wall flow considerably increases the elementary separation effect of a centrifuge. In this way, the stage (centrifuge) separation factor attains a range of 1.2 to 1.5. For the production of low enriched uranium of about 3% U-235 enrichment, therefore, only 12 stages are needed. A schematic representation of an ultracentrifuge with an internal countercurrent flow is shown in Fig. 3-10.

The gaseous uranium hexafluoride, UF$_6$, feed is introduced into the centrifuge through a tube on the axis. Enriched UF$_6$ is extracted from the upper part near the wall through a scoop resembling a small Pitot tube arranged perpendicular to the rotor axis. On the enriched side, this scoop is enclosed in a small chamber, which permits the extraction of the enriched gas through a screen while preventing any interaction of the scoop with the internal gas flow. The depleted UF$_6$ gas is extracted by a similar scoop at the lower end of the rotor. This scoop not only removes the depleted UF$_6$, but also interacts with the

Fig. 3-11. View of the gas ultracentrifuges of the German demonstration plant at Almelo, Netherlands (URANIT)

spinning gas. This interaction generates the countercurrent flow, which ascends along the inner axis of the centrifuge and descends along the wall. A similar effect can also be generated by a linear temperature profile along the axial solid boundary of the centrifuge (slight heating of the bottom combined with some slight cooling of the top). The rotor of such high speed centrifuges requires a material of high strength and low specific weight. Aluminum, which allows a maximum peripheral speed of about 400 m/s to be used, is more and more being replaced by other materials like titanium, special steels and composite materials, such as glass and carbon fibers as well as aromatic polyamides of the nylon family (peripheral speed, 600–1100 m/s).

Since 1977, two 200 te SWU/a centrifuge enrichment plants have been operated at Capenhurst, UK, and Almelo, Netherlands. In the USA and Japan, similar demonstration facilities are being run. Fig. 3-11 presents a view of the centrifuge enrichment plant of Almelo. The total number of centrifuges for 200 te SWU/a capacity is about 40,000. An increase in centrifuge enrichment capacity is planned for the time period of 1985–1990 (see Table 3-6). Some 20,000 te SWU/a or more additional separative work capacity could be available from 1990/95 on.

While the investment costs of centrifuge enrichment plants are higher, the specific energy consumption is in the range of 100–400 kWh/kg SWU, i.e., an order of magnitude lower than in gaseous diffusion plants.

3.4.5 Aerodynamic Methods

Two aerodynamic separation techniques were developed in Germany (separation nozzle) and South Africa (advanced vortex tube). The separation nozzle

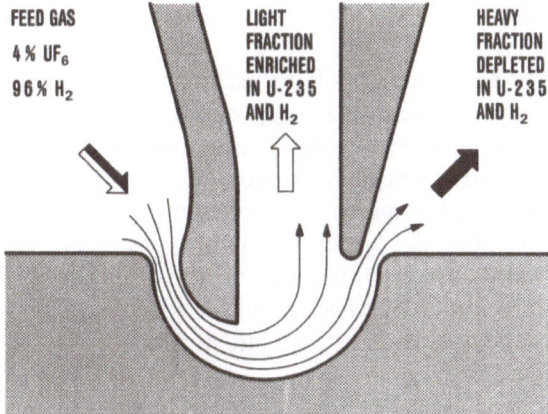

FEED GAS

4 % UF$_6$

96 % H$_2$

LIGHT
FRACTION
ENRICHED
IN U-235
AND H$_2$

HEAVY
FRACTION
DEPLETED
IN U-235
AND H$_2$

Fig. 3-12. Schematic of the separation nozzle process (W. Ehrfeld)

process is based on the centrifugal force in a fast curved jet flow (Fig. 3-12). The process gas is a mixture of about 96% H$_2$ or He and about 4% UF$_6$. The principle of the advanced vortex tube process is based on a vortex system for the separation of uranium isotopes contained in a hydrogen carrier gas employing the Helikon cascade technique. The first step towards commercial implementation of the separation nozzle process will be the construction of a demonstration plant in Brazil with an annual capacity of about 300 te SWU. South Africa is planning the construction of a small pilot plant with a capacity of 80–90 te SWU/a.

For both aerodynamic processes, the specific investment costs and the specific power consumptions of commercial size plants are estimated to be close to those of gaseous diffusion plants.

3.4.6 Advanced Separation Processes

Laser separation processes based on selective excitation with subsequent ionization, dissociation or chemical combination of uranium atoms or uranium hexafluoride molecules by laser beams are being investigated extensively in several countries. Chemical exchange has been studied in France for some time in order to develop a technically feasible uranium separation process. Some research effort is also being devoted to electromagnetic separation methods based on plasma rotation and selective ion cyclotron resonance heating of uranium plasmas.

3.4.7 Effects of Tails Assay and Economic Optimum

A 1 GW(e) LWR requires 84 te of enriched uranium as the initial load. This means that, with a tails assay of 0.2% of U-235, 367 te of natural uranium and 257 te SWU are needed. Including all reloads at 3.15% of U-235 enrichment over a life of 30 years of the reactor plant, a total of 4220 te of natural uranium and 3320 te SWU is necessary. If the tails assay were decreased, the required amount of natural uranium would be reduced, but the amount of separative work would increase.

Both natural uranium savings as a function of decreasing tails assay and increased separative work lead to cost variations with opposed functional dependencies. Thus, there is an optimum tails assay with minimum costs, which is only dependent on the costs of separative work and the cost of natural uranium including uranium conversion to UF_6. For given costs of natural uranium of US $ 100/kg and costs of $ 100/kg SWU, the optimum tails assay is determined as 0.218%. Present enrichment plants are operated around this optimum tails assay. Tails assays below 0.20% are technically feasible but, under these conditions, would not guarantee economic operation of enrichment plants.

3.5 Fuel Fabrication

Uranium is now used in the reactor core almost exclusively in the form of UO_2 pellets (LWR, HWR, AGR). Only in very few cases are uranium oxide or carbide particles used (HTGR and HTR). The source material for the production of enriched UO_2 pellets is uranium hexafluoride gas (UF_6), which is supplied by enrichment plants in 1.5 te gas cyclinders. Conversion into uranium oxide can follow a number of procedures. One process used especially in the Federal Republic of Germany is the *ammonium uranyl carbonate* (AUC) process (Fig. 3-13). It involves mixing of the UF_6 gas with water in a reaction column, thus producing uranyl fluoride (UO_2F_2). Further mixing with ammonia (NH_3) and carbon dioxide (CO_2) generates ammonium uranyl carbonate, which is precipitated from the solution. The suspension is passed through a rotary filter, washed and then put into a fluidized bed furnace where ammonia and carbon dioxide is split off by thermal decomposition. This produces uranium trioxide (UO_3), which is reduced to uranium dioxide (UO_2) by means of hydrogen at temperatures around 500 °C. Fluoride residues in the uranium dioxide powder are decreased to 110 ppm by water vapor at 650 °C.

The production of UO_2 pellets follows powder metallurgical processes. First the UO_2 powder is homogenized and a suitable grain size distribution is achieved by crushing and screening. After adding binders and lubricants, pellets are pressed to a density of approx. 5.5 g/cm³. In a pusher-type sintering furnace,

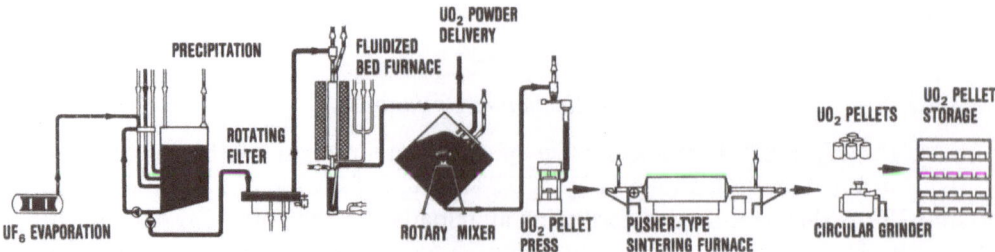

Fig. 3-13. Flowsheet of UO_2 pellet production (C. Keller)

the binders and lubricants as well as fluorine are first expelled at 500–1000 °C and hyperstoichometric uranium oxide is reduced to UO_2 by hydrogen. This is followed by the sintering process, which takes 2–3 hours at 1600–1750 °C, achieving a density of the UO_2 pellets of 10.3–10.5 g/cm^3. Afterwards, the pellets are ground on centerless circular grinders to the required geometric dimensions and tolerances.

The UO_2 pellets then pass quality control, where the geometric dimensions, density, surface quality, moisture content and the O/U ratio are checked. After this quality control, the UO_2 pellets are stored ready for fuel fabrication. In PWR, BWR and CANDU fuel elements, zircaloy tubes are filled with the UO_2 pellets and welded tight. These fuel rods are then assembled into fuel elements (see e.g. Fig. 4-3 and 4-4).

The fabrication of uranium and thorium particles with ceramic coatings for HTGR's will be described in Section 7.3.3. The design of prismatic or spherical fuel elements with such dispersed fuel particles will be addressed in Section 4.2.2. Similarly, the fabrication of PuO_2/UO_2 pellets or particles for plutonium recycling and LMFBR fuel elements are described in Section 7.2.2 and 7.4.5.

Selected Literature

Uranium resources and requirement

Best, F.R., Driscoll, M.J.: The Prospects for Uranium Recovery from Seawater. In: 1980 Annual Meeting, Las Vegas, Nev., June 9–12, 1980. American Nuclear Society Transactions *34*, 380–381 (1980).

Connolly, T.J., *et al.*: World Nuclear Energy Paths. New York-London: The Rockefeller Foundation/The Royal Institute of International Affairs. 1979.

Evaluation of Uranium Resources. Proceedings of an Advisory Group Meeting. Vienna: International Atomic Energy Agency. 1979.

International Nuclear Fuel Cycle Evaluation, Fuel and Heavy Water Availability. Report of INFCE Working Group 1. Vienna: International Atomic Energy Agency. 1980.

International Uranium Resources Evaluation Project (IUREP), Report on Phase 1. San Francisco, Cal.: Miller Freeman. 1980.

Problems of US Uranium Resources and Supply to the Year 2010, National Research Council, Supporting Paper No. 1. Committee on Nuclear and Alternative Energy Systems (CONAES), Uranium Resource Group. Washington: National Academy of Science. 1978.

Uranium: Resources, Production and Demand. A Joint Report by the OECD Nuclear Energy Agency and the International Atomic Energy Agency. Paris: Organisation for Economic Cooperation and Development. 1979 and 1982.

Uranium enrichment

Becker, E.W., *et al.*: Uranium Enrichment by the Separation Nozzle Method Within the Framework of German/Brazilian Cooperation. Nuclear Technology *52*, 105–114 (1981).

Cohen, K.: The Theory of Isotope Separation as Applied to the Large-Scale Production of U-235. New York: McGraw-Hill. 1951.

Ehrfeld, W., Ehrfeld, U.: Anreicherung von U-235. In: Gmelin Handbuch der Anorganischen Chemie, Uran, Ergänzungsband 2A, Isotope. Berlin-Heidelberg-New York: Springer. 1980.

International Nuclear Fuel Cycle Evaluation, Enrichment Availability. Report of INFCE Working Group 2. Vienna: International Atomic Energy Agency. 1980.

Uranium Enrichment (Villani, S., ed.). Topics in Applied Physics, Vol. 35. Berlin-Heidelberg-New York: Springer. 1979.

Technology of uranium fuel

Benedict, M., *et al.*: Nuclear Chemical Engineering. New York: McGraw-Hill. 1981.

Brandberg, S.G.: The Conversion of Uranium Hexafluoride to Uranium Dioxide. Nuclear Technology *18*, 177–184 (1973).

Hackstein, K.G., Plöger, F.: Neue Anlage zur Erzeugung von UO_2-Pulver aus UF_6. Atomwirtschaft/Atomtechnik *12*, 524–526 (1967).

Hardy, C.J.: The Chemistry of Uranium Milling. Radiochimica Acta *25*, 121–134 (1978).

Kernbrennstoffkreislauf, Band I (Keller, C., Möllinger, H., eds.). Heidelberg: Dr. Alfred Hüthig. 1978.

Schneider, V.W., Plöger, F.: Herstellung von Brennelementen. In: Chemie der nuklearen Entsorgung, Teil I (Baumgärtner, F., ed.), pp. 115–138. München: Karl Thiemig. 1978.

Seidel, D.C.: Extracting Uranium from Its Ores. International Atomic Energy Agency Bulletin *23* (2), 24–28 (1981).

4 Converter Reactors with a Thermal Neutron Spectrum

4.1 Light Water Reactors

At present, nuclear power generation is mainly based on *light water reactors* (LWR's) designed as *pressurized water reactors* (PWR's) or *boiling water reactors* (BWR's) (see Table 1-1). LWR's use enriched uranium fuel, which makes for greater flexibility in the choice of reactor core materials, especially allowing normal (light) water to be used as a coolant and moderator. PWR's deliver the heat generated in their reactor core to water circulating under high pressure in primary coolant circuits. From here the heat is transferred to a secondary coolant system via a steam generator to produce steam driving a turbo-generator system. In BWR's with direct cycle, the steam for the turbo-generator system is generated right in the reactor core and sent directly to the turbo-generator. PWR's and BWR's have been advanced to a high level of technical maturity in more than twenty years of development. They are built in unit sizes up to 1300 MW(e). The most important design data of large PWR and BWR power plants are listed in Table 4-1.

4.1.1 Pressurized Water Reactors

In the early fifties, PWR's were developed in the USA particularly as power plants for nuclear submarines ("N.S. Nautilus," 1954). The successful application of the PWR concept in naval reactors then resulted in the construction of the first non-military experimental nuclear power plant, Shippingport (60 MW(e)), which was commissioned in 1957. Present PWR's are built basically according to the same technical principles by a number of manufacturers in various countries of the world. In this Chapter, chiefly the standard PWR of 1300 MW(e) will be described as built by Kraftwerk Union in the Federal Republic of Germany. PWR designs by other manufacturers have some technical differences relative to this concept but these are not relevant to understanding the basic principles of operation.

Fig. 4-1 shows the main design principles of a 1300 MW(e) PWR. The heat generated by nuclear fission in the reactor core is transferred from the fuel elements to the coolant in the primary system. The highly pressurized water (158 bar) is circulated by coolant pumps and heated in the reactor core from 292 °C (inlet) to 326 °C (outlet). It flows to four steam generators, where it

Table 4-1. *Design characteristics of large PWR and BWR power plants (Kraftwerk Union, General Electric)*

		PWR (Kraftwerk Union)	BWR (General Electric)
Reactor power			
Thermal	MW(th)	3780	3580
Electrical gross	MW(e)	1300	1270
Electrical net	MW(e)	1240	1220
Plant efficiency (net)	%	32.8	34.1
Reactor core			
Equivalent core diameter	m	3.6	4.9
Active core height	m	3.9	3.8
Specific core power	kW(th)/l	95	54
Density	kW(th)/kg U	36.5	23
Number of fuel elements		193	748
Total amount of fuel	kg U	103,500	136,200
Fuel element and control element			
Fuel		UO_2	UO_2
U-235 fuel enrichment			
– initial	w/o	1.9/2.5/3.2	2–3
– reloadings	w/o	3.4	2.4–3
Cladding material		Zircaloy-4	Zircaloy-2
Cladding outer diameter	mm	10.75	12.3
Cladding thickness	mm	0.725	0.81
Fuel rod spacing	mm	14.3	16.2
Av. specific fuel rod power	W/cm	208	203
Av. heat flux on fuel element surface	W/cm^2	61	52.5
Fuel assembly array		16 × 16	8 × 8
Control/absorber rod type		20 rods inserted in fuel element	Cruciform control element inserted from the bottom between set of 4 assemblies
Number of control elements		61	177

Table 4-1 (continued)

		PWR (Kraftwerk Union)	BWR (General Electric)
Heat transfer system			
Primary system			
Total coolant flow rate (core flow)	te/s	18.8	13.1
Coolant pressure	bar	158	68
Coolant inlet temp.	°C	292	216
Coolant outlet temp.	°C	326	285
Steam supply system			
Steam generation	te/s	2	2
Steam pressure	bar	68	70
Steam temperature	°C	285	286
Fuel cycle			
Average fuel burnup	MWd(th)/te	35,000	28,400
Refueling sequence		1/3 per year	1/4 per year 1/3 per 18 months
Average fissile fraction in spent fuel			
– U-235	%	0.82	0.8
– Pu-fiss.	%	0.66	0.6

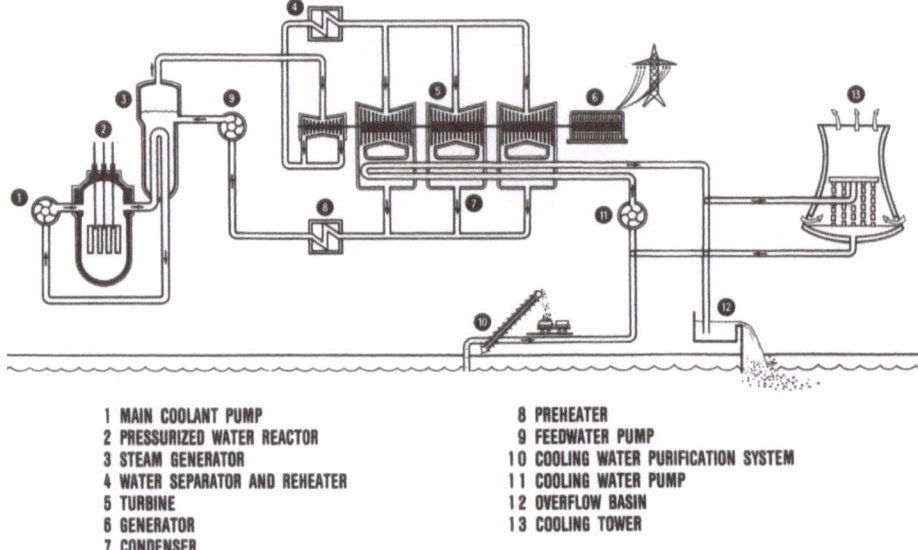

1 MAIN COOLANT PUMP
2 PRESSURIZED WATER REACTOR
3 STEAM GENERATOR
4 WATER SEPARATOR AND REHEATER
5 TURBINE
6 GENERATOR
7 CONDENSER

8 PREHEATER
9 FEEDWATER PUMP
10 COOLING WATER PURIFICATION SYSTEM
11 COOLING WATER PUMP
12 OVERFLOW BASIN
13 COOLING TOWER

Fig. 4-1. Functional design diagram of a pressurized water reactor power plant (Kraftwerk Union)

transfers its heat to the secondary steam system. In the secondary system, steam of 68 bar and 285 °C is generated. The steam drives the turbine and generator. The steam exhausted by the turbine is precipitated in the condenser, and the condensate water is pumped back into the steam generators. Waste heat delivered to the condenser is discharged into the environment either through river, lake or sea water or through a cooling tower.

4.1.1.1 Core

The core initially contains fuel elements with three different levels of U-235 enrichment (1.9%, 2.5% and 3.2%). The more highly enriched fuel elements are arranged at the core periphery, the less enriched fuel elements are distributed throughout the interior of the core (Fig. 4-2). This provides for a relatively flat power distribution over the core. In core reloadings, fuel elements with different burnups also may be arranged in a similar pattern. The average power density in the core is about 95 kW(th)/l or 36 kW(th)/kg uranium. At present, the fuel attains a maximum average burnup of about 35,000 MWd(th)/te during an irradiation period in the core of about three years. A one-year unloading and reloading cycle can be reached by unloading one third of the fuel elements at maximum burnup, reloading 3.4% enriched fresh fuel, and reshuffling fuel elements with medium or low burnups.

Uranium dioxide (UO_2) is used as a core fuel. The UO_2 powder is pressed and sintered into pellets of 10 mm height and diameter with an average density of 10.40 g/cm^3. The pellets along with pressurized helium are placed into tubes (cladding) made of Zircaloy-4 with an active core length of 3.9 m. This cladding

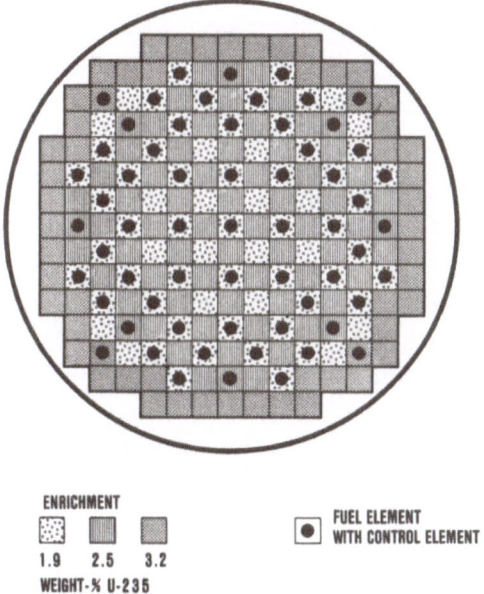

ENRICHMENT

1.9   2.5   3.2

WEIGHT-% U-235

⊙ FUEL ELEMENT
     WITH CONTROL ELEMENT

Fig. 4-2. Arrangement of fuel and control elements in the core of a 1300 MW(e) PWR (Kraftwerk Union)

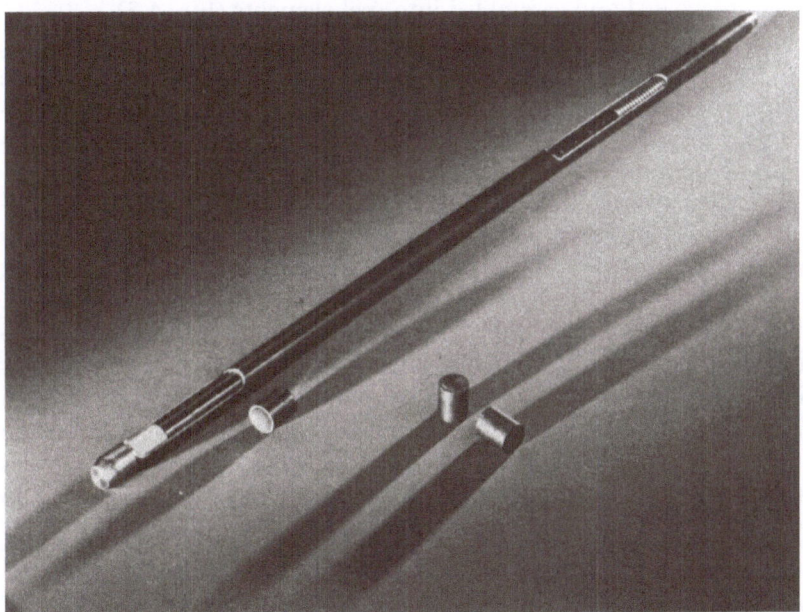

Fig. 4-3. PWR fuel rod (Kraftwerk Union)

Fig. 4-4. Fuel element and control element of a 1300 MW(e) PWR (Kraftwerk Union)

material is chosen for its low neutron absorption and good mechanical properties. The tubes are welded and assembled into fuel elements (Fig. 4-3).

A fuel element of a 1300 MW(e) reactor contains 236 fuel rods (Fig. 4-4). The core has 193 fuel elements with a total uranium mass of 103,500 kg. Some of the fuel elements contain control elements with twenty axially movable absorber rods. These absorber rods are filled with boron carbide or silver-indium-cadmium as neutron absorbing materials.

The fuel elements, control elements and core monitoring instruments are contained in a large pressure vessel designed to withstand the operating pressures at operating temperatures (Fig. 4-5). The vessel has an inner diameter of 5 m, a thickness of the cylindrical wall of 250 mm, and a height of approx. 13.2 m. The top head of the vessel, which holds all the control element drives, is removable for refueling. The water flowing at a rate of 18,800 kg/s enters the reactor vessel through four inlet nozzles close to the top and flows downward through the annulus between the vessel and the thermal shield to the core inlet at the bottom. It then returns upward through the core, leaving the four outlet nozzles of the pressure vessel. The reactor vessel is made of 22NiMoCr37 steel and its inner surface is plated with austenitic steel.

4.1.1.2 Coolant System

The water leaving the outlet nozzles of the pressure vessel transports the heat generated in the reactor core through four identical primary coolant circuits to four steam generators and is then recirculated to the pressure vessel. The inner diameter of the primary system pipes is 750 mm. Each primary coolant

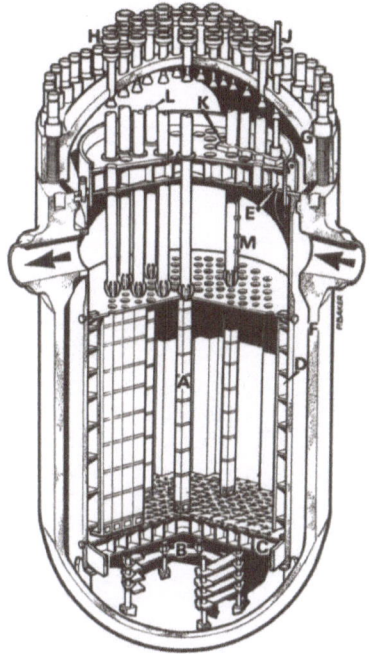

A   REACTOR CORE
B   SUPPORT STRUCTURE
C   LOWER GRID
D   CORE TANK, THERMAL SHIELD
E   UPPER GRID
F   PRESSURE VESSEL
G   REACTOR PRESSURE VESSEL HEAD
H   NOZZLE FOR CONTROL ELEMENT DRIVES
J   NOZZLE FOR CORE INSTRUMENTATION
K   INSTRUMENTATION LANCE
L   CONTROL ELEMENT GUIDE TUBE
M   CORE INSTRUMENTATION GUIDE TUBE

Fig. 4-5. Pressure vessel of a 1300 MW(e) PWR (Kraftwerk Union)

pump has a pressure head of 8 bar and consumes 5.4 MW(e) of power. The whole primary system is also plated with austenitic steel. The pressure of 158 bar in the primary coolant system is maintained by a pressurizer filled partly with water and partly with steam. It has heaters for boiling water and sprayers for condensing steam to keep the pressure within specified operating limits.

In the steam generators heat is transferred through a large number of tubes (Fig. 4-6). Feed water at a pressure of about 70 bar is evaporated here and steam is passed through separators to remove water droplets and attain a moisture content of less than 0.25%. In the steam generator, the radioactive primary coolant is separated from the secondary coolant by the tubes. The steam flows through the turbine valves into the high pressure section and, after reheating, into the low pressure section of the turbine. The expanded steam is precipitated in the condenser and pumped as water by the main condensate pumps into the feedwater tank. The main feedwater pumps move the pressurized water

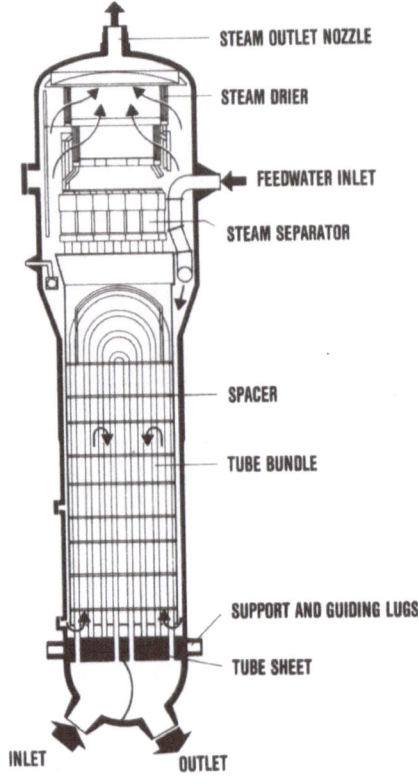

STEAM OUTLET NOZZLE

STEAM DRIER

FEEDWATER INLET

STEAM SEPARATOR

SPACER

TUBE BUNDLE

SUPPORT AND GUIDING LUGS

TUBE SHEET

INLET   OUTLET

REACTOR COOLANT

Fig. 4-6. Steam generator of a 1300 MW(e) PWR (Kraftwerk Union)

from the feedwater tank through four main feedwater pipes into the four steam generators.

In case the turbine has to be suddenly disconnected from the grid as a result of some fault condition, steam can be directly passed into the condenser by means of bypass valves. If the condenser should not be available due to some failure, the steam can be blown off by means of blowdown valves and safety valves.

A number of supporting systems are required for operation of the coolant circuit systems. The volume control system offsets changes in volume of the reactor cooling system resulting from temperature and operational influences. It is controlled by the water level in the pressurizer. Part of the cooling water is extracted continuously and purified in ion exchangers. At the same time, corrosion products and radioactive products are removed.

4.1.1.3 Containment

The reactor pressure vessel, the coolant pumps, steam generators, emergency and afterheat cooling systems as well as the vault for fresh and spent fuel elements are arranged within the reactor building, which is enclosed in a spherical double containment (Fig. 4-7). The double containment is made up of the

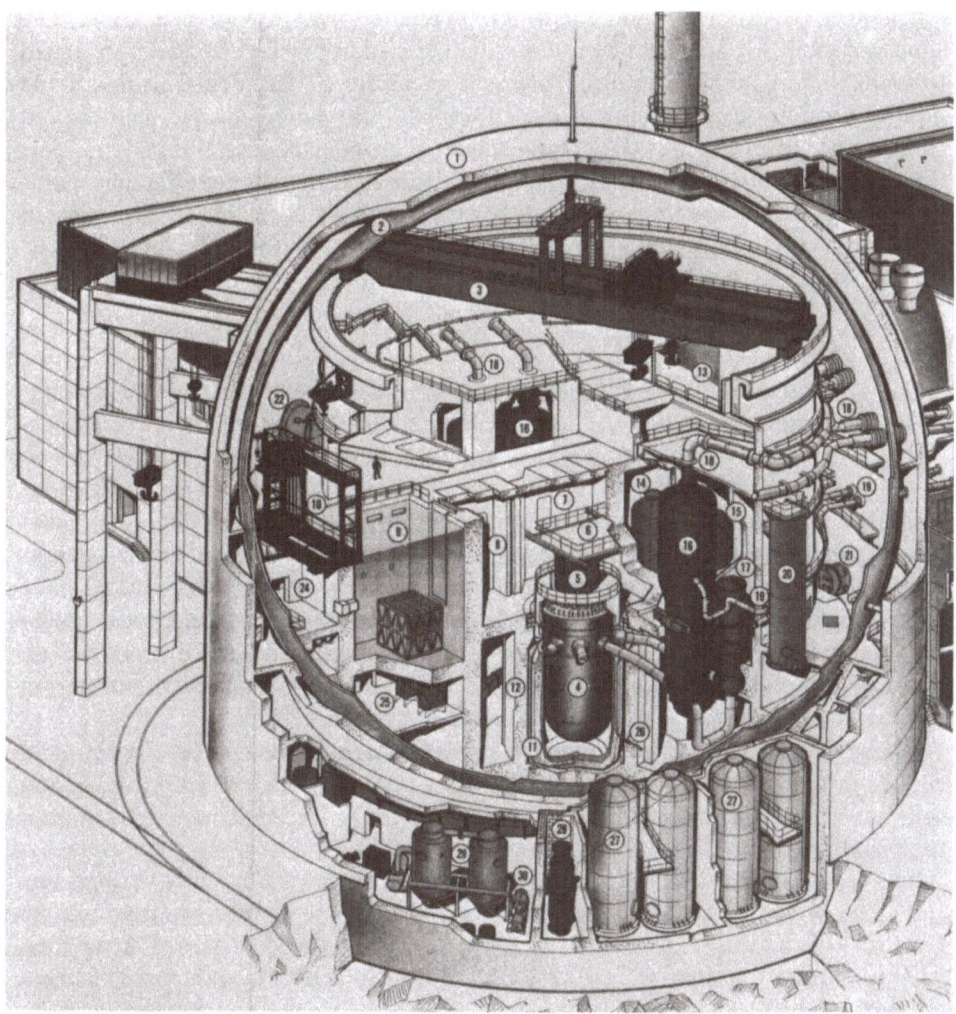

1	CONCRETE SHIELD	16	STEAM GENERATOR
2	STEEL CONTAINMENT	17	MAIN COOLANT PUMPS
3	REACTOR POLAR CRANE	18	MAIN STEAM PIPES
4	REACTOR PRESSURE VESSEL	19	FEEDWATER PIPES
5	CONTROL ROD DRIVES	20	PRESSURE ACCUMULATOR
6	CABLE BRIDGE	21	PERSONNEL LOCK
7	REACTOR COMPARTMENT	22	MATERIALS LOCK
8	CORE INTERNALS LAY-DOWN LOCATION	23	SEMI-PORTAL STRUCTURE WITH TROLLEY
9	FUEL ELEMENT STORAGE POOL	24	STORAGE FACILITY FOR FRESH FUEL ELEMENTS
10	FUELING MACHINE	25	TRANSDUCER ROOM
11	INNER SHIELD (BIOLOGICAL SHIELD)	26	MEASURING CHAMBER GUIDE TUBES FOR EXTERNAL
12	SUPPORT SHIELD (BIOLOGICAL SHIELD)		NEUTRON FLUX MEASUREMENT
13	VESSEL HEAD LAY-DOWN LOCATION	27	FLOODING TANK
14	PRESSURIZER	28	RESIDUAL HEAT EXCHANGER
15	PRESSURIZER RELIEF TANK	29	COMPONENT COOLING HEAT EXCHANGERS
		30	SAFETY FEED PUMPS

Fig. 4-7. Cutaway view of the containment of a 1300 MW(e) PWR (Kraftwerk Union)

inner steel containment and an outer concrete shield up to 2 m thick. The spherical inner steel containment has a diameter of approx. 50 m and is designed to a maximum possible internal pressure of about 5 bar. Penetrations of the piping through the containment are equipped with vented double bellows and can be checked for leaks. The outer steel reinforced concrete shield protects the reactor against external impacts and shields the environment against radiation exposure in case of accidents. External impacts considered as a design basis for the containment are earthquakes, floods, storms, airplane crashes, and pressure waves generated by chemical explosions.

4.1.1.4 Control Systems

At nominal power, PWR's have negative reactivity coefficients of coolant temperature and power (see Section 2.8). Any reduction in core coolant temperature therefore will result in an increase in reactivity and power. Any increase in the core coolant temperature will cause a decrease of reactivity and power. If higher loads are demanded by the generator and the turbine, more heat must be extracted from the primary cooling system through the steam generators. This is done by opening the turbine governor valve, which causes the primary coolant temperature in the reactor core to drop and the reactor power to rise; the power automatically balances out at a slightly higher level. However, in order to prevent the steam temperature and the steam pressure from dropping too far, the control elements are also moved at the same time.

Reactivity changes in the core are balanced by axial movements of the control elements; slow changes of the kind brought about by fuel burnup and fission product buildup are controlled by changing the boric acid concentration of the primary coolant.

The coolant pressure is kept at the systems pressure (158 bar) by the pressurizer. The water level in the pressurizer is controlled by the volume control system. The addition of feedwater to the steam generator must be matched by the feedwater control system as a function of the amount of steam extracted for the turbine.

4.1.1.5 Protection System

The PWR protection system (Fig. 4-8) processes the main measured data important for plant safety, and automatically initiates, among other steps, the following actions as soon as certain set points are exceeded:
– fast shutdown of the reactor core with turbine trip (reactor scram),
– emergency power supply,
– afterheat removal and emergency water injection,
– closure of reactor containment.

4.1.1.5.1 Reactor Scram. In a reactor scram, the absorber rods are dropped into the reactor core by gravity. The reactor goes subcritical and the reactor power drops to the level of afterheat generation.

4.1.1.5.2 Emergency Power Supply. During normal reactor operation, the plant is connected to the public grid system. In the case of a breakdown of the

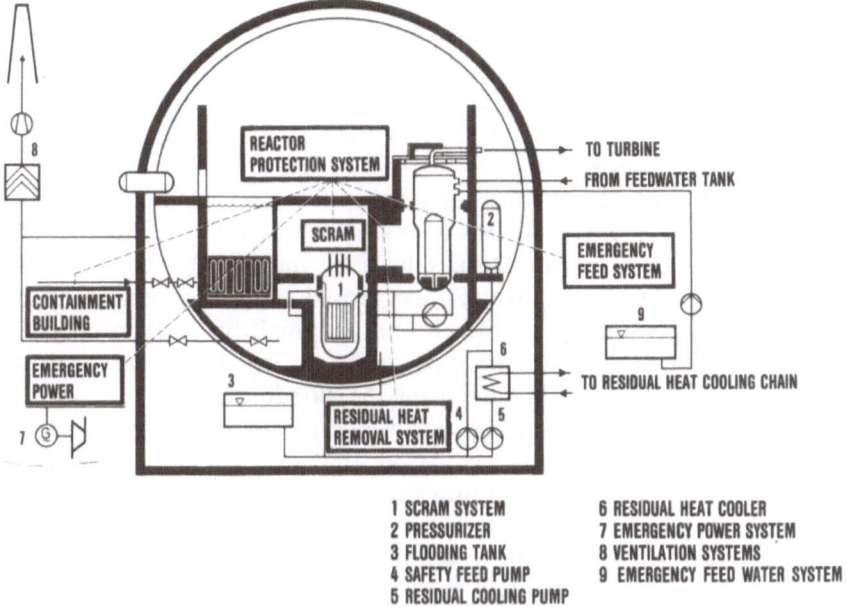

1 SCRAM SYSTEM
2 PRESSURIZER
3 FLOODING TANK
4 SAFETY FEED PUMP
5 RESIDUAL COOLING PUMP

6 RESIDUAL HEAT COOLER
7 EMERGENCY POWER SYSTEM
8 VENTILATION SYSTEMS
9 EMERGENCY FEED WATER SYSTEM

Fig. 4-8. Reactor protection system of a PWR (Kraftwerk Union)

public grid, the plant is disconnected from the grid and the output of the turbo-generator and the power generation of the reactor is reduced to the level of the plant requirement. The plant generates its own supply in isolated operation and is maintained operational. If also the isolated mode of operation fails, only the safety systems will be supplied by emergency power. Emergency power is supplied in a redundant layout by diesel generators and battery systems.

4.1.1.5.3 Emergency Feedwater System. The emergency feedwater system supplies feedwater to the steam generators, if the main feedwater pumps can no longer do so. They are supplied by emergency power in case the main power supply were to fail. The emergency feedwater system is equipped with fourfold redundancy, having water reserves (feedwater tank and demineralized water tank) which allow the removal of afterheat to be maintained for 10–15 hours. Removal of the afterheat from the reactor core is ensured first by natural convection of the primary cooling water flowing through the core and steam generators. Natural convection is enforced by installing the steam generators at a higher level than the reactor core. After the coolant pressure in the reactor cooling system has dropped to a sufficiently low level the afterheat removal system takes over this function.

4.1.1.5.4 Emergency Cooling and Afterheat Removal Systems. This cooling system, which is also called the residual heat, or afterheat, removal system, serves for both operational and safety related purposes (see Fig. 4-8):
– When the power of the PWR is reduced, the emergency cooling and afterheat removal system is started up automatically, after the pressure and the tempera-

ture in the primary systems have dropped to sufficiently low levels. The system then removes the afterheat and continues to cool the reactor core and coolant circuits.

– In a loss-of-coolant accident, the emergency cooling and afterheat removal system has to maintain the coolant level in the reactor pressure vessel and ensure cooling of the reactor core.

The emergency cooling and afterheat removal system has fourfold redundancy and is supplied by emergency power. It can feed water both into the cold (inlet) and the hot (outlet) main coolant lines by means of a high pressure and a low pressure feed system. Major leaks would cause the pressure in the reactor cooling system to drop so quickly that the high pressure feed system would not be started up. In that case, borated water will be injected into the primary main cooling system directly from the flooding tanks and through the low pressure injection systems. If the flooding tanks have run dry, the low pressure injection system feeds water from the reactor building sump into the primary systems. The building sump can collect leakage water from the primary system (sump recirculation operation).

Smaller leaks cause the pressure to drop only gradually, so that initially only the high pressure feed system will start to function. However, the pressure and temperature drop in the main coolant system is supported by a temperature drop of 100 °C/h on the secondary side. This is done automatically. After the coolant pressure and temperature have dropped sufficiently the low pressure feed systems will be started up.

Finally, the afterheat is discharged into river, lake or sea water, a cooling pond or special cooling towers by way of intermediate heat exchangers (secondary parts of the emergency cooling and afterheat removal system), which are also fourfold redundant and driven by emergency power.

4.1.1.5.5 Closure of the Reactor Containment. If there is an accident, the reactor protection system automatically closes the building and all piping penetrations. The annulus between the inner steel containment and the outer steel reinforced concrete shielding is kept at a pressure lower than atmospheric. In this way, radioactivity escaping from minor leaks can be extracted and, if necessary, discharged into the exhaust air stack in a controlled way through filters (see Fig. 4-8).

4.1.2 Boiling Water Reactors

The development of commercial BWR's began in 1956 when the Vallecitos BWR received the first US power reactor license. The first prototype or demonstration nuclear power plant to be equipped with a BWR, Dresden (180 MW(e)), was commissioned in 1960. This was followed by Oyster Creek, a BWR committed by a utility in the USA in late 1963 against the competition of fossil fueled power plants and without any government support, thus marking the beginning of the commercial utilization of nuclear power.

The BWR's built today by a number of manufacturers all over the world are characterized by almost identical technical designs. This Chapter will mainly

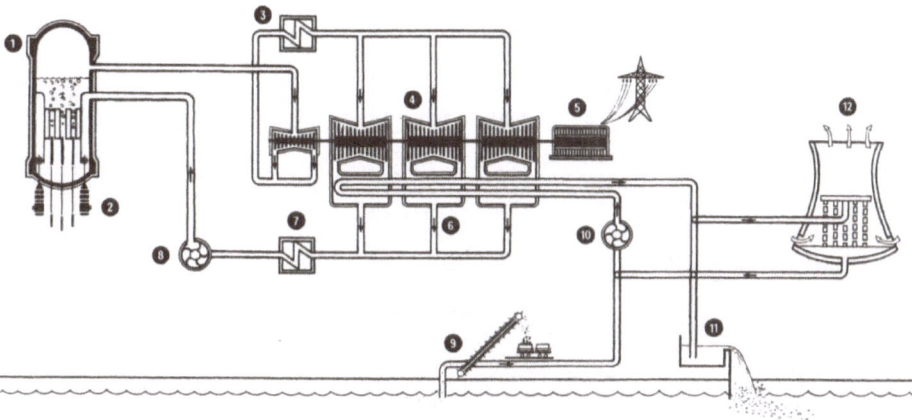

1 BOILING WATER REACTOR	7 PREHEATER
2 RECIRCULATION PUMP	8 FEEDWATER PUMP
3 WATER SEPARATOR AND REHEATER	9 COOLING WATER PURIFICATION SYSTEM
4 TURBINE	10 COOLING WATER PUMP
5 GENERATOR	11 OVERFLOW BASIN
6 CONDENSER	12 COOLING TOWER

Fig. 4-9. Functional design diagram of a boiling water reactor power plant (Kraftwerk Union)

deal with the standard BWR of 1220 MW(e) as designed and built by General Electric, USA. Fig. 4-9 shows the functional design diagram of a BWR.

4.1.2.1 Core, Pressure Vessel and Cooling System

The reactor core consists of a square array of fuel elements about 3.6 m long. The fuel elements each contain 8×8 fuel rods with outer diameters of 1.23 cm in a closed square box called a fuel channel (Fig. 4-10). For moderation of the neutrons and cooling of the core, water flows through the core and is allowed to boil in the upper part of the core. Cruciform absorber rods, containing boron carbide as the absorber material, are installed in between the square fuel channels. The absorber rods are moved hydraulically into and out of the reactor core from below. The fuel rods have claddings of Zircaloy-2 and contain UO_2 pellets with enrichments of about 2–3% U-235. The fuel is unloaded after a maximum burnup of 28,400 MWd(th)/te. Roughly one quarter to one third of the fuel elements are unloaded after 12 and 18 months, respectively, and replaced by fresh fuel elements with enrichments of approx. 2.4–3% of U-235. Fuel elements which have not attained their maximum burnups at that time are reshuffled.

Some fuel rods contain gadolinium as a burnable absorber to compensate for the burnup of fissile material and the buildup of fission products in the fuel during reactor operation. In addition to fuel rods, the fuel elements also contain water rods for power flattening across the fuel elements. The average power generation density in the core is some 54 kW(th)/l or 23 kW(th)/kg uranium. The water inlet temperature in the core is 216 °C, the outlet temperature is 286 °C, which corresponds to a saturation steam pressure of roughly 70 bar.

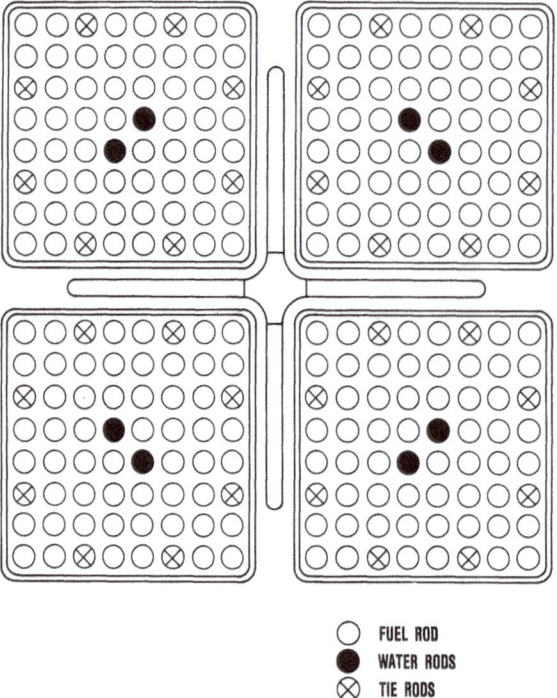

○ FUEL ROD
● WATER RODS
⊗ TIE RODS

Fig. 4-10. Four bundle fuel module of a BWR (General Electric)

The steam is generated by the water boiling in the reactor core, during which process some of the water flow is evaporated. To provide sufficient core flow for ample heat transfer and yet reduce the number of external recirculation loops required, the BWR employs internal jet pumps which accept recirculation flow from the two recirculation loops and increase the core flow rate (Fig. 4-11). The core with the absorber rods is contained in a large steel pressure vessel (Fig. 4-12). Above the core, there are the steam separators and steam driers. The reactor vessel head can be removed for loading and unloading fuel elements. A typical BWR pressure vessel has a diameter of 6 m, a wall thickness of roughly 150 mm, and a height of 22 m. It is made of manganese-molybdenum-nickel steel, the inside being plated with austenitic stainless steel.

The saturated steam flows from the reactor pressure vessel directly to the turbo-generator system and is pumped back from the condenser to the pressure vessel (see Fig. 4-9). A question of particular interest in BWR's with direct cooling systems is the radioactivity of the cooling water. The amount of radioactivity is determined by the impurities contained in the water and by an (n,p)-reaction of oxygen producing nitrogen, N-16, with a halflife of 7.2 s. Many years of experience in operating boiling water reactors have shown that, because of the short halflife of the N-16, maintenance work on the turbine, the condenser and the feedwater pumps is not impaired by radioactivity.

To run the jet pumps in the annulus between the reactor core and the inner wall of the reactor pressure vessel, some 30% of the recirculating water is extract-

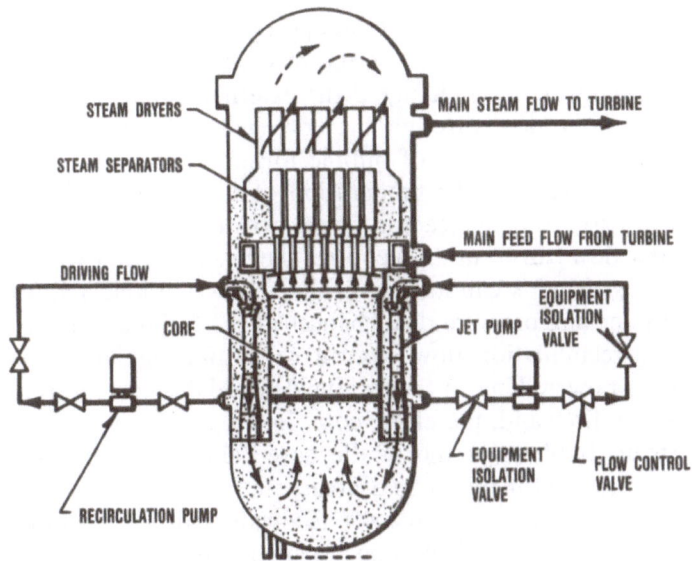

Fig. 4-11. BWR pressure vessel with recirculation system and high and low pressure spray and injection systems (General Electric)

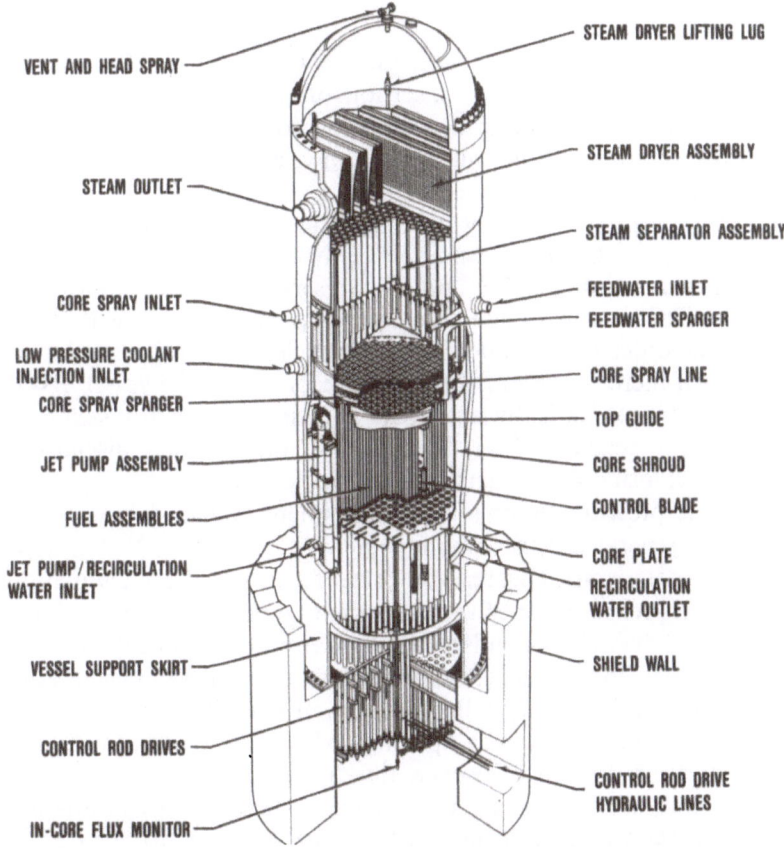

Fig. 4-12. Pressure vessel of a 1220 MW(e) BWR (General Electric)

ed from the annulus and forced into the jet pumps by external pumps (see Fig. 4-11). In this way, the water circulation in the reactor pressure vessel and through the core can be controlled by changing the external water circulation flow rate. Reduction of water flow through the core will result in a higher evaporation rate and in a larger volume of bubble formation. Increasing the volume of steam in the core reduces the moderation of the neutrons, and as a consequence, the reactivity and the reactor power will be reduced. In this way, changes in the water flow can be used to control the reactor power without movement of control rods. BWR's can automatically follow the load require-ments of the turbine, by sensing pressure disturbances at the turbine, transmit-ting these signals to the recirculation flow control valve and regulating core flow and therefore reactor power. Thus, if the turbine demand decreases, turbine pressure rises, recirculation flow and, therefore, core flow is caused to decrease, which increases the steam bubble volume in the core. This chain of events causes the reactivity and the reactor power to decrease.

In order to ensure high quality of the reactor feedwater, all the feedwater recirculated from the turbine condenser is pumped through filters (demineralizer units) and cleared of any corrosion products and other impurities.

4.1.2.2 Safety Systems

If there are reactivity perturbations or losses of coolant flow, the reactor is shut down in a short time by rapid insertion of the absorber rods. As a backup shutdown system, the BWR can poison the coolant (moderator) with a neutron absorbing boron solution and, in this way, also quench the nuclear reaction and shut down the reactor.

Ruptures in any pipe in the primary cooling system (e.g., the recirculation system) cause the main steam pipes and the feedwater pipes to be blocked by two series connected isolation valves and by two series connected non-return valves, respectively. This action isolates the reactor pressure vessel within the containment from the outer turbine and condenser cycles. When these isolation valves are closed, or if there is overpressure in the reactor pressure vessel, BWR safety/relief valves are automatically actuated, allowing a path for steam to be discharged from the reactor vessel. In this case, the steam is discharged into a large water pool inside the pressure suppression containment.

Fig. 4-13 shows the pressure suppression containment system demonstrated by the example of a General Electric boiling water reactor of the BWR-6 type with a Mark III containment. The reactor pressure vessel, the recirculation system and the pressure relief valves of the main steam pipes are accommodated in a chamber called the drywell, which isolates them from the rest of the reactor containment. The drywell communicates with a pressure suppression water pool which completely surrounds the drywell. The drywell consists of a reinforced concrete structure with horizontal vent openings which allow communication between the pool and the drywell. However, the drywell is kept dry by a weir wall. Above the drywell there is an upper water pool which serves as radiation protection during operation and refueling and for fuel element storage during shutdown/refueling.

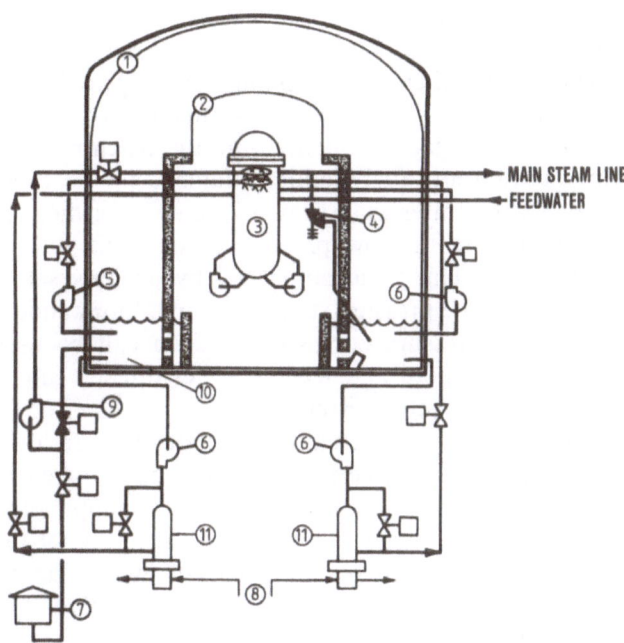

→ MAIN STEAM LINE
← FEEDWATER

1	SHIELD BUILDING AND STEEL CONTAINMENT	6	LOW PRESSURE COOLING INJECTION FUNCTION OF RESIDUAL HEAT REMOVAL SYSTEM
2	DRYWELL	7	CONDENSATE STORAGE TANK
3	REACTOR PRESSURE VESSEL	8	SERVICE WATER
4	SAFETY PRESSURE RELIEF VALVE DEPRESSURIZATION	9	HIGH PRESSURE CORE SPRAY SYSTEM
5	LOW PRESSURE SPRAY SYSTEM	10	PRESSURE SUPPRESSION POOL
		11	HEAT EXCHANGER

Fig. 4-13. BWR Mark III containment with emergency cooling and residual heat removal systems (General Electric)

If primary coolant is released through a leak in the primary system pipes, steam will enter into the drywell and will be channeled into the pressure suppression pool, where it will condense. A steel containment surrounds the drywell pressure suppression system and all reactor equipment. It is designed to withstand temperatures and pressures that could be caused by a loss-of-coolant accident and is designed to retain fission products which could potentially be released from the reactor system during an accident.

4.1.2.3 Emergency Cooling and Afterheat Removal Systems

If the water level in the reactor pressure vessel drops, or if there is a leak (loss-of-coolant accident), water is automatically added to the reactor by the following systems (see Fig. 4-13):
- a high pressure core spray system,
- a low pressure core spray system,
- a low pressure coolant injection system, and
- a reactor core isolation cooling system.

The pressure pump of the high pressure core spray system initially takes water from tanks that store condensate water or, if necessary, takes water from the pressure suppression pool. The water is then directly pumped to a spray header inside the pressure vessel. Its function is to supply large quantities of water to the core in the case of a loss-of-coolant accident while the reactor is still in a high pressure condition. It prevents fuel cladding damage in the event the core becomes uncovered by jetting water as a spray over the fuel elements.

In case it becomes necessary to use the low pressure system, the reactor pressure vessel can be depressurized. This is accomplished by opening safety relief valves and discharging steam to the pressure suppression pool for condensation. After this is done, the low pressure core spray system can be used by taking water from the suppression pool and feeding water directly into a spray header above the core inside the reactor vessel.

The low pressure coolant injection system is actually one mode of the residual heat removal system. The system employs three pumps, two pumps linked to heat exchanger loops while one communicates directly between the suppression pool and the reactor. The two heat exchanger loops can be used to cool the suppression pool or the heat exchangers can be bypassed and water can be injected at low pressure directly into the pressure vessel to cool the core. The third low pressure injection pump has no heat exchanger loops as it takes water from the pressure suppression pool and feeds it directly into the pressure vessel.

An additional function of the residual heat removal system is to remove and condense any steam bypassing the drywell. Water is taken from the pressure suppression pool and supplied to the containment spray headers where spray injected into the containment free volume would remove decay heat and over-pressurization of the containment would be prevented. The combined application of these systems guarantees cooling of the reactor core, the suppression pool and the containment for afterheat removal and in case of a loss-of-coolant accident.

A reinforced concrete structure encloses and protects the steel containment against such external events as were described for the PWR in Section 4.1.1.3 above. The annulus between the outer concrete shield building and the steel containment is maintained at a pressure lower than atmospheric so that any radioactive gases leaking into this annulus can be filtered prior to the release to the environment.

4.2 Gas Cooled Thermal Reactors

Graphite is a good neutron moderator with a relatively low absorption of neutrons. Thermal reactors with graphite moderators and gas (carbon dioxide or helium) as a coolant can therefore be operated on natural uranium. Because of the low concentration of fissile material, the attainable burnup of the fuel is low. For this reason, advanced gas cooled reactors use U-235 enrichment of the fuel. The lower capacity for moderating fission neutrons to thermal neutron energies results in a relatively large moderator volume, if graphite is used, which makes for a large reactor core and very low power density.

Gas cooled and graphite moderated natural uranium reactors were developed in the United Kingdom and in France and built in the fifties and sixties (MAGNOX reactors). The world's first nuclear power plant, Calder Hall (4×40 MW(e), commissioned in 1956) belongs in this category of reactors. MAGNOX reactors are no longer built today and, for this reason, will not be covered in more detail in this Section.

In Britain, further technical development then led to the *advanced gas-cooled reactor* (AGR). In the USA and in Germany, the *high temperature gas cooled reactor* line (HTGR) has been developed more recently. These advanced gas cooled reactors are attractive, above all, because of their high gas outlet temperatures and the resultant high thermal efficiencies. The high gas outlet temperatures, moreover, allow such plants to be used as sources of industrial high temperature process heat (see Section 1.2.3).

4.2.1 Advanced Gas Cooled Reactors

The AGR line has so far been built in unit sizes up to 625 MW(e). It also uses graphite as a moderator and carbon dioxide (CO_2) as a coolant gas. The primary cooling system operates at a gas pressure of 42 bar and coolant gas outlet temperatures of 670 °C. This allows steam temperatures to be reached in the secondary cooling system of 540 °C at a steam pressure of 170 bar. The overall plant efficiency of 41% (with sea water cooling) is correspondingly high. The whole primary cooling system, i.e., the reactor core, the gas circulators and the steam generators, are arranged in a prestressed concrete reactor pressure vessel. Eight gas circulators ensure circulation of the coolant gas between the core and four steam generators. The fuel element consists of 36 rods with UO_2 pellets enclosed in steel claddings. The UO_2 pellets contain slightly enriched uranium fuel (2% U-235). The fuel elements are arranged in vertical cooling channels in a graphite structure acting as a moderator. On-load fueling is possible. The main data of an AGR with 625 MW(e) power are listed in Table 4-2.

4.2.2 High Temperature Gas Cooled Reactors

High temperature gas cooled reactors also use graphite as a moderator, but helium as a coolant. Helium, being an inert gas, will not react with graphite at high temperatures and, consequently, allows even higher coolant outlet temperatures to be reached than in the AGR line. Two prototype high temperature gas cooled reactors will be briefly described below, which mainly differ in the shapes of their fuel elements and, accordingly, in the arrangements of their reactor cores, namely the *high temperature gas cooled reactor* (HTGR) with prismatic fuel elements developed in the USA and the *high temperature reactor* (HTR) with spherical fuel elements developed in the Federal Republic of Germany.

4.2.2.1 HTGR with Prismatic Fuel Elements

The Fort St. Vrain HTGR prototype reactor was built in the USA by General Atomic in a unit size of 300 MW(e) after successful operation of the smaller

Table 4-2. *Design characteristics of large gas cooled thermal power reactor plants (UKAEA, General Atomic, HRB)*

		AGR (Hinkley Point B)	HTGR	THTR-300
Reactor power				
Thermal	MW (th)	1493	3000	750
Electrical gross	MW (e)	665	1175	310
Electrical net	MW (e)	621	1160	300
Plant efficiency (net)	%	41.6	38.6	40
Reactor core				
Equivalent core diameter	m	9.1	8.4	5.6
Active core height	m	8.3	6.3	5.1
Specific core power	kW (th)/l	2.76	8.6	6
Density	kW (th)/kg fuel	13.1	76.5	115
Number of fuel elements		308	3944	675,000
Total amount of fuel	kg	114,000 U	1725 U + 37,500 Th	330 U + 6220 Th
Fuel element		UO$_2$ fuel (2.0–2.55% enriched) hollow pellets 5.1 mm i.d. 14.5 mm o.d. Cladding: stainless steel	Th/U-235 (93% enriched) coated particles in cylindrical rods, 15.6 mm diameter	U–Th oxide kernels, 0.4 mm diameter, coated with pyrolytic carbon in a spherical graphite element of 6 cm diam., 0.96 g U-235, (93% enriched), 9.62 g Th per fuel element
Control/Absorber rod		44 control rods (boron inserts in stainless steel claddings) 21 override regulating rods	73 pairs, hollow cylindrical, B$_4$C/graphite elements	36 control rods operating in holes in the reflector around the core, 42 safety rod moving into the pebble bed core

Table 4-2 (continued)

		AGR (Hinkley Point B)	HTGR	THTR-300
Heat transfer system				
Primary system				
Coolant		CO_2	He	He
Total coolant flow rate (core flow)	te/h	13,250	5080	1080
Coolant pressure	bar	43	51	40
Coolant inlet temp.	°C	292	316	260
Coolant outlet temp.	°C	645	741	750
Steam supply system				
Steam generation	te/h	2,200	3,900	930
Steam pressure	bar	160	169	180
Steam temperature	°C	538	510	530
Fuel cycle				
Average fuel burnup	MWd(th)/te	18,000	98,000	113,000
Refueling sequence		on-load, continuously 3 channels per week	off-load	on-load, continuously

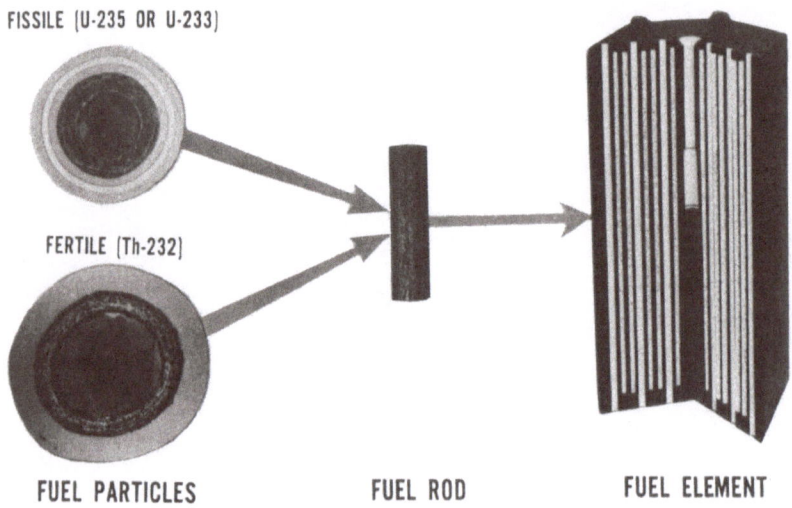

FISSILE (U-235 OR U-233)

FERTILE (Th-232)

FUEL PARTICLES   FUEL ROD   FUEL ELEMENT

Fig. 4-14. HTGR fuel components (General Atomic)

prototype reactors, Peach Bottom in the USA and Dragon in the UK. Larger plants of 1160 MW(e) were planned in the early seventies. Table 4-2 lists the main design characteristics of an 1160 MW(e) HTGR power plant.

The HTGR runs in the thorium/uranium fuel cycle. The fuel consists of highly enriched uranium particles (fissile) and thorium particles (fertile) with ceramic coatings. The fissile particles have diameters of 200–800 µm and contain either highly enriched U-235 (93%) or recycled U-233. They are coated with pyrolytic carbon and silicon carbide layers of 150–200 µm thickness. The fertile particles contain Th-232 as ThO_2 and are coated with pyrolytic carbon only. The particles are dispersed in a graphite matrix to form a fuel rod. Fig. 4-14 shows the fuel elements of an HTGR core as used for the Fort St. Vrain reactor plant and later for larger plants of 1160 MW(e) size. The fuel rods are again incorporated in a hexagonal graphite block to form a hexagonal fuel element. These hexagonal fuel elements are arranged in groups of seven elements to make up the core block. The core block is composed of several hundred hexagonal graphite elements, each consisting of three bottom reflector graphite blocks, eight fuel element blocks, and three upper reflector graphite blocks. The helium coolant gas flows downward through vertical holes in the hexagonal fuel elements.

The reactor core of an 1160 MW(e) HTGR plant has a diameter of 8.5 m and a height of 6.3 m. The core power density is 8.4 kW(th)/l, which is considerably lower than in LWR's. The helium transmits its heat to the secondary system while flowing upward in the steam generators. The core, the steam generators and the circulators of the primary cooling system are contained in cavities of a prestressed concrete reactor vessel (PCRV) (Fig. 4-15). These cavities are steel lined for sealing the high pressure coolant system. A thermal barrier protects the prestressed concrete from the high temperature. The helium coolant

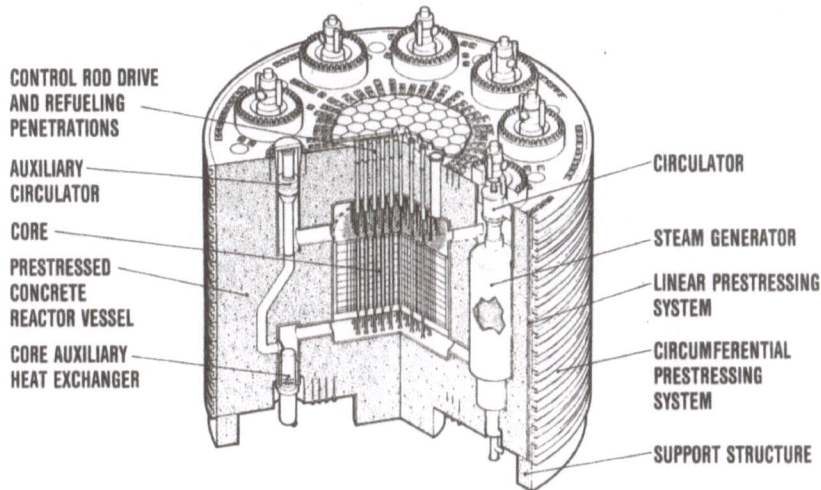

CONTROL ROD DRIVE
AND REFUELING
PENETRATIONS

AUXILIARY
CIRCULATOR

CORE

PRESTRESSED
CONCRETE
REACTOR VESSEL

CORE AUXILIARY
HEAT EXCHANGER

CIRCULATOR

STEAM GENERATOR

LINEAR PRESTRESSING
SYSTEM

CIRCUMFERENTIAL
PRESTRESSING
SYSTEM

SUPPORT STRUCTURE

Fig. 4-15. HTGR prestressed concrete reactor vessel arrangement (General Atomic)

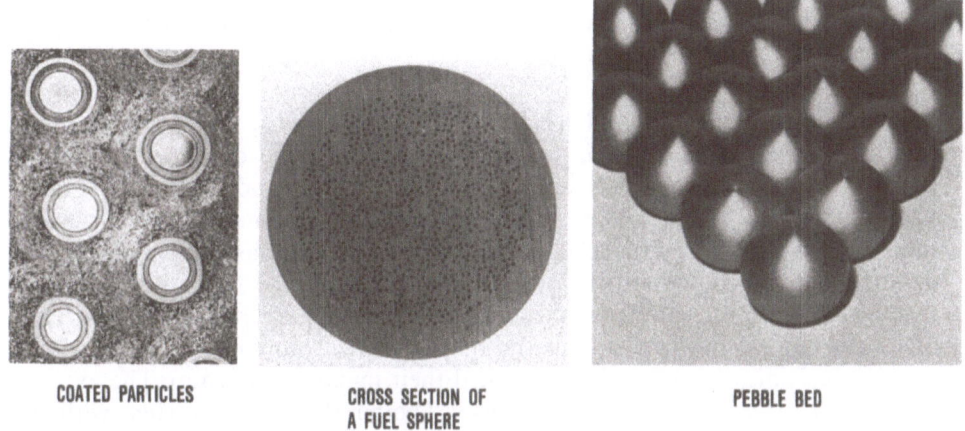

COATED PARTICLES

CROSS SECTION OF
A FUEL SPHERE

PEBBLE BED

Fig. 4-16. THTR 300 spherical fuel element (NUKEM)

gas is at a pressure of 51 bar and leaves the core at a temperature of about 740 °C. This leads to an overall efficiency of the plant of about 39%.

4.2.2.2 HTR with Spherical Fuel Elements

Following the construction and operation of a small experimental test reactor (AVR, 15 MW(e)) in Jülich, Federal Republic of Germany, the 300 MW(e) *thorium high temperature reactor*, THTR 300, has been under construction in Germany since 1972. As the HTGR line in the USA, it will be operated in the U-235/thorium/U-233 cycle. Except for the core and the fuel elements, its design is similar to that of the HTGR. The reactor core consists of 675,000 spherical fuel elements of 6 cm diameter each (Fig. 4-16). These spherical fuel

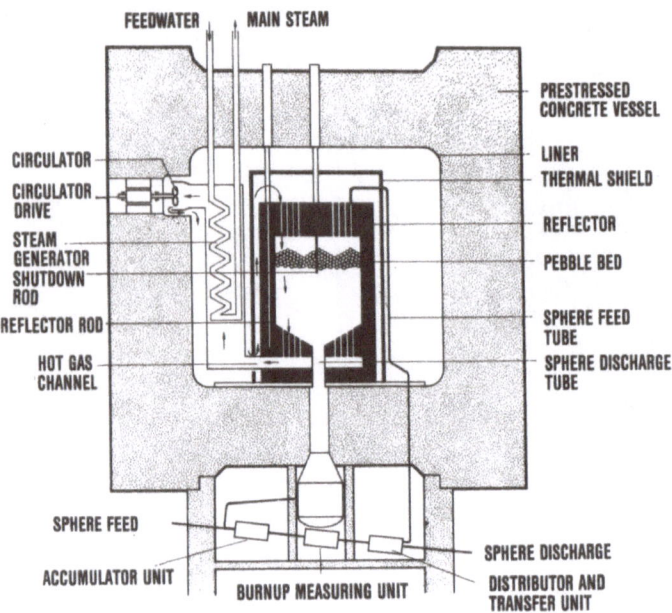

Fig. 4-17. Schematic diagram of a pebble bed high temperature reactor (HRB)

elements contain the fuel and the fertile material as UO_2 and ThO_2 particles which, as in the HTGR fuel, are coated with pyrolytic carbon. The pebble bed is enclosed in a graphite structure of 5.6 m diameter and 6 m height (Fig. 4-17 and Table 4-2). The bottom reflector is conical, terminating in a discharge tube for the spheres. The fuel elements are loaded continuously during operation into the core through refueling tubes located above the core. After passing through the core, they are removed continuously during operation. The spherical fuel elements are passed through the reactor core from top to bottom about six or seven times until they have reached their maximum burnup. The average core power density is around 6 kW(th)/l. The whole primary system with the coolant circulators and steam generators is integrated in a prestressed concrete reactor vessel. 42 absorber rods are inserted vertically into the pebble bed for reactor shutdown. 36 control rods are moved vertically within the graphite reflector for reactor control and shutdown. The coolant pressure in the primary system is 40 bar. The core structure is surrounded by six steam generators with coolant circulators. The coolant passes through the pebble bed in a downward flow and is heated to a core outlet temperature of 750 °C. It then passes through holes in the bottom reflector into the plenum below the core and is circulated from there through hot gas ducts to the six steam generators. In the steam generators the heat is transferred to a secondary steam-water system with a turbo-generator unit. The overall efficiency of the plant is 40%.

The *very high temperature gas cooled reactor* (VHTR) is intended to achieve temperatures of the helium coolant of 950–1000 °C so as to be able to extend the application of process heat to coal gasification and thermochemical splitting of water (see Section 1.2).

4.2.2.3 General Safety Considerations of HTGR's and HTR's

Since the cores of HTGR's and HTR's are basically made up of graphite and ceramic fuel, the melting and sublimation points of the fuel elements are very high. The solid core (power density of 6–8 kW(th)/l) has a relatively high heat capacity, which is important in case of failure of the core heat removal system. In normal operation the reactor shows high stability and very good selfregulating properties. The high heat capacity of the graphite moderator and the comparatively low power density retard all temperature transients. Even considerable increases in temperature do not result in abrupt or irreversible changes of the physical properties of core components, such as melting or evaporation.

The helium coolant is chemically inert. Because of possible chemical reactions of graphite with water and air after defects in a steam generator or air ingress, detailed analyses of graphite corrosion have been carried out. The results show that even with high combined ingress rates of air and water, the reactor can be safely cooled down and carbon burn-off of the fuel elements is limited to the fuel-free zone.

4.2.2.3.1 Control and Shutdown Systems. HTGR's with cylindrical fuel elements use rod type absorbers. The central fuel element of each group of seven fuel elements has two adjacent absorber rod channels (see Fig. 4-14). In these two channels a pair of absorber rods driven by a common drive can be inserted. An 1160 MW(e) HTGR plant has 98 such rod pairs. There is a third channel adjacent to the two absorber rod guide channels. This third channel enables small boron carbide granules, which are stored in hoppers between the control rod drives and the upper thermal shield, to be introduced in the core. These hoppers can be pressurized to break their rupture disks, thus allowing the absorber balls to fall through a guide tube into their respective channels in the core. This absorber system acts as a standby shutdown system.

In the THTR 300 pebble bed core, scram and hot shutdown is achieved by 36 absorber rods. These absorber rods are inserted in bores in the radial reflector. Reactor control and secondary shutdown is provided by a bank of 42 rods freely inserted under pressure into the pebble bed. For large pebble bed reactors rotating rods and helical absorbers are proposed as an alternative design. In addition, small neutron poison granules penetrating into the gaps between the pebbles are suggested as a standby shutdown system.

4.2.2.3.2 Afterheat removal and Emergency Cooling. Under normal operating conditions, the afterheat of the reactor after shutdown is accommodated by the primary systems with steam generators. Auxiliary cooling loops, which are also contained in the prestressed concrete reactor vessel, are designed to remove the afterheat in case of failure of the main coolant loops after reactor shutdown.

4.2.2.3.3 Design Base Accidents. In the safety analyses of high temperature gas cooled reactor plants a number of accidents have to be considered, e.g., uncontrolled control rod withdrawal, steam/water leaks into the primary system, loss of forced circulation of the helium coolant, and primary system depressurization into the reactor building, with a potential for air ingress.

To limit graphite corrosion in case of a steam generator tube rupture, the amount of water entering the core must be limited and the reactor core temperature must be cooled to a level below 700 °C. After steam or water has been detected in the coolant system, the defective loop is isolated and the reactor is shut down. Flow restrictors at penetrations through the prestressed concrete reactor vessel reduce the helium loss, if the vessel integrity is violated. In case of a leak in the penetration of the prestressed concrete pressure vessel the reactor must be shut down and long term cooling of the afterheat of the reactor core must be assured.

The purpose of the outer containment is to protect the environment against severe internal accidents with radioactivity releases and to protect the reactor against external events (earthquakes, airplane crashes, gas cloud explosions).

4.3 Heavy Water Reactors

Heavy water (D_2O) is an excellent moderator, absorbing far fewer neutrons than either light water (H_2O) or graphite. Reactors using heavy water as a moderator and coolant can therefore be run on natural uranium fuel. However, this requires a ratio of the D_2O/UO_2 volumes of almost 20, which means relatively wide fuel rod lattices or relatively low power densities. Variants of the *heavy water reactor* (HWR) design also use light water or CO_2 as coolants besides heavy water as a moderator. In that case, slight enrichment of the fuel is required.

Reactors with heavy water as the moderator have been developed in Canada, Europe and Japan. The *steam generating heavy water reactor* (SGHWR) developed in the United Kingdom uses light water in pressure tubes surrounded by the D_2O moderator. The coolant can boil in these pressure tubes. In the Atucha heavy water reactor developed in the Federal Republic of Germany, the heavy water coolant in the cooling channels and the surrounding heavy water moderator in the reactor pressure vessel are at the same pressure levels. The D_2O coolant under high pressure will not reach the boiling point, however. In France, the CO_2 cooled, D_2O moderated pressure tube reactor has been developed.

In this Chapter, the 600 MW(e) standard CANDU reactor type developed in Canada will be explained as an example. Table 4-3 shows some characteristic design data of a 600 MW(e) CANDU nuclear power plant.

Table 4-3. *Design characteristics of large HWR CANDU power plants (AECL)*

		CANDU-PHWR
Reactor power		
Thermal	MW(th)	2156
Electrical gross	MW(e)	680
Electrical net	MW(e)	633
Plant efficiency (net)	%	29.4

Table 4-3 (continued)

		CANDU-PHWR
Reactor core		
Equivalent core diameter	m	6.28
Active core height	m	5.94
Specific core power	kW(th)/l	11
Density	kW(th)/kg U	24
Number of fuel elements		380
Total amount of fuel	kg U	86,000
Fuel element and control rod		
Fuel		UO_2
U-235 fuel enrichment		
– initial	%	0.72 (natural)
– reloadings	%	0.72 (natural)
Cladding material		Zircaloy
Cladding outer diameter	mm	13.1
Cladding thickness	mm	0.38
Fuel channel spacing	cm	28.6
Nominal fuel element power		
– outer ring	W/cm	508
– intermediate ring	W/cm	417
– inner ring	W/cm	365
Reactivity devices		
Control system		
– light water compartments		14
– Cd absorber rods		4
– stainless steel adjuster rods		21
Safety systems		
– Cd shutoff units		28
– Gd injection nozzles		6
Heat transfer system		
Primary system		
Total coolant flow		
rate (core flow)	te/s	7.6
Coolant pressure	bar	100
Coolant inlet temp.	°C	267
Coolant outlet temp.	°C	310
Steam supply system		
Steam generation	te/s	1.05
Steam pressure	bar	47
Steam temperature	°C	260
Fuel cycle		
Average fuel burnup	MW d(th)/te	7,000
Refueling sequence		continuously, on-load
Average fissile fraction		
in spent fuel		
– U-235	%	0.2
– Pu-fiss.	%	0.3

4.3.1 CANDU Pressurized Heavy Water Reactor

In the CANDU PHWR (*Can*ada *d*euterium *u*ranium *p*ressurized *h*eavy *w*ater *r*eactor) D_2O is used as the coolant and moderator. So far, this line has only been built in unit sizes up to 750 MW(e). Design studies of 1200 MW(e) have been completed. In the primary system, the heavy water coolant flows through individual pressure tubes of the reactor core, thus cooling the fuel bundles. The coolant is kept at a pressure of roughly 100 bar and at temperatures of 267 °C at the inlet and 310 °C at the outlet of the cooling channel. For thermal insulation, the pressure tubes are surrounded by another gas filled annular gap. Some 380 of these tubes are arranged horizontally in the reactor vessel

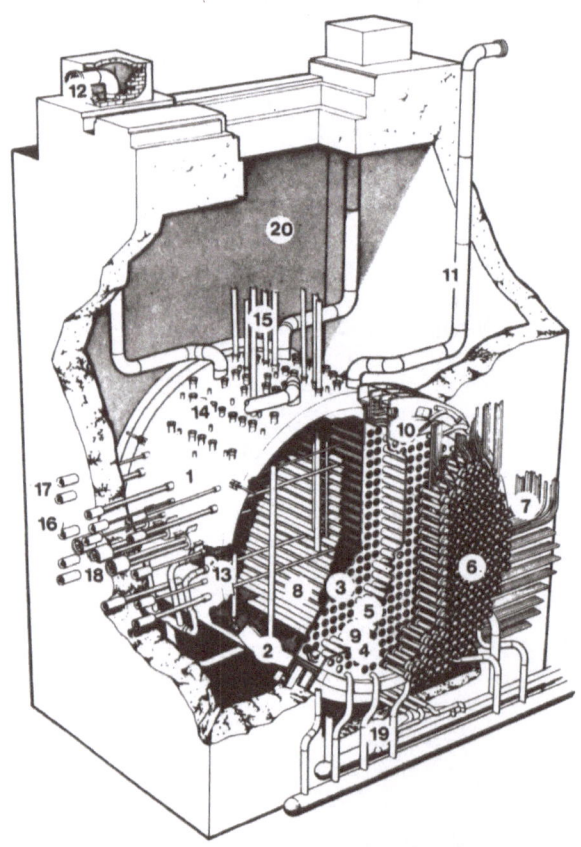

1 CALANDRIA	11 PRESSURE RELIEF PIPES
2 CALANDRIA SHELL	12 RUPTURE DISC
3 CALANDRIA SIDE TUBE SHEET	13 MODERATOR INLETS (4 EACH SIDE)
4 FUELING MACHINE SIDE TUBE SHEET	14 REACTIVITY CONTROL NOZZLES
5 LATTICE TUBES	15 REACTIVITY CONTROL DEVICES
6 END FITTINGS	16 HORIZONTAL FLUX DETECTORS (9)
7 FEEDERS	17 POISON INJECTOR NOZZLES (6)
8 CALANDRIA TUBES	18 ION CHAMBER HOUSING (0 EACH SIDE)
9 STEEL BALL SHIELDING	19 END SHIELD COOLING PIPING
10 ANNULAR SHIELDING SLAB	20 CALANDRIA VAULT (LIGHT WATER SHIELD)

Fig. 4-18. Cutaway view of a 600 MW(e) CANDU reactor core (AECL)

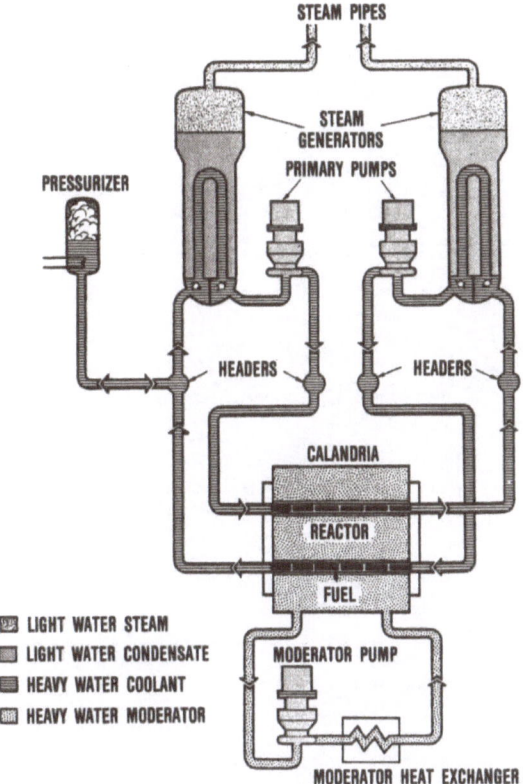

STEAM PIPES

STEAM GENERATORS

PRIMARY PUMPS

PRESSURIZER

HEADERS   HEADERS

CALANDRIA

REACTOR

FUEL

LIGHT WATER STEAM
LIGHT WATER CONDENSATE
HEAVY WATER COOLANT
HEAVY WATER MODERATOR

MODERATOR PUMP

MODERATOR HEAT EXCHANGER

Fig. 4-19. CANDU reactor simplified flow diagram (AECL)

filled with D_2O, which is kept at a pressure close to atmospheric (Fig. 4-18). Each pressure tube is connected via distribution headers to the pumps and steam generators of the primary system (Fig. 4-19). A pressurizer in the primary system designed similar to the device used in pressurized water reactors maintains the coolant pressure. The heavy water moderator in the calandria has its own cooling system and is kept at a temperature of approx. 70 °C and a pressure close to atmospheric. The calandria in a 600 MW(e) plant has a diameter of 7.6 m and an inside (core) length of 6 m. Because of the low internal pressure, its wall thickness is only 29 mm. The reactor power is removed by two cooling circuits shaped like the figure eight, using pumps and steam generators. In the secondary system, light water serves as the coolant to generate steam to drive a turbogenerator system. The overall thermal efficiency of the CANDU reactor line is 29.4%. Fig. 4-20 shows an overall view of the calandria and the primary systems. The reactor building is a prestressed concrete structure.

4.3.1.1 Fuel Elements

The CANDU reactor uses natural uranium (0.72% U-235) as a fuel in UO_2 pellets welded into Zircaloy tubes of 13.1 mm outer diameter. A fuel bundle

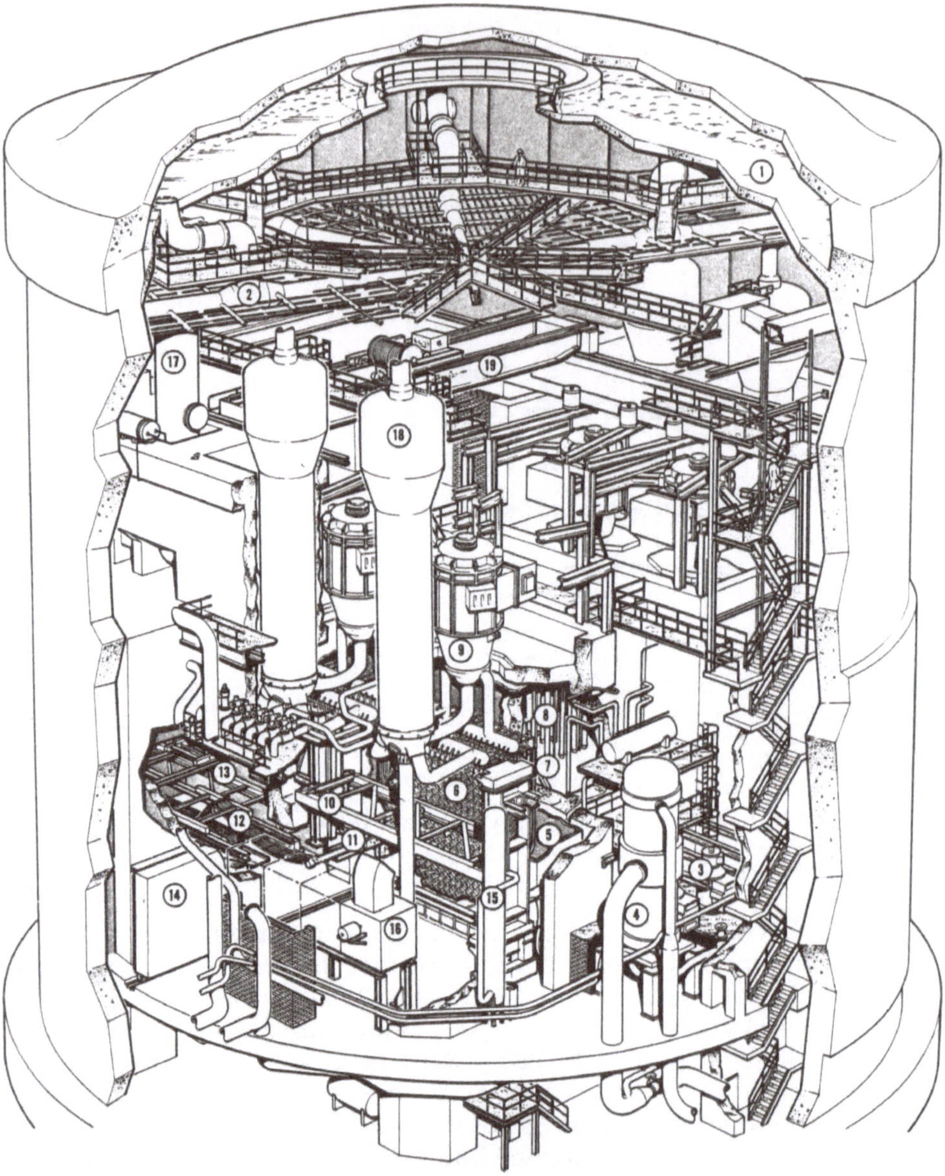

1	DOUSING WATER TANK	10	FUELING MACHINE BRIDGE
2	DOUSING WATER VALVES	11	FUELING MACHINE CARRIAGE
3	MODERATOR PUMP	12	FUELING MACHINE CATENARY
4	MODERATOR HEAT EXCHANGER	13	FUELING MACHINE MAINTENANCE LOCK
5	FEEDER CABINETS	14	FUELING MACHINE MAINTENANCE LOCK DOOR
6	REACTOR FACE	15	END SHIELD COOLING WATER DELAY TANK
7	REACTOR	16	VAULT COOLER
8	REACTIVITY MECHANISM	17	PRESSURIZER
9	PRIMARY HEAT TRANSPORT SYSTEM PUMP	18	STEAM GENERATOR
		19	STEAM GENERATOR ROOM CRANE

Fig. 4-20. 600 MW(e) CANDU reactor building cutaway view (AECL)

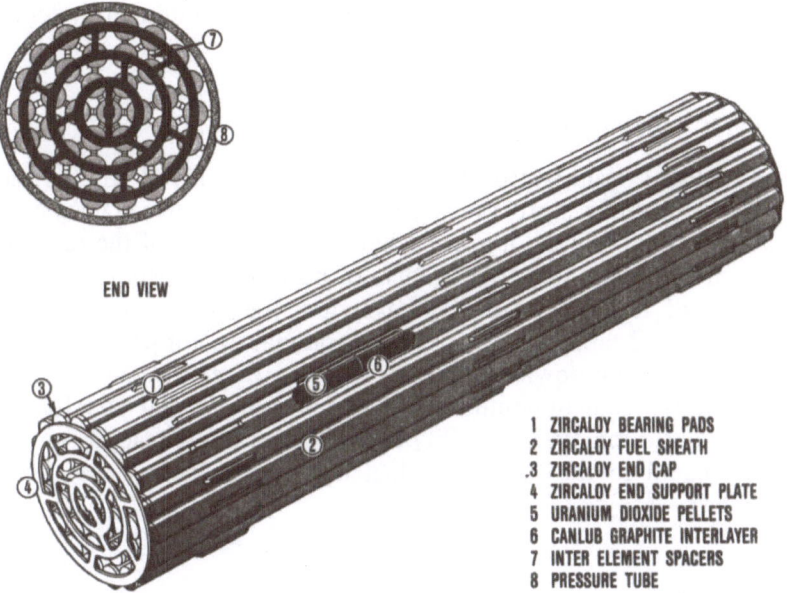

END VIEW

1 ZIRCALOY BEARING PADS
2 ZIRCALOY FUEL SHEATH
.3 ZIRCALOY END CAP
4 ZIRCALOY END SUPPORT PLATE
5 URANIUM DIOXIDE PELLETS
6 CANLUB GRAPHITE INTERLAYER
7 INTER ELEMENT SPACERS
8 PRESSURE TUBE

Fig. 4-21. 37-element fuel bundle of a 600 MW(e) CANDU reactor (AECL)

contains 37 of these fuel elements combined into a cylindrical cluster (Fig. 4-21). Twelve of these fuel bundles are loaded in the pressure tube in series.

The reactor core contains 4560 fuel bundles and a total of 95 te of UO_2 fuel. It has a diameter 6.3 m and a length of 5.9 m, attaining a power density of 11 kW(th)/l. In CANDU reactors, the fuel bundles can be exchanged during operation without requiring the reactor to be shut down. Over about 200 full load days, the fuel attains a burnup of 7000 MWd(th)/te, still containing some 0.2% U-235 and 0.3% plutonium when unloaded.

Refueling is done by two refueling machines. Fuel bundles are pushed into the reactor channel by a remotely operated fueling machine. Spent fuel bundles are discharged simultaneously into another fueling machine at the opposite end of the reactor. The fuel is then transferred to a water filled storage bay.

4.3.1.2 Reactivity Control

Long term control of burnup and fission products is achieved by on-load refueling. Minor reactivity variations are compensated by a liquid zone control system. This consists of 14 compartments of variable amounts of light water (acting as neutron absorbers). Bulk reactivity control is achieved by variation of the average amount of light water in these compartments, whereas spatial (stability) control is achieved by differential filling or draining of these compartments. Core power distribution flattening is achieved with a set of 21 stainless steel adjuster rods. Fast, controlled (as distinct from scram) power reductions are accomplished with four cadmium loaded absorber rods. Dissolved moderator poisons (both boron and gadolinium) are also used for reactivity trim.

Reactor scrams are initiated by one of two special safety shutdown systems, as discussed in Section 4.3.1.4.

4.3.1.3 Shutdown Cooling System

A shutdown cooling system is provided in the CANDU PHWR to cool the fuel and the primary heat transport system after a reactor shutdown. The system consists essentially of a pump and a heat exchanger at each end of the reactor, connected between the inlet and outlet headers of both heat transport circuits. Under normal use of the shutdown cooling system (i.e., at reduced system pressure) its heat removal capability is $\approx 1.2\%$ of nominal reactor power. Under accident conditions the shutdown cooling system can be connected into the primary heat transport system at nominal operating pressure (≈ 100 bar). Under these conditions the heat removal capability is $\approx 7\%$ of nominal reactor power, which is roughly equal to the fission product decay heat immediately after a reactor shutdown.

4.3.1.4 Safety Systems

CANDU PHWR's have four special safety systems:
- shutdown system 1,
- shutdown system 2,
- emergency core cooling system,
- containment system.

These systems are independent of each other and of any of the process or regulating systems.

Shutdown system 1 makes use of 28 cadmium loaded shutoff units. These enter the core vertically.

Shutdown system 2 injects a liquid neutron poison (gadolinium nitrate) into the moderator. This is done through the use of six horizontally placed, perforated injection nozzles, each connected to its own poison tank.

The emergency core cooling system would inject cooling water into the core after a loss-of-coolant accident. The system has a high pressure, intermediate pressure, and low pressure stage. During the high pressure stage water is injected from a high pressure, external tank. The intermediate stage takes water from a tank located under the dome of the reactor building. The low pressure stage recirculates water collected in the reactor building sump. In the CANDU PHWR design complete failure of the emergency core cooling system can be tolerated because in such cases the heavy water moderator, which is independent of the primary coolant, and is therefore not lost during a loss-of-coolant accident, becomes the ultimate heat sink.

The containment system in single-unit stations consists of a concrete structure with a plastic liner. Reactor building air coolers and a dousing tank act as energy suppression systems during loss-of-coolant accidents. Multi-unit stations use a common vacuum building to contain potential radioactivity releases. In these stations each unit reactor building is connected via a duct to this vacuum building.

4.4 Near Breeder and Thermal Breeder Reactors

In a thermal neutron spectrum, the neutron yield, η, of U-233 is considerably higher than that of Pu-239 or U-235. This value of η for U-233 opens up a potential of high conversion ratios or even breeding ratios around 1.0 (see Section 2.5). Reactors able to attain breeding ratios of 1–1.03 with U-233/Th fuel and a thermal neutron spectrum are called "thermal breeders." Converter reactors attaining conversion ratios between 0.94 and 1.0 are sometimes called "near breeders." In such reactor cores, parasitic neutron absorption and neutron leakage losses must be kept extremely low. Fission product poisoning and removal of Pa-233 with a high neutron absorption cross section and a halflife of 27.4 days are then two of the key problems. Enhanced fission product removal necessitates either frequent refueling or the introduction of the homogeneous reactor concept with continuous fission product removal. Pa-233 poisoning can be avoided by separating the fissile U-233 from the fertile Th-232 fuel containing Pa-233.

4.4.1 Homogeneous Core Thermal Breeders

A reactor concept of the homogeneous core type is the *molten salt breeder reactor* (MSBR), a small test reactor developed, built and operated at Oak Ridge, USA, in the sixties.

The MSBR is fueled with a homogeneous fluid salt containing both the fissile uranium and the fertile thorium fuels. The fuel carrier salt is a mixture of the fluorides of lithium, beryllium and thorium ($LiF-BeF_2-ThF_4$). The fissile material is contained as UF_4. Either U-235 or U-233 can be used. The fuel carrier salt is pumped through a core structure of bare graphite. Pa-233 and some of the neutron absorbing fission products are removed continuously by a purification system and on-site reprocessing.

The heat of the core fuel salt is transferred in a heat exchanger to a secondary coolant (molten salt), which then passes through steam generators to generate steam. All metal surfaces contacting fuel salt are made of Hastelloy-N, a nickel base alloy.

4.4.2 Light Water Breeder Reactors (LWBR's)

In principle, thermal converter reactors can attain high conversion ratios ≥ 0.9 in the Th/U-233 cycle (see Chapter 6). Further improvement in the neutron economy of the core and short time refueling can raise the conversion ratio even further. However, short time refueling adds to the fuel cycle costs, because the fuel remains in the core for power production only for short periods of time and must pass through reprocessing and refabrication more often.

One particularly interesting variant of a high converting thermal reactor is the so-called seed-and-blanked concept in a light water reactor. In the seed-and-blanket reactor concept, the core is subdivided into a number of modules each containing a seed and a blanket region. The reactor core is water cooled. Seed and blanket have U-233/Th fuel of different enrichments and different neutron spectra. The seed region contains mainly fissile material, whereas the blanket region is made mainly of fertile material. Because of the concentration of reactivity in the seed region, the reactor can be controlled by axial movement of the seed regions. Axial movement of the seed changes the leakage of neutrons from the fissile region (seed) to the fertile region (blanket). Moreover, the fuel and moderator volume fractions are chosen in such a way that a thermal spectrum dominates in the seed zone. Consequently, the advantages of the high thermal η-value of U-233 can be utilized. The blanket region contains less moderator. As a consequence, the neutron spectrum has a high average neutron energy, which increases the fast fission effect in Th-232 and the fraction of (n,2n)-reactions. All these measures support the breeding effect. When cooled with light water, this reactor can attain a breeding ratio of 1.01. The use of heavy water would raise the breeding ratio even further.

It is remarkable that this type of reactor retains all design features of a PWR, except for the reactor core. A first demonstration program of the seed-and-blanket reactor utilizing the Th/U-233 cycle was started at Shippingport in the USA at the end of 1977.

Selected Literature

Light water reactors

Druckwasserreaktor. Erlangen: Kraftwerk Union AG, K/10567 (1981).
General Description of a Boiling Water Reactor. San Jose, Cal.: General Electric Company. 1978.
Lamarsh, J.R.: Introduction to Nuclear Engineering. Reading, Mass.: Addison-Wesley. 1975.
Lewis, E.E.: Nuclear Power Reactor Safety. New York: John Wiley. 1977.
Märkl, H.: Core Engineering and Performance of KWU Pressurized Water Reactors. Erlangen: Kraftwerk Union AG. 1976.
Nero, A.V.: A Guidebook to Nuclear Reactors. Berkeley-Los Angeles-London: University of California Press. 1979.
Druckwasserreaktoren für Kernkraftwerke (Oldekop, W., ed.). München: Karl Thiemig. 1974.
Sauer, A.: Siedewasserreaktoren für Kernkraftwerke. Berlin: AEG Telefunken, Handbücher Band 10. 1969.
Smidt, D.: Reaktor-Sicherheitstechnik. Berlin-Heidelberg-New York: Springer. 1979.
Technical Evaluation of Bids for Nuclear Power Plants – A Guide Book. Vienna: International Atomic Energy Agency, TR 204 (1981).

Gas cooled reactors

Boyer, V.S., *et al.*: High Temperature Reactors. In: Nuclear Energy Maturity, Proc. European Nuclear Conference, Paris, 21–25 April, 1975 (Zaleski, P., ed.), Vol. 2, pp. 191–217. Oxford: Pergamon Press. 1976.
Dahlberg *et al.*: HTGR Fuel and Fuel Cycle Summary Description. San Diego, Cal.: General Atomic Company, GA-A 12801 (1974).
Development Status and Operational Features of the High Temperature Gas Cooled Reactor. Palo Alto, Cal.: Electric Power Research Institute, EPRI NP-142 (1976).

Engelmann, P., Oehme, H.: Ziele der HTR-Entwicklung für Stromerzeugung und Prozeß-wärme. Atomwirtschaft/Atomtechnik *22*, 484–490 (1977).

Fassbender, J., *et al.*: Zur Störfalltopologie des Hochtemperaturreaktors. Atomkernener-gie/Kerntechnik *37*, 81–86 (1981).

Harder, H., *et al.*: Das 300 MWe Thorium-Hochtemperatur-Kernkraftwerk (THTR). Atomwirtschaft/Atomtechnik *16*, 238–245 (1971).

Schulten, R., *et al.*: The Pebble-bed High Temperature Reactor as a Source of Nuclear Process Heat. Kernforschungsanlage Jülich, Jül 1115-RG (1974).

Wessmann, G.L., Mofette, T.R.: Safety Design Bases for HTGR. Nuclear Safety *14*, 618–634 (1973).

Heavy water reactors

Foster, J.S., Critoph, E.: The Status of the Canadian Nuclear Programme and Possible Future Strategies. Annals of Nuclear Energy, Vol. 2, pp. 689–703. Oxford: Pergamon Press. 1975.

McIntyre, H.C.: Natural Uranium Heavy Water Reactors. Scientific American *233* (4), 17–27 (1975).

Smith, H.A.: A Review of the Development Status of CANDU Nuclear Power Plants. In: Nuclear Energy Maturity, Proc. European Nuclear Conference, Paris, 21–25 April, 1975 (Zaleski, P., ed.), Vol. 2, pp. 38–44. Oxford: Pergamon Press. 1976.

Study of the Development Status and Operational Features of Heavy Water Reactors, Palo Alto, Cal.: Electric Power Research Institute, EPRI NP-365 (1977).

Woodhead, L.W., Ingolfsrud, L.J.: Performance of Canadian Commercial Nuclear Units and Heavy Water Plants. In: Nuclear Energy Maturity, Proc. European Nuclear Con-ference, Paris, 21–25 April, 1975 (Zaleski, P., ed.), Vol. 2, pp. 103–112. Oxford: Perga-mon Press. 1976.

Near breeders and thermal breeder reactors

Conceptual Design Study of a Single Fluid Molten Salt Breeder Reactor. Oak Ridge National Laboratory, ORNL-4541 (1971).

The Development Status of Molten Salt Breeder Reactors. Oak Ridge National Laborato-ry, ORNL-4812 (1972).

Perry, A.M., Weinberg, A.M.: Thermal Breeder Reactors. In: Annual Review of Nuclear Science, Vol. 22, pp. 317–354. Palo Alto, Cal.: Annual Reviews Inc. 1972.

Status and Prospects of Thermal Breeders and Their Effect on Fuel Utilization. Vienna: International Atomic Energy Agency, IAEA Technical Report Series No. 195 (1979).

5 Breeder Reactors with a Fast Neutron Spectrum

5.1 The Potential Role of Breeder Reactors
with a Fast Neutron Spectrum

Breeder reactors with a fast neutron spectrum have a sufficiently high breeding ratio to allow independence of any external supply of fissile material in practical operation. The U-238/Pu-239 or the Th-232/U-233 conversion processes enable this type of reactor to utilize natural uranium or thorium theoretically with 100% efficiency and practically, including losses in the fuel cycle, with an efficiency in excess of 60%. This is a factor of 100 higher than the fuel utilization in present standard LWR's without reprocessing, and approximately a factor of 25 to 50 higher than in converter reactors operating with high conversion ratios and in the recycling mode (see Section 2.6 and Chapter 6).

For a specific energy release of 0.92 MWd(th) per gram of uranium or plutonium undergoing fission (see Section 2.1.5), this results in $0.92 \times 0.60 = 0.55$ MWd(th) of energy extracted from 1 gram of natural uranium. Conversely, this allows the uranium consumption of a *fast breeder reactor* (FBR) to be determined as

$$U_{FBR} = \frac{1}{0.55} \cdot 365 \cdot \frac{1}{0.40} \left[\frac{g\,U}{MWd(th)} \cdot \frac{d}{a} \cdot \frac{MW(th)}{MW(e)} \right] = 1660\,(g\,U/MW(e) \cdot a)$$

According to this formula, FBR's annually consume 1.66 te of natural uranium or depleted uranium fuel per 1000 MW(e) (at 40% thermal efficiency and 100% plant load factor). Because of their low uranium consumption, FBR's are quite insensitive to the price of uranium.

If the estimated world nuclear energy requirement listed in Fig. 1-1 is included in these assessments, breeder reactors with a fast neutron spectrum are found to open up an energy potential with the existing uranium reserves, which can be good for several thousand years. FBR's can utilize for nuclear fission not only fertile U-238, but also fertile Th-232. This would further add to the energy potential referred to above. Accordingly, with the use of FBR's, the supplies of nuclear fuel can be considered inexhaustible far beyond any time scale of conceivable planning interest. This is comparable with the energy potential that is hoped to be tapped by fusion reactors operating on the D-T cycle with lithium as the breeding material. However, fusion reactors are still in their infancy of development compared to FBR's.

5.2 Brief History of the Development of Fast Breeder Reactors

The principle of breeding had been recognized at the very beginning of the development of nuclear fission reactors. Construction of the first reactors with a fast neutron spectrum was begun in the United States of America before 1950. In the UK and the Soviet Union, this development started in the early fifties. These first-generation breeder reactors, however, mainly served for studies of fast neutron physics (CLEMENTINE, EBR-I, BR-1, BR-2). In addition, they were to demonstrate the feasibility of the technical solutions adopted (EBR-II, EFFBR, DFR). Consequently, some of them had rather low thermal power levels. In accordance with the state of the art at that time, they were equipped with fuel elements of enriched uranium or plutonium metals. Because of the high power per unit volume in the core, liquid metals such as mercury, sodium-potassium or sodium were used as coolants at relatively moderate coolant temperatures (*l*iquid *m*etal cooled *f*ast *b*reeder *r*eactors, LMFBR's).

Along with the development of LWR's it became apparent around 1960 that LMFBR cores would be able to attain the required high burnup of 100,000 MWd(th)/te at relatively low fuel cycle costs if fueled with ceramic fuel (PuO_2/UO_2). The use of PuO_2/UO_2 mixed oxide fuels and higher volume fractions of the sodium coolant in large LMFBR cores led to higher neutron moderation than in the initial small LMFBR cores. As a consequence, reactivity coefficients, such as the instantaneous Doppler coefficient and the sodium void coefficient, became subjects of detailed investigations (see Section 2.8.2). LMFBR test reactors up to powers of 60 MW(th) were therefore built in an interim phase between 1960 and 1970, which mainly served to demonstrate properties of ceramic fuels up to high burnups and the safe operation of this type of reactor (SEFOR, BR-5, BOR 60, RAPSODIE). All of these reactors had PuO_2/UO_2 mixed oxide fuels and were cooled with sodium. Coolant temperatures were chosen to allow steam conditions for high thermal plant efficiencies (Table 5-1).

The proven good predictability of the physics and safety characteristics and the good operating experience accumulated in this first generation of test reactors have been the basis for construction of the second generation of breeder power plants with electric powers around 250–300 MW(e) (Table 5-2). The technical data and the design features of these second generation reactors are already geared to the characteristics of commercial size LMFBR's. They have PuO_2/UO_2 mixed oxide fuels with target burnups of 70,000–100,000 MWd(th)/te and sodium as a coolant. The core outlet temperature of the coolant is approx. 550 °C, permitting steam conditions to be reached in the turbogenerator system with thermal efficiencies around 40%. The breeding ratios of these LMFBR's are around 1.15–1.20. Regarding the primary cooling systems, two design alternatives are applied: the loop and the pool type systems (see Section 5.4).

Three prototype reactors in this power category have already accumulated several years of operating experience. The French PHENIX prototype LMFBR has been in operation since 1973. The Russian BN 350 prototype LMFBR was first connected to the grid in 1973. The British PFR *prototype fast reactor* reached criticality in 1974 and has delivered power into the public grid system

Table 5-1. Second-generation experimental fast reactors (KfK)

		USA		USSR	France	Germany	Italy	Japan
		SEFOR	FFTF	BOR 60	RAPSODIE [a]	KNK-II	PEC	JOYO
Reactor power								
Thermal	MW (th)	20	400	60	40	58	130	100
Electrical	MW (e)	0	0	12	0	20	0	0
Core								
Fuel		PuO_2/UO_2	PuO_2/UO_2	PuO_2/UO_2 or UO_2	PuO_2/UO_2	PuO_2/UO_2 $+UO_2$	UO_2	PuO_2/UO_2
Core volume	liter	500	1030	53	45	320	420	280
Linear rod-power (max.)	W/cm	650	500	590	400	430	400	430
Neutron-flux (max.)	n/cm$^2 \cdot$s	6×10^{14}	7.2×10^{15}		3×10^{15}	2.3×10^{15}	2.8×10^{15}	4×10^{15}
Primary heat-transfer system								
Type		Loop	Loop	Loop	Loop	Loop	Loop	Loop
Coolant		Na	Na	Na	Na	Na	Na	Na
Number of coolant loops		1	6	2	2	2	2	2
Coolant temperature at								
Core inlet	°C	370	320	360 to 450	410	360	375	370
Core outlet	°C	430	480	600	530	550	525	500
Date of operation		1969 [b]	1980	1969	1970	1978	1978	1978

[a] Upgraded version.   [b] Shut down 1972.

since 1975. For all three prototype plants, the original design characteristics were confirmed in terms of fast reactor core physics, control and safety engineering parameters, and performance of the primary cooling system. The control and safety performance of these prototypes has been tested not only in normal operation, but also under simulated emergency cooling conditions, and has fully met expectations. Initial difficulties in running the large sodium components (heat exchangers and steam generators) have meanwhile been overcome. In the operation of PHENIX, a breeding ratio of 1.16 was demonstrated, while a large proportion of the irradiated fuel assemblies have already been reprocessed. The FFTF reactor in the USA, a large sodium cooled test reactor of 400 MW(th), began operation in 1979/80 after extensive pretest programs. It will mainly be used for fuels and materials testing. The Russian BN 600, the first LMFBR with an electric power of 600 MW(e), went into operation in 1980. It represents an intermediate step between the prototypes of the 250–300 MW(e) class and commercial size LMFBR's. In the Federal Republic of Germany, the small sodium cooled test reactor, KNK-II, 20 MW(e), is in operation and a 300 MW(e) prototype LMFBR, SNR 300, is under construction. Japan plans the construction of a 250 MW(e) prototype LMFBR, MONJU, after sufficient operating experience with its JOYO test reactor becomes available. In the USA, the design of CRBR and fabrication of its main components is in an advanced state.

The phase of commercial size demonstration power plants was begun in France in 1977 with the construction of SUPERPHENIX. The plant has a net electric power of 1200 MW(e), a thermal efficiency of 40% and, like PHENIX, is a pool type LMFBR. It is to be operated at full power in 1983. In the UK, the commercial size demonstration power plant, CDFR, is in its detailed planning phase. A decision about construction of the plant is expected to be taken in the next few years. In the USSR, plans for an 800 MW(e) and a 1600 MW(e) demonstration LMFBR have been completed. These are to serve as prototypes for later commercial size facilities. Similar studies on commercial size LMFBR's are underway in the USA, the Federal Republic of Germany and Japan.

5.3 The Physics of LMFBR Cores

5.3.1 LMFBR Core Design

LMFBR cores mainly consist of a cylindrical arrangement of hexagonal fuel elements surrounded radially by hexagonal blanket elements. The hexagonal fuel elements are designed in such a way that the inner core is also axially surrounded by an upper and a lower breeding blanket (Fig. 5-1). The blanket elements initially contain depleted uranium as uranium dioxide (UO_2). By contrast, the fuel elements of the core are filled with PuO_2/UO_2 mixed oxide as the fissile material. Most cores contain two radial zones of different enrichments in order to make the radial neutron flux and power distributions as flat as possible.

Table 5-2. *Fast breeder prototype and demonstration reactors (KfK)*

		France	
		PHENIX	SUPERPHENIX
Reactor power			
Thermal	MW(th)	568	3000
Electrical net	MW(e)	250	1200
Primary circuit		Pool	Pool
Number of primary circuits		3	4
Primary/secondary coolant		Na/Na	Na/Na
Coolant temperature at			
Core inlet	°C	385	395
Core outlet	°C	552	545
Steam conditions turbine inlet			
Pressure	bar	168	177
Temperature	°C	510	487
Diameter of reactor vessel	m	11.8	21
Core dimension			
Eq. diameter	cm	139	366
Height	cm	85	100
Fuel		UO_2/PuO_2	UO_2/PuO_2
Cladding material		316 SS	316 SS
Pin diameter	mm	6.6	8.5
Number of pins per fuel element		271	271
Core power density	kW(th)/l	406	280
Max. linear rod power	W/cm	450	450
Fuel burnup			
Average	MWd(th)/te	40,000	70,000
Maximum	MWd(th)/te	72,000	100,000
Breeding ratio		1.16	1.18
Status		Beginning of operation 1973	Expected beginning of operation 1983

In an LMFBR core, such as SUPERPHENIX with a net power generation of 1200 MW(e), the inner radial core zone contains 196 hexagonal fuel elements with 14% Pu enrichment, and the outer radial core zone comprises 171 hexagonal fuel elements with 18% Pu enrichment. The core is surrounded by 233 hexagonal blanket elements. It also incorporates 24 positions for hexagonal absorber elements containing boron carbide (B_4C) as an absorber. Insertion or withdrawal of these absorber elements regulates the criticality and power of the reactor and guarantees safe control and shutdown conditions. The core has a diameter of 3.66 m, a height of 1 m and a volume of 10.8 m³. This corre-

FRG	Japan	UK		USA	USSR	
SNR 300	MONJU	PFR	CDFR	CRBR	BN 350	BN 600
762	714	600	3230	975	1000	1470
312	300	250	1250	350	350[a]	600
Loop	Loop	Pool	Pool	Loop	Loop	Pool
3	3	3	6	3	6	3
Na/Na	Na/Na	Na/Na	Na/Na	Na/Na	Na/Na	Na/Na
377	397	394	370	388	300	380
546	529	550	540	535	500	550
160	125	128	160	100	49	142
495	483	513	486	462	435	505
6.7	7.0	12.2	19.2	6.2	6.0	12.8
178	179	147	290	188	158	206
95	93	91	100	91	106	75
UO_2/PuO_2	UO_2/PuO_2	UO_2/PuO_2	UO_2/PuO_2	UO_2/PuO_2	UO_2	(UO_2/PuO_2)
SS (1.4970)	316 SS	316 SS	316 SS	316 SS	SS	SS
6.1	6.5	5.8	5.8	5.8	6.1	6.9
169	169	325	325	217	169	127
290	300	380	410	380	430	550
380	360	480	400	500	440	530
57,000	80,000	75,000	100,000	50,000		
87,000	100,000			80,000	50,000	100,000
(1.0)	1.2	1.2	1.25	1.23	(1.4)	(1.3)
Expected beginning of operation 1986	Planned	Beginning of operation 1975	Planned	Planned	Beginning of operation 1973	Beginning of operation 1980

[a] $\cong 150$ MW(e) $+ 120,000$ m^3/d desalinated water

sponds to an average power per unit volume of the core of 280 kW(th)/l and a maximum power per unit volume of 435 kW(th)/l. The radial blanket has a thickness of 50 cm, the axial blanket a height of 30 cm.

The fuel assembly is made up of 271 fuel rods held in position by spacers. The fuel rods have outside diameters of 8.5 mm and are filled in the core region with cylindrical pellets made of PuO_2/UO_2 mixed oxide. The fuel rod cladding is made of Ti-stabilized austenitic steel with a wall thickness of approx. 0.7 mm. After fuel rod fabrication, the rods are filled with helium and welded tight at the ends. The fissile zone of a fuel rod is 1 m long. It is followed on both

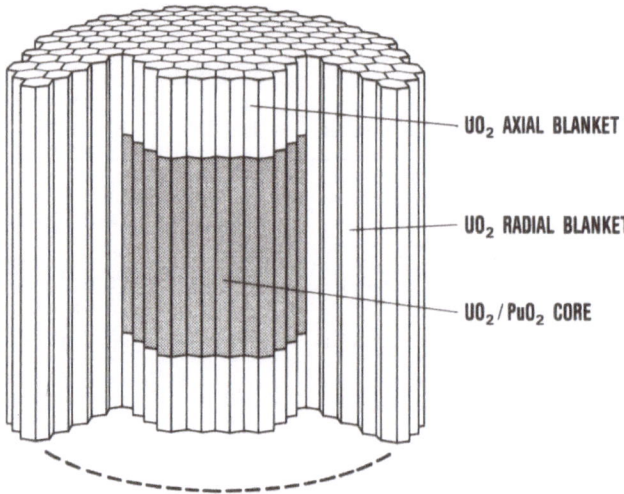

UO$_2$ AXIAL BLANKET

UO$_2$ RADIAL BLANKET

UO$_2$ / PuO$_2$ CORE

Fig. 5-1. Schematic of a fast breeder reactor core

ends by a fertile zone each of 30 cm length. Below the lower axial blanket there is a gas plenum 850 mm long, in which gaseous fission products accumulate. Sodium flows through the fuel elements at a rate of approx. 3–6 m/s. On its way through the fuel elements, it is heated from 395 °C to 545 °C. The fuel elements are surrounded by a hexagonal can made of austenitic or ferritic steel with a wall thickness of 4.6 mm. The complete fuel element has a length of 5.4 m and is 16.6 cm across the hexagonal flats. The blanket rods have diameters of 16.3 mm, which are larger than the diameters of the fuel rods; the blanket elements accordingly only contain 91 rods. The maximum fuel rod power in the core is roughly 450 W/cm of fuel rod length. A 1200 MW(e) core contains some 4.6 te of Pu-239 and Pu-241. This corresponds to 3.7 te of (Pu-239 and Pu-241)/GW(e). The core contains a total of 37 te of PuO$_2$/UO$_2$, the axial breeding blanket has 22 te of UO$_2$, and the radial blanket has 52 te of UO$_2$. Each absorber element consists of 18 absorber rods with an outside diameter of 18.8 mm, filled with B$_4$C pellets. B$_4$C is enriched to about 93% in its B-10 isotope.

5.3.2 Energy Spectrum and Neutron Flux Distribution

The cores of LMFBR's in the 1200 MW(e) power category contain some 34–37% of PuO$_2$/UO$_2$ mixed oxide fuel, some 39–41% of sodium, and 22–27% of austenitic steel. This results in a neutron energy spectrum (see Fig. 2-5), whose mean energy is in the range of 200 keV. The low energy part of the neutron energy spectrum extends into the resonance ranges of the fission cross sections of the fuel and the capture cross sections of the fuel, steel and sodium. In this region, the resonances of these cross sections in part overlap very strongly. This must be taken into account in determining the temperature coeffi-

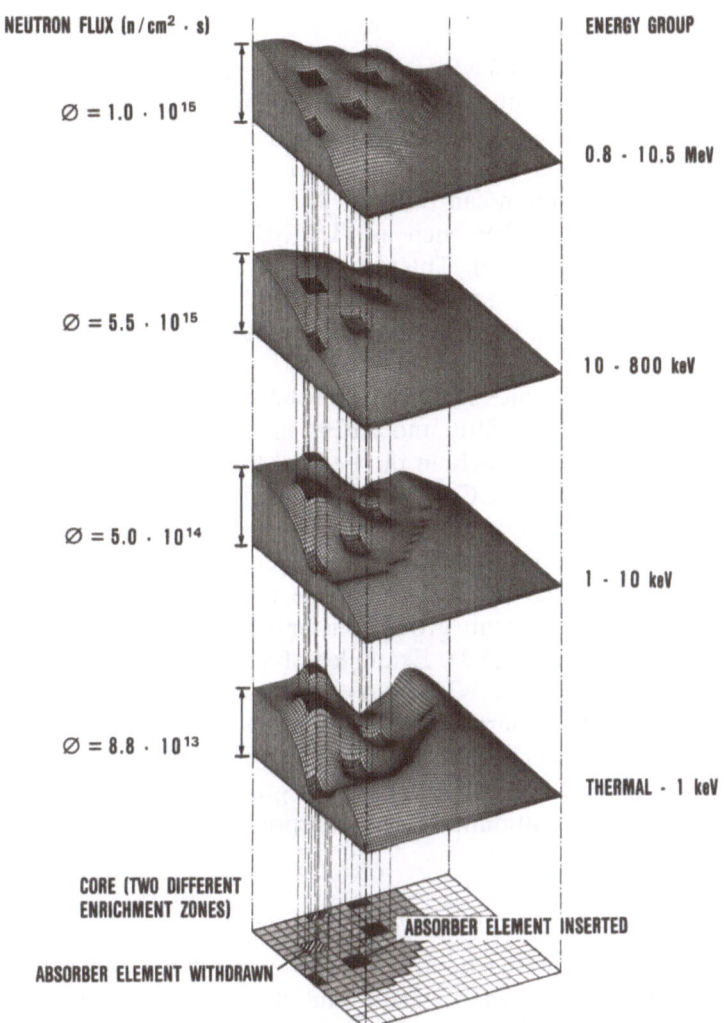

NEUTRON FLUX (n/cm² · s)

∅ = 1.0 · 10¹⁵

∅ = 5.5 · 10¹⁵

∅ = 5.0 · 10¹⁴

∅ = 8.8 · 10¹³

CORE (TWO DIFFERENT
ENRICHMENT ZONES)

ABSORBER ELEMENT WITHDRAWN

ENERGY GROUP

0.8 · 10.5 MeV

10 · 800 keV

1 · 10 keV

THERMAL · 1 keV

ABSORBER ELEMENT INSERTED

Fig. 5-2. Neutron flux distribution in a fast reactor core (KfK)

cient of reactivity (Doppler coefficient). The high energy part of the neutron energy spectrum extends into the region of the fission cross sections of U-238, where it contributes to a pronounced fast fission effect (see Section 2.5).

Fig. 5-2 shows the spatial distribution of the neutron flux in such an LMFBR core for a rough subdivision into only four groups of the neutron energy spectrum. The neutron flux drops markedly towards the edge of the core. Neutron flux and power distribution attain their peaks in the center of the core. At maximum power generation, 450 W/cm, the neutron flux in the center of the core reaches approximately 7×10^{15} n/cm²s, integrated over the entire energy spectrum. Neutrons leaving the core are captured in the U-238 blanket material of the radial and axial blankets.

5.3.3 Breeding Ratio

The total breeding ratio of a 1200 MW(e) core, such as SUPERPHENIX, is 1.18. It is composed of contributions by the core of 0.8 and by the axial and radial blankets of 0.38. The annual excess production in that case is 83 kg of (Pu-239 and Pu-241)/GW(e)·a at a load factor of 0.75 and 2% losses in the fuel cycle. This excess plutonium can be saved for construction of another large LMFBR core. For an LMFBR such as SUPERPHENIX, this would take some forty years (doubling time). This doubling time is relatively long. However, it plays no major role during the commercial introduction phase of LMFBR's since the plutonium is predominantly made available externally by operating LWR's.

The physics of large LMFBR cores has been thoroughly investigated in numerous critical assemblies, such as ZPR and ZPPR in the USA, ZEBRA in the UK, MASURCA in France, SNEAK in the Federal Republic of Germany, FCA in Japan, and BES in the USSR. Computation techniques were developed to account for energy self-shielding in the resonance range of the cross sections and for heterogeneity effects in fast reactor cores. Microscopic cross section data libraries, e.g., the *E*valuated *N*uclear *D*ata *F*ile (ENDF), allow special computer programs to be used to prepare group constants as inputs for extended multigroup calculations (see Section 2.3). Examples of such programs are the MC2-2SDX program developed at ANL (USA), the GALAXY program of the UKAEA or the MIGROS program of KfK (Germany). A few hundred up to 2000 neutron energy group calculations are run to calculate the neutron energy spectra within the LMFBR core. Condensed group constant sets are developed and finally used for multidimensional transport or diffusion calculations.

5.3.4 Reactivity Coefficients and Control Stability

Changes in k_{eff} resulting from the buildup of fission products and higher actinides and the changes in k_{eff} during startup and shutdown of the reactor determine the design of control and shutdown systems. Because of the relatively small cross sections in the important range of a fast reactor neutron energy spectrum, these changes in k_{eff} are smaller in LMFBR's than in thermal converter reactors (see Section 2.8.3). In a 1200 MW(e) core, such as SUPERPHENIX, the buildup of fission products and actinides causes only a 3% change of k_{eff} over a power cycle of roughly one year. The feedback reactivities (Doppler coefficient, density changes and core expansion) to be covered as a result of temperature changes during startup from zero power to full power only amount to roughly a 1.5% change of k_{eff}. As a consequence, sufficient absorber material for a control system and two shutdown systems can be accommodated in about 24 positions of LMFBR cores, such as SUPERPHENIX. The aggregate negative reactivity of these control and shutdown systems is about -13% of k_{eff}. Since the capture cross sections for absorbers are relatively small in the important range of the fast reactor neutron energy spectrum, the absorber material B_4C has to be enriched in its B-10 content in order to provide the desired amount of negative reactivity within a given volume available for absorber elements.

For SUPERPHENIX, the reactivity coefficient for increasing power (power coefficient) is $-0.19 \times 10^{-5}/\mathrm{MW}(\mathrm{th})$, at reference power level. This means that increases in power are counteracted by strong negative feedback reactivities. Accordingly, large LMFBR cores can be controlled in an inherently stable mode.

5.3.5 The Doppler Coefficient

The negative Doppler coefficient makes the greatest contribution to the negative power coefficient. It instantaneously generates a negative reactivity effect whenever the power and the fuel temperature increase. It results from the Doppler broadening of resonances of the capture and fission cross sections. In large cores, its main contribution comes from the range of the neutron energy spectrum below 75 keV. Unlike thermal reactors, where the main contributions to the Doppler coefficient arise from resolved resonances of the cross sections, some 50% of the contributions in fast reactors result from the range of statistical resonances where the details of nuclear data are not known precisely. The negative contributions by the capture resonance cross sections of U-238 by far outweigh the small positive contributions by fission resonance cross sections of Pu-239 (see Section 2.8.2).

The Doppler coefficient in fast reactor cores was measured in a number of critical assemblies, such as ZPPR, ZEBRA, SNEAK etc., and in the SEFOR reactor. In large LMFBR cores, there is at present only an uncertainty of about $\pm 15\%$ between theoretical calculations of the Doppler coefficient and its experimental verification. The Doppler coefficient under conditions of rising fuel temperature, T_f, in large LMFBR cores is in the range of

$$\frac{1}{k_{\mathrm{eff}}} \cdot \frac{\partial k_D}{\partial T_f} = -\frac{0.008}{T_f^x}$$

For normal operating conditions, the exponent x is between 0.8 and 1.0. The Doppler coefficient can be influenced by the composition of the core:
– Within the composition of fuel isotopes, Pu-241 raises the Doppler coefficient, whereas Pu-240 lowers it.
– The fractions of sodium and steel in the core influence the contribution of the neutron energy spectrum below 75 keV. Voiding of the sodium from the core (boiling) may reduce the Doppler effect by up to 50%.
– Adding solid moderator, such as BeO, increases the fraction of the neutron spectrum below 75 keV and raises the Doppler coefficient.

5.3.6 The Coolant Temperature Coefficient

Increases in temperature of the sodium coolant decrease its density. Increases in temperature of the cladding and fuel result in radial expansion of the fuel rod and, at the same time, expulsion of sodium from the core. This leads to a reduction of the sodium density per cm^3 of core volume. Three different individual effects must be distinguished. The reduction in sodium density

– increases neutron leakage. This effect is negative (reducing k_{eff}), dominating mainly in the outer core regions at high spatial gradients of the neutron flux. It depends on the size of the reactor core;
– changes moderation. The neutron energy spectrum is shifted towards higher energies. This effect is positive (increasing k_{eff}), because the shift towards higher energies slightly increases the average η-value of fissile isotopes and also increases fast fission of U-238;
– reduces neutron absorption. This usually relatively small reactivity effect is positive throughout the core (adding to k_{eff}).

While the leakage term is influenced by the size and shape of the reactor core, the spectral term depends on the plutonium enrichment, the content of higher Pu isotopes, and the composition of the core, e.g., the solid moderator fraction. In a cylindrical two zone LMFBR core with 1200 MW(e) power, the overall coolant temperature coefficient of reactivity is 5×10^{-6} per °C of coolant temperature increase. Special core designs can reduce the coolant temperature coefficient. For instance, for a so-called heterogeneous core (see Section 5.7) containing also blanket elements in the fuel zone, it can be reduced to about 2×10^{-6} per °C of coolant temperature increase. However, in this case also the negative Doppler coefficient is reduced.

The positive coolant temperature coefficient is dominated in the normal operating power range of LMFBR cores by the quantitatively higher negative contributions of the Doppler coefficient, by fuel expansion and by structural expansion of the core. Only if in extremely improbable accident situations the central regions of the core were to be voided as a result of sodium boiling, would the resulting coolant density changes lead to overall positive reactivity contributions.

5.3.7 Fuel and Structural Temperature Coefficients

Temperature increases of the fuel cause the fuel rod to expand axially and decrease its density, thus reducing k_{eff}. In a large LMFBR core, the fuel temperature coefficient of reactivity due to axial expansion is approximately -2×10^{-6} per °C of fuel temperature increase.

The radial expansion of the core structure resulting from temperature increases in the steel causes the core to expand and thereby reduces the fuel density. This again results in a negative temperature coefficient of reactivity (structural expansion coefficient). In a large LMFBR core, such as SUPER-PHENIX, the structural expansion coefficient is determined by two effects. These are the expansion of the core grid plate and bowing effects of core fuel elements caused by radial temperature gradients. Radial clamping and the structural design of the core prevent bowing fuel elements from leading to significant overall positive bowing reactivity contributions. Early difficulties experienced with a positive bowing coefficient in the EBR-I are avoided in today's LMFBR's. The expansion coefficient of the grid plate of large LMFBR cores is in the range of -10^{-5} per °C of coolant inlet temperature variation. The bowing coefficient caused by radial temperature gradients is in the range of -5×10^{-7} per °C of temperature difference between core inlet and outlet.

5.3.8 Delayed Neutron Characteristics and Prompt Neutron Lifetime

In addition to the feedback reactivity coefficients, the following set of characteristics govern the dynamics of LMFBR cores, as with any other fission reactor cores (see Section 2.1):
- the effective fraction of delayed neutrons, β_{eff},
- the average decay constant of delayed neutron precursors, λ,
- the lifetime of prompt neutrons, l_{eff}.

These characteristics are compared in Table 5-3 with those of a PWR representing the class of thermal converter reactors. This comparison is made to show differences between thermal converter reactors and LMFBR cores and to discuss whether LMFBR cores show specific peculiarities in their dynamic behavior compared to thermal converters. As can be seen from Table 5-3, the average decay constants of the precursors of delayed neutrons are similar for U-235 fueled thermal spectrum PWR's and Pu-239 fueled fast spectrum LMFBR's. The effective delayed neutron fraction, β_{eff}, of a plutonium fueled LMFBR is smaller by almost a factor of 2 than β_{eff} in PWR's (β_{eff} includes the fast fission contribution of U-238 in LMFBR cores as well as the thermal fission of Pu-239 in PWR cores). However, β_{eff} is not the only decisive quantity. As is apparent from Section 2.8, the dynamic behavior of a reactor core is determined by $(\rho-\beta_{eff})$ as well as by λ and l_{eff}. The prompt neutron lifetime, l_{eff}, shows the greatest difference (see Table 5-3). For LMFBR cores, l_{eff} is

Table 5-3. *Comparison of design characteristics (approximate values) of control and shutdown systems of PWR's and LMFBR's (KfK)*

		PWR (1300 MW(e))	Fast breeder SNR 300
Fuel		U-235	Pu-239
Prompt neutron lifetime	$l_{eff}(s)$	2.5×10^{-5}	4.5×10^{-7}
Effective fraction of delayed neutrons	$\beta_{eff} \cong 1$ $	0.005–0.0065	0.0035
Average decay constant of delayed neutron precursors	λ (1/s)	0.077	0.065
Speed of control-rod movements	mm/s 10^{-2} $/s	1 2.5	1.2 $\leqq 4$
Speed of shutdown rod movements	cm/s	156	85–190
Delay time prior to reaction of shutdown system	s	0.2	0.2
Time span for full insertion of shutdown rods in core	s	2.5	0.6
Reactivity of shutdown system	Δk $	11	≈ 10
Reactivity of shimrod system (burnup reactivity)	Δk $	19	≈ 8

REACTOR POWER (MW(e))

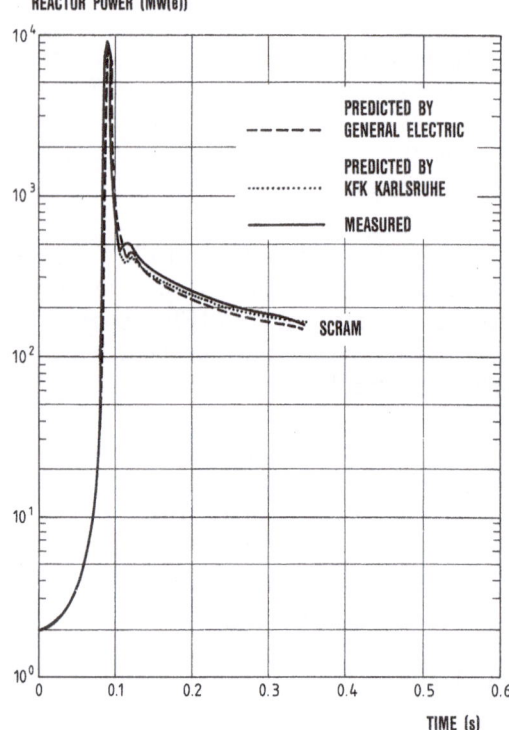

Fig. 5-3. SEFOR core I superprompt power transient (KfK)

about two orders of magnitude smaller than in a PWR core. As explained in Section 2.8. the whole non-steady state regimes of reactor cores can be divided into two subregimes, i.e., the delayed critical regime, in which all normal control and power rise operations take place, and the prompt supercritical regime, which can become important for the class of extremely unlikely severe reactivity accidents. Differences in the quantitative values for (ρ-β_{eff}), λ and l_{eff} of LMFBR cores as against PWR or other thermal converter cores must be analyzed in detail for these two different dynamic regimes.

In the delayed critical regime, i.e., as long as reactivity increases, ρ, remain essentially below β_{eff}, the dynamic behavior of any reactor core is determined by the average decay constant, λ, of the precursors of the delayed neutrons. As can be seen from Table 5-3, the decay constants of plutonium fueled FBR cores and U-235 fueled PWR's are very similar. Consequently, the control behavior of LMFBR cores in the delayed critical range for control and power rise or shutdown operations is not different from that of PWR or other thermal converter cores. The designs of control and shutdown systems are very similar to those of PWR's or other thermal converter reactors. This is apparent from Table 5-3.

Under prompt and prompt supercritical conditions, i.e., when $\rho \geq \beta_{eff}$, the short neutron lifetime is not a fundamental problem as long as the power coefficient of reactivity or the dominant fuel temperature coefficient of reactivity is negative. With a negative power coefficient, as represented by the negative Doppler coefficient, LMFBR cores are subjected to sharply limited narrow power bursts. Such prompt supercritical power bursts were intentionally produced within the SEFOR experimental program, as is shown in Fig. 5-3. The SEFOR core was safely shut down even after such a prompt supercritical power burst. The energy released during such a power burst even decreases with decreasing neutron lifetime in the presence of a highly negative Doppler coefficient.

In conclusion, it can be stated that the differences in β_{eff} and l_{eff} of LMFBR cores do not lead to significant differences in control behavior. Therefore, there is no need for electromechanical designs of control and shutdown (scram) devices to be significantly different from those in thermal converters.

5.4 Technical Aspects of Sodium Cooled FBR's

Sodium as a coolant of fast reactor cores at present clearly dominates fast reactor research, development and demonstration programs all over the world. Helium cooled fast reactor concepts were pursued as an alternative coolant concept, but no helium cooled fast test or demonstration reactor has been built so far.

The design concept of liquid sodium cooled fast breeder reactors is mainly determined by the thermal and nuclear properties of the coolant: good heat transfer, small moderating effect and low neutron capture cross section. Capture processes in sodium lead to the formation of Na-24 with a halflife of 15 hours. This activates the primary sodium flowing through the core. Sodium has a high melting point (98 °C) and a high boiling point (892 °C at atmospheric pressure and 900–1000 °C at coolant pressures within the core). It has a high specific heat and very good thermal conductivity.

The high melting point of sodium requires preheating of pipes and components of the cooling system before sodium filling and after certain maintenance periods. The high boiling point allows high coolant temperature conditions to be maintained at very low system pressures (6 to 10 bar). This results in high thermal efficiencies around 40% for the whole LMFBR plant. The relatively high specific heat permits moderate coolant flow rates of 2 to 6 m/s in the fuel elements and low pumping powers, while the good thermal conductivity, coupled with other thermal properties, leads to very good natural convection conditions in the core and the cooling system in case of a pump failure. However, these excellent thermal properties also give rise to special design and operating requirements. Thermoshock problems at pipe nozzles and valves in the reactor vessel must be avoided during short term power reductions or reactor scram conditions.

Although sodium is not very corrosive to stainless steel, its impurities, mainly oxygen and carbon, must be held at acceptably low contents (5 to 10 ppm for O_2, < 50 ppm for carbon). High impurity contents cause radioactive corrosion products to be detached from the surfaces of fuel claddings and then transported to low temperature parts of the primary coolant system (heat exchangers). Such undesired concentrations of corrosion products must be avoided

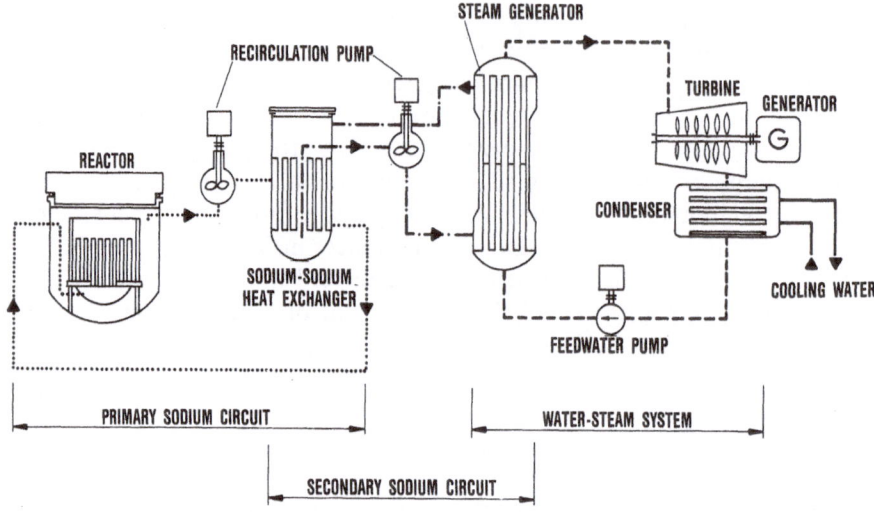

Fig. 5-4. Heat transfer system of a sodium cooled fast breeder reactor (LMFBR) (INTER-ATOM)

because of possible maintenance and repair difficulties, which could arise after several years of plant operation. It was observed that Mn-54 produced by the (n,p)-reaction of Fe-54 is rapidly transferred from the hot core regions to low temperature areas.

The opacity of sodium affects the design of the refueling systems and requires ultrasonic devices to be used for supervision of refueling and repair processes.

Major design consequences arise from the potential of sodium to enter into chemical reactions with water and air. This property, together with the fact that sodium becomes radioactive under neutron irradiation in the core, leads to a plant design with
– a primary cooling system containing the radioactive sodium heated up in the core;
– a secondary cooling system coupled with the primary system by intermediate heat exchangers;
– a tertiary water system producing steam for electricity generation in the turbo-generator system (Fig. 5-4).

Within the primary cooling system, radioactive sodium is protected against air by steel barriers and cells filled with argon or nitrogen. Radioactive sodium of the primary cooling system is separated from the non-radioactive sodium of the secondary system by the steel tubes of the intermediate heat exchangers.

So far, two principal design concepts have been used for LMFBR's, i.e., the pool and the loop type concepts. In the pool type concept, all primary system components with the core, primary pumps and intermediate heat exchangers are built into the pool tank filled with sodium. This concept was used in PFR, PHENIX, SUPERPHENIX and BN 600. In the loop type concept by contrast, only the reactor core is in the reactor vessel, the primary sodium being pumped to the intermediate heat exchanger through a piping system.

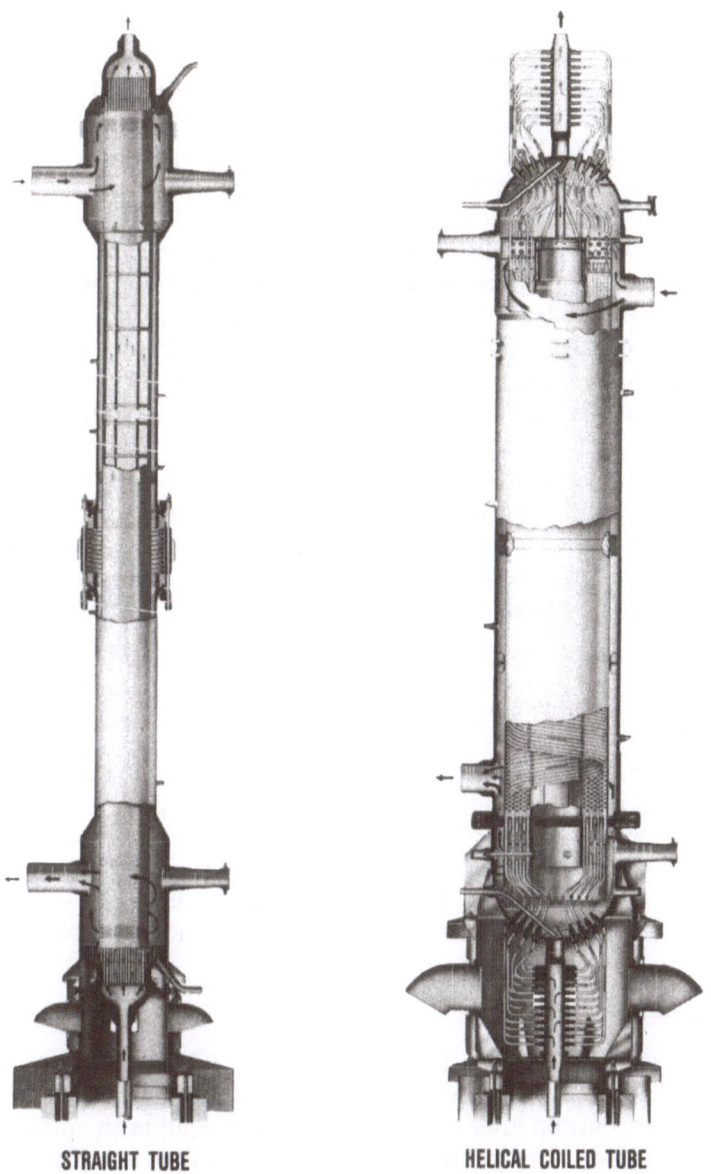

STRAIGHT TUBE   HELICAL COILED TUBE

Fig. 5-5. Steam generators of the SNR 300 LMFBR demonstration plant (NERATOOM)

This design concept was chosen in BN 350, SNR 300, FFTF and MONJU. Both design concepts have a number of advantages and disadvantages, which depend on design details and roughly balance out. Only future experience in the operation and licensing of large LMFBR's can tell whether one of these design concepts must be preferred to the other.

Much care is required in the design and construction of sodium heated steam generators. LMFBR steam generators contain non-radioactive secondary-system sodium and water or steam, both separated from each other only by the tube walls. As is well known sodium and water develop a violent chemical reaction when brought into contact, due to their different physical and chemical properties. Many design aspects, therefore, have to be taken into account, such as fabrication, safety, operational availability, leak detection, inspectability, corrosion effects, repair, etc. Consequently, much research within LMFBR programs has been devoted to steam generator development and testing in full scale test rigs. Fig. 5-5 shows two different concepts of steam generators, a straight tube steam generator and a helical tube steam generator, both belonging to the SNR 300 plant. A number of other design concepts are employed in other fast reactor projects. Over the past decades, sodium-water interactions and design optimization of pressure relief systems for LMFBR steam generators have been the subject of intensive research. There is no doubt that large LMFBR steam generators can be built upon a sound and safe technological concept. Analysis of the difficulties encountered with steam generators in the early phases of BN350 and PFR operation has shown that no major technological problems will be encountered in LMFBR steam generators. However, the proper choice of tube steel (intergranular corrosion, weldability, etc.) and quality assurance during fabrication are very important factors.

5.5 SUPERPHENIX – A Commercial Size Demonstration LMFBR

SUPERPHENIX has a thermal power of 3000 MW and an electric net power of 1200 MW with a net plant efficiency of 40%. The reactor core is cooled by sodium. The design of the primary heat transfer systems is based on the pool principle, i.e., the primary coolant systems are contained in the pool tank. The design of the reactor core and the blanket of SUPERPHENIX has been described in Section 5.3.1. The technical data are summarized in Table 5-4.

5.5.1 Reactor Core and Blankets

The fuel elements remain in the core for approximately 2.2 years, attaining a burnup of 70,000–100,000 MWd(th)/te. Half the fuel elements are unloaded annually at this maximum burnup after having reached their equilibrium cycles. The blanket elements have a longer residence time, the inner rows of elements being unloaded earlier and more frequently than the outer ones. By the time they reach their maximum burnups of 100,000 MWd(th)/te, the fuel elements have been exposed to severe materials stresses and radiation damage. Fuel rod claddings attain maximum temperatures at their inner surfaces of 650 °C and, towards the end of their burnup periods in the core, must sustain internal pressures of about 50 bar. This internal pressure results from the gaseous fission products building up in the fuel rod. At high burnup and high specific power, the PuO_2/UO_2 mixed oxide fuel considerably changes its physico-mechanical and chemical behavior. The fuel rod cladding must withstand these loads and, in addition, considerable thermal stresses and still maintain its good mechanical properties. Fuel rod claddings and fuel element cans are also exposed to high

Table 5-4. *Design characteristics of SUPERPHENIX (NOVATOME)*

Reactor Power		
Thermal	MW(th)	3000
Electrical gross	MW(e)	1240
Electrical net	MW(e)	1200
Plant efficiency	%	40
Reactor Core		
Fuel		UO_2/PuO_2
Core outer diameter	cm	360
Core height	cm	100
Pu eq. enrichment		
Inner core zone	%	14
Outer core zone	%	18
Pu eq. mass	te	4.8
Total UO_2/PuO_2 mass in core	te	36.9
Fuel rod outer diameter	mm	8.5
Core power density		
Average	kW(th)/l	280
Maximum	kW(th)/l	435
Max. fuel rod power	W/cm	450
Max. burnup	MWd(th)/te	100,000
Blankets		
Fuel		UO_2
Axial thickness	cm	30
Radial thickness	cm	50
Total UO_2 mass	te	74
Fertile rod outer diameter	cm	15.8
Total breeding ratio		1.18
Fissile Fuel Bundles		
Number of bundles		364
Number of pins per bundle		271
Pin total length	m	2.7
Bundle total length	m	5.4
Cladding material		stainless steel
Cladding maximum rated temperature	°C	620
Fertile Fuel Bundles		
Number of bundles		233
Number of pins per bundle		91
Pin total length	m	1.94
Bundle total length	m	5.4
Cladding material		stainless steel
Control Bundles		
Main shutdown system:		
Number of bundles		21
Number of absorber elements per bundle		31
Pin length	m	1.3
Cladding material		stainless steel

Table 5-4 (continued)

Control Bundles (continued)		
Standby shutdown system:		
Number of bundles		3
Number of absorber elements per bundle		3
Cladding material		stainless steel
Main Reactor Vessel		
Shape		cylindrical with torispherical bottom head
Inner diameter	m	21
Height	m	19.5
Material		stainless steel
Primary System		
Coolant		sodium
Primary Na mass	te	3250
Rated flow	te/s	4×4.1
IHX sodium outlet temperature	°C	392
Core sodium inlet temperature	°C	395
Core sodium outlet temperature	°C	545
IHX sodium inlet temperature	°C	542
Secondary System		
Coolant		sodium
Secondary Na mass	te	1500
Rated flow	te/s	4×3.3
SG sodium outlet temperature	°C	345
IHX sodium inlet temperature	°C	345
IHX sodium outlet temperature	°C	525
SG sodium inlet temperature	°C	525
Water-Steam System		
SG water inlet temperature	°C	237
Turbine steam inlet temperature	°C	487
SG water inlet pressure	bar	218
Turbine steam inlet pressure	bar	177
Rated flow	kg/s	4×340

temperatures and radiation damage brought about by fast neutrons. The maximum neutron fluence reaches approximately 2×10^{23} n/cm^2. As a consequence, volume swelling, creep effects and high temperature embrittlement occur in the steel of the fuel rod claddings and the fuel element cans. Austenitic steel, such as Ti-stabilized alloys, and ferritic steel developed more recently are used under these severe conditions.

5.5.2 Reactor Tank and Primary Coolant Circuits

The core and the radial blanket are supported on a core diagrid plate resting in a support structure in the pool tank (Fig. 5-6). Sodium enters this double

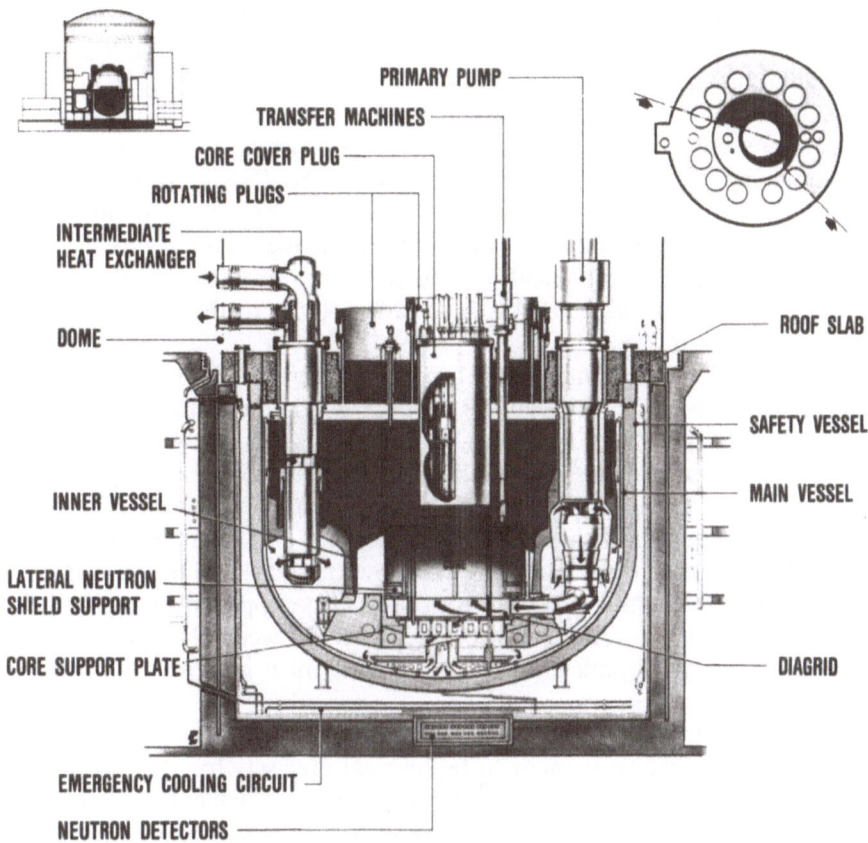

Fig. 5-6. Core and primary cooling system of SUPERPHENIX (NOVATOME)

bottom core support plate at a temperature of 395 °C from the primary sodium pumps and flows into the fuel elements through openings in the fuel element feet. It flows through the core and radial blanket from bottom to top and is heated to a temperature of 545 °C. The core and the radial blanket are surrounded by a radial steel reflector and the neutron shield. The hot sodium now flows upward within the inner vessel structure into the large sodium pool. It moves on into the inlet openings of eight *i*ntermediate *h*eat e*x*changers (IHX) arranged radially around the core inside the pool tank. Sodium passes through these IHX's from top to bottom and is cooled to 392 °C by secondary sodium moving in a countercurrent flow. The cooled sodium is now taken in by four primary sodium pumps and forced back into the core diagrid plate. The four primary sodium pumps are also installed radially, each between two IHX's in the pool tank.

The internal tank structure serves to separate the hot sodium leaving the core at 545 °C from the sodium cooled to 392 °C. For this purpose, it contains the openings for the eight IHX's and the four primary sodium pumps. The sodium chamber, which is at a temperature of 545 °C, contains approx. 2100 m^3 of sodium, while the chamber holding the sodium cooled to 392 °C contains a volume of roughly 1900 m^3 of sodium. Each of the four primary sodium pumps moves 4.1 te/s, which means that the sodium flow through the core and the radial blanket is roughly 16.5 te/s.

The sodium contained in the pool tank is kept at atmospheric pressure. The primary pumps only generate a fairly low pressure, which is mainly necessary to overcome the drag forces in the core and in the IHX's. The large mass of sodium in the pool tank ensures that the whole primary cooling system will react only slowly to increases in power or in the outlet temperature of the core. The radial neutron shield around the core prevents activation of the secondary sodium in the eight IHX's. The radioactivity of the primary sodium is approximately 18 mCi/cm^3.

The pool has an inner diameter of 21 m and a height of 19 m. It is a double walled structure, the space between the primary main vessel and the secondary safety vessel being filled with nitrogen. The wall thickness of the double walled pool tank is 25–60 mm. The pool tank is made of austenitic steel welded on site. It is closed at the top by a roof slab up to 3 m thick, which also carries the two rotating plugs with the fuel element transfer and loading machine and the core cover plug with the control and shutdown systems and the instrumentation. On its outer circumference, the roof slab also supports the eight IHX's and the four primary sodium pumps. The space between the open primary sodium level in the pool tank and the roof slab is filled with argon as a cover gas.

Two eccentric rotating plugs allow the positions of each fuel element, blanket element or radial shielding element to be reached precisely for the loading and unloading procedures. The fuel elements and blanket elements are removed from the core or the radial blanket by means of the fuel element transfer machine and brought into a transfer cask in a loading position. At this point, they are grabbed, pulled up over a ramp at an inclined angle and taken into the fuel element storage facility beside the pool tank. During the loading and unloading processes, the absorber rod drives are uncoupled. All absorber rods are in the core. The fuel element transfer machine can move freely above the core.

The whole pool tank is contained in a cavity of reinforced concrete lined with a leaktight steel liner and, on the inner surface, equipped with a cooling system for emergency cooling conditions. The upper roof slab also contains an integrated sodium purification system controlling all impurities in the primary sodium (e.g., oxygen and carbon contents). It is fitted with pipe connections for the cover gas feed (argon) and its purification. The roof slab with all penetrations, pump motors, IHX's, absorber rod drive systems and fuel element transfer machine is enclosed in a dome made of steel.

5.5.3 Secondary Coolant Circuits and Steam Generators

SUPERPHENIX has four independent heat transfer systems. One of them is shown in Fig. 5-7. It consists of the primary system described above, with a primary pump and two IHX's, the secondary sodium system, and the tertiary water-steam system. The secondary sodium leaves the two IHX's at a tempera-

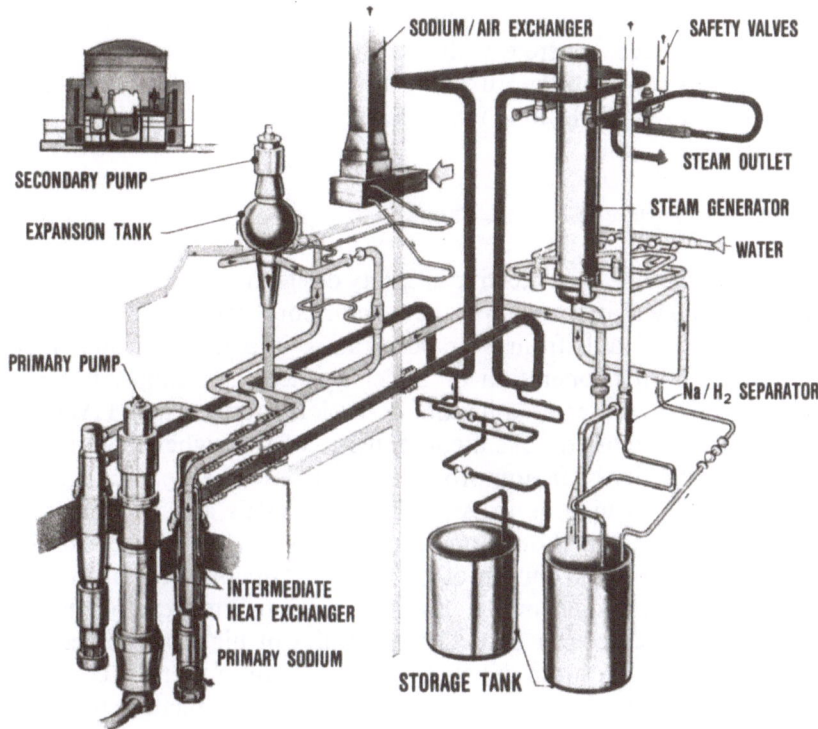

Fig. 5-7. Main secondary sodium circuits and steam generators of SUPERPHENIX (NO-VATOME)

ture of 525 °C. It penetrates the steel dome through two pipes with flexible joints and flows to a steam generator, where it delivers its heat of 750 MW(th) to the water-steam system, reducing its temperature to 345 °C. From the steam generator, the secondary sodium flows back to a large spherical expansion tank. The secondary sodium pump is immersed in this expansion tank. It forces the secondary sodium back into the two IHX's through two separate pipes. The inlet temperature into the IHX of the secondary sodium is 345 °C. In the steam generators, steam of 218 bar and 487 °C is produced. The steam is fed to the turbogenerator system.

The afterheat of the reactor following shutdown is removed by the four systems. After shutdown, the pumps are kept running at a few percent of their full speed by small pony motors. These pony motors are driven by auxiliary diesel engines. The decay heat of the reactor core can also be removed by natural convection. In case of non-availability of the decay heat removal system through the steam generators, the decay heat can be removed by hot sodium extracted from the expansion tank and transporting of the heat to sodium-air heat exchangers. The sodium is cooled there and passed back to the return pipes of the IHX. In the sodium-air heat exchangers, the heat can be removed either through forced convection by means of air or by natural convection through an exhaust air stack.

The outer reactor building has a height of 80 m, an inner diameter of 64 m and is made of reinforced concrete with a wall thickness of about 1 m. The reactor building also contains the auxiliary argon systems and storage tanks and rooms for handling and cleaning irradiated fuel elements.

5.6 Safety Design Aspects of LMFBR Plants

As in all nuclear reactors, the objective of safety design measures in LMFBR's is to prevent the radioactivity present in the reactor core, primary heat transfer system and fuel storage tank from ever penetrating into the environment of the plant, either in normal operation or under accident conditions. LMFBR plants produce roughly the same amounts of fission products per GW(th)·a as thermal converter reactors do. As a consequence of the higher thermal efficiency (40%), this quantity of fission products, relative to the electricity generated, is approximately as high as in AGR's and HTGR's and slightly lower than in LWR's and HWR's. Although plutonium fission by fast neutrons clearly dominates, the percentage distribution of the different fission product isotopes differs only slightly from that encountered in thermal converter reactors usually fueled with uranium (see Section 2.1). The quantities of higher actinides produced in LMFBR's per GW(th)·a are slightly larger than in thermal converter reactors. The absolute quantity of fissile plutonium differs relative to U-235 and U-233 fueled reactors and is only reached approximately by converter reactors recycling plutonium. Also the total fuel inventory in the core and the blankets, which is 111 te of fuel at 1200 MW(e), can be compared, e.g., with 104 te of U ($\cong$ 122 te UO_2) in a 1240 MW(e) PWR (see Table 4-1). However, in LMFBR's, roughly 90% of the power is produced in the core, which constitutes only about half of the fuel volume and fuel weight.

5.6.1 The Multiple Barrier Principle

One of the most important safety design principles applied in LMFBR's, as in any other reactor concept, is the principle of multiple safety barriers between the fuel generating the energy, and thus also the radioactive fission products, and the environment. This principle is explained by the example of the SNR 300 which, unlike SUPERPHENIX, is a loop type reactor.

The first solid barrier against the fuel is the tightly welded steel cladding of the fuel rod. Solid fission products remain enclosed in the crystalline structure of the fuel. Gaseous fission products are collected in the fission product gas plenum of the fuel rod. Leaks in the cladding are detected by means of fission product gas monitors in the cover gas or by delayed neutron monitors in the primary sodium. If necessary, the reactor will be shut down so that the element containing the defective fuel rod can be detected and replaced.

The second barrier is constituted by the reactor tank and the primary system. The reactor tank is limited at the top by the rotating shield plug. Argon as a cover gas is contained in the space between the free sodium level and the rotating shield plug. The IHX's and the pumps of the primary systems are

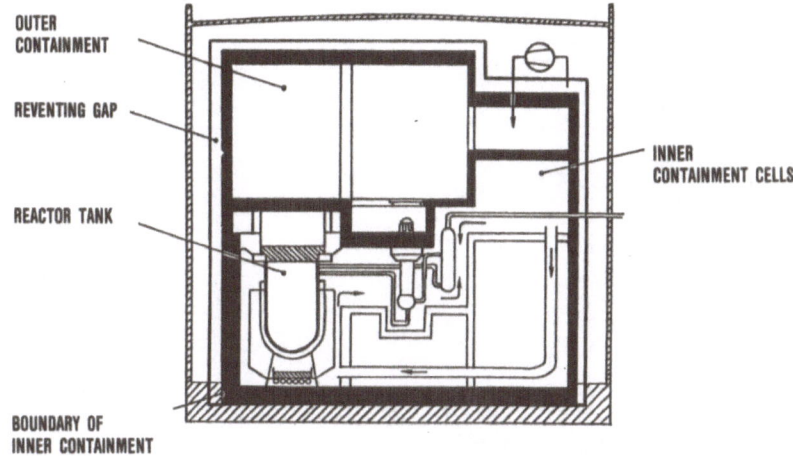

Fig. 5-8. Containment system of the SNR 300 (INTERATOM)

contained in so-called inner containment cells. The interior of these inner containment cells is lined with steel plates and filled with nitrogen. The pipings, pumps and IHX's are installed in such a way that, in case of a leak in the primary pipe, the primary sodium will run into special cavities and a certain height of the sodium level in the reactor tank will not be underrun. In addition, crack growth in the thin walled pipings of austenitic steel guarantees that small leaks can be detected by sodium leak detectors long before large scale cracks could occur.

The last barrier is the outer containment filled with air. In the SNR 300, it consists of reinforced concrete with a wall thickness of approximately 1.2 m, protecting the plant against such external impacts as airplane crashes, exploding gas clouds, earthquakes, etc. It is completely surrounded by a steel liner and has a reventing gap between the concrete walls and the liner (Fig. 5-8). Radioactive aerosols leaking into this gap can be revented by blowers into the outer containment. This reventing procedure can be kept up for a period of days, if needed. After that time period, the containment atmosphere can be released through filters into the stack. During this procedure, radioactive aerosols are almost totally trapped in the filters. Similar containment concepts are planned for commercial size loop type reactors.

5.6.2 Control and Shutdown Systems

LMFBR cores are controlled by means of absorber rods containing B_4C as an absorber material. Also shutdown can be brought about by means of B_4C absorbers. After being released, the absorber rods can be introduced into the core by dropping under gravity within 0.7–0.8 s. To add to the reliability of this shutdown concept, both the SNR 300 and other LMFBR's have two completely independent, diverse shutdown systems. In this way, a failure threshold for the shutdown systems of $<10^{-6}$ failures/a can be attained. By way of example, Fig. 5-9 shows the design principle of two independent shutdown sys-

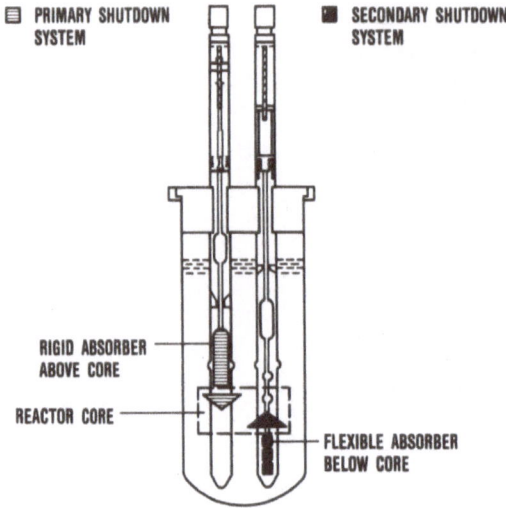

PRIMARY SHUTDOWN
SYSTEM

SECONDARY SHUTDOWN
SYSTEM

RIGID ABSORBER
ABOVE CORE

REACTOR CORE

FLEXIBLE ABSORBER
BELOW CORE

Fig. 5-9. Schematic of the SNR 300 shutdown systems (INTERATOM)

tems in the SNR 300. The primary shutdown system drops absorber rods into
the core. The secondary shutdown system pulls a flexible absorber chain into
the core from below. Both systems have diverse electronic channels. The magnetic
release of the absorbers is direct and indirect, respectively. In SUPERPHENIX,
some of the absorber rods are additionally released by inherent mechanisms
as soon as the sodium temperature in the core rises above a certain threshold.

5.6.3 Afterheat Removal and Emergency Cooling of LMFBR Cores

Even after shutdown of the reactor power, the decay heat of an LMFBR core
must still be removed safely. Under normal conditions, it is carried to the
steam generators by the main heat transfer systems in such a way that the
pumps are driven by pony motors at low speed. However, sodium has such
excellent natural circulation characteristics that the afterheat can also be dissi-
pated from the reactor core through the main systems by way of natural circulation. This has been proven experimentally both for pool type reactors, such
as PHENIX and PFR, and for loop type reactors in SEFOR and FFTF. In
case the main heat transfer systems are not available, SUPERPHENIX has
additional systems, which will then be operated to transport the heat from
the secondary system to sodium-air coolers (see Section 5.5.3). In the SNR
300, six immersion coolers arranged in the reactor tank below the emergency
sodium level transfer the decay heat to sodium-air coolers. Even if all active
components of the emergency cooling systems were to fail, enough heat would
still be delivered through the surfaces of the pipes of the primary system in
the SNR 300 to leave the maximum temperature in the reactor tank at <700 °C.
Unlike water or gas cooled thermal converter reactors, which need active emergency cooling systems, LMFBR's thus have the potential to remove decay heat
by natural convection without any active systems.

5.6.4 Core Instrumentation and Protection Against Fault Propagation

In addition to the high reliability required of the safety shutdown systems, it must also be assured that the initiation of faults, which could develop, e.g., from local blockages in fuel elements, is counteracted from the beginning. Local blockages followed by local boiling and fuel pin failure were considered to have the potential to lead to partial fuel element destruction and, possibly, severer consequences to the core. However, recent in-pile research and development programs show that such accident developments can be excluded. Nevertheless, the cores of LMFBR's are equipped with individual subassembly instruments (thermocouples and, in some cases, flowmeters) to detect coolant anomalies. In addition, delayed neutron monitors and covergas monitors are able to detect fuel rod failures.

Other possible faults in loop type reactors are counteracted by inherent design measures or additional detectors. The SNR 300 is equipped with a gas bubble separator underneath the core diagrid plate to prevent gas from passing through the core. This is achieved by directing entrained gas bubbles to the radial blanket and reflector regions where they can be tolerated. Piping integrity is ensured and surveyed by in-service inspection methods, sodium level detectors in the reactor vessel and expansion tanks, and by leak detection systems.

5.6.5 Design Bases of the Primary System and Containment

While the *loss-of-coolant* accident (LOCA) must be regarded as a design base accident in LWR's, it is a different class of extremely improbable accidents, which play the major role in the safety analysis of present LMFBR's. In LMFBR's, core destruction can occur only if the pumps are shut down after scram initiation while the two shutdown systems fail at the same time. In this so-called unprotected *loss-of-f*low (LOF) accident, the reactor power remains constant, while the coolant pumps will be shut off and, hence, sodium cooling will cease. Accidents initiated by unexpected positive reactivity insertions are less probable than LOF accidents, but would have very similar consequences. Failure of the two redundant and diverse shutdown systems is supposed to occur only with a probability of $< 10^{-6}$ per annum. Accordingly, also the unprotected LOF accident must be classified in this region.

Safety analyses of the unprotected LOF accident in the SNR 300 show that there will be a rapid temperature increase in the sodium coolant in the reactor core after a few seconds, because of the imbalance between power generation and cooling. When the boiling temperature of sodium has been reached, which is 900–1000 °C, sodium begins to boil in the core. Initially, because of the positive sodium void coefficient, there will be further power increases counteracted by negative temperature reactivity coefficients, e.g., the Doppler coefficient and the fuel expansion coefficient. However, the temperatures eventually reached will be high enough for steel, fission products and PuO_2/UO_2 mixed oxide fuel to melt, boil and evaporate in the center of the core. For a few milliseconds, the very hot mixture will be under a pressure of a few 10 bar locally and will expand rapidly. This will make the reactor core subcritical and cause it

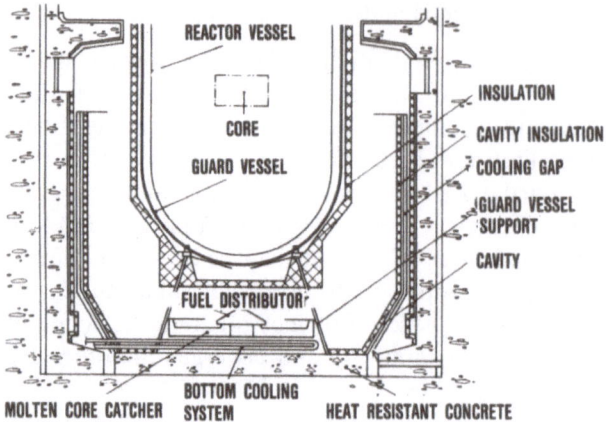

Fig. 5-10. SNR 300 molten core catcher (INTERATOM)

to shut itself down automatically. In the SNR 300 core, the isentropic work potential of this mixture of hot core material was calculated to be below 100 MJ. This is well below the design load limit of 370 MJ specified by the licensing authorities for the mechanical integrity of the reactor tank system.

For a commercial size LMFBR, such as SUPERPHENIX, a similar accident sequence was analyzed and a mechanical work potential of a few hundred MJ was found for the meltdown or disassembly of the core.

Most of the LMFBR core will be molten after this accident. Cooling of this molten core material must be ensured also for this extremely unlikely case. For the SNR 300, theoretical and experimental analyses show that cooling in the reactor tank is ensured and molten fuel would not penetrate through the lower tank bottom. In other demonstration LMFBR's and in SUPER-PHENIX, similar conclusions were reached. In the SNR 300, however, there is an additional safety barrier, which is an external molten core cooling device for long term cooling in the unlikely event that molten fuel were to penetrate through the tank bottom. This molten core cooling device is explained in Fig. 5-10. It can collect the sodium contained in the reactor vessel. The molten core debris is caught by a layer of uranium oxide. A NaK cooling system under-neath this layer serves to cool the heat developed by the core debris, thereby preventing the heat from reaching the concrete of the containment foundation plate.

In case the reactor tank system were to develop a leak, the remaining barriers would still retain their sealing functions against releases of radioactivity even in this severe accident under discussion. Sources of radioactivity to be considered would primarily be gaseous and volatile fission products and sodium aerosols. In SUPERPHENIX, aerosols would reach the dome above the reactor tank cover plug. From here they could be released through filters and delay lines. Aerosols also overcoming the inner containment barrier would enter the outer containment, from where they would be released to the environment after filtration.

The retention capability of the containment (barriers) and the filter systems is designed on the basis of aerosol physics. Most of the aerosols entering the outer containment would plate out on the walls and the bottom by coagulation, thermophoresis and sedimentation processes. Only plutonium aerosol concentrations in the mg/m^3 range would be found to remain airborne in the outer containment of the SNR 300. After a few days it would be necessary to release some of the contents of the outer containment over sandbed carbon filters through a stack into the environment. The resultant irradiation dose at the fence of the reactor plant would remain below the legally permissible limits.

For commercial size loop type LMFBR's similar design solutions for the outer containment are presently under discussion. The favored solution is a leakproof cylindrical steel containment, which will be surrounded by a reinforced concrete containment for protection against external impacts.

5.6.6 Sodium Fires

Sodium fires in the primary cooling systems containing radioactive sodium are prevented by enclosing those systems in cells filled with nitrogen. Free sodium surfaces in tanks are covered with argon. Sodium leak detection systems survey the leaktightness of the cooling system. The secondary systems are usually surrounded by air.

Sodium leaking out of the cooling systems will be directly collected in special catch pans and containers underneath the pipes and pumps. Access of oxygen to the hot sodium will largely be prevented in those installations. Fire extinguishing systems are also available. Extensive experimental data on sodium burning rates, burning temperatures and sodium aerosol formation are at hand from large out-of-pile test rigs in the USA and Europe. In addition, extensive operating experience with demonstration LMFBR's has proved the high standard of experience of sodium technology.

5.6.7 Sodium-Water Interactions in Steam Generators

Only in the steam generators non-radioactive sodium could contact water as a result of a ruptured pipe. In that case, a high pressure water jet would penetrate into the sodium and a violent Na-H$_2$O interaction with temperatures of about 1300 °C in the reaction zone would develop. Hydrogen and sodium hydroxide are generated with peak pressures of 90–130 bar in the reaction zone. Steam generators are therefore fitted with special pressure relief systems (Fig. 5-11). A rupture disk would break on the sodium side, and the hydrogen, sodium and sodium hydroxide would pass through a release pipe into a cyclone where the hydrogen would be separated from the sodium and sodium hydroxide. Hydrogen would be vented to the air where it would ignite spontaneously and burn. Hydrogen monitors and other detectors in the pressure relief system produce clear signals of such an accident, and the steam generator would then be shut down, taken out of operation and repaired.

Extensive experimental data on temperature and pressure buildups and pressure wave propagation through the steam generator bundle are available to

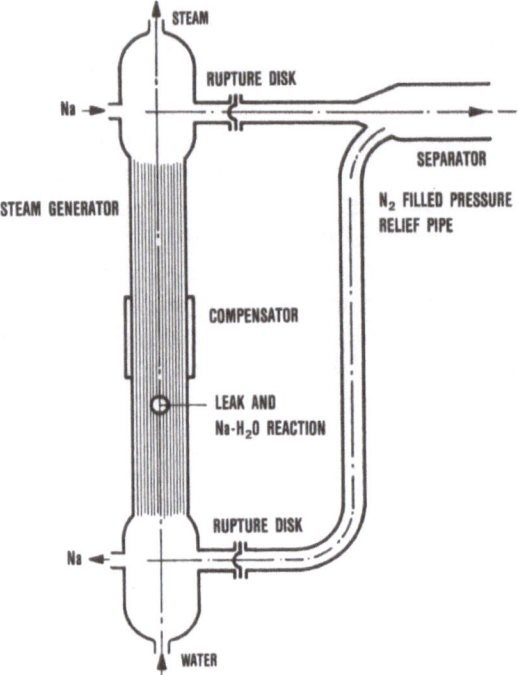

Fig. 5-11. Test model of the pressure relief system of a Na-H$_2$O steam generator (INTER-ATOM)

each group developing LMFBR's in the US, Europe, Japan and the USSR. Hydrogen detectors detecting small leaks in water or steam pipes have been developed and are being used in LMFBR steam generators. They allow timely detection of small leaks before these can develop or propagate into large leaks or pipe ruptures.

5.7 Heterogeneous Core Designs of LMFBR's

The main core performance characteristics of LMFBR's can be influenced to some degree by the design of the core and the geometrical arrangement of fissile and fertile fuel elements. Several alternative core designs have been proposed in the course of LMFBR development. These were, e.g., relatively flat cores (pancakes) or modular core arrangements. In recent years, in particular so-called heterogeneous core designs were discussed and investigated. Heterogeneous core designs contain blanket elements also in the core zone. Possible advantages of heterogeneous versus conventional two zone-cores are, e.g., higher breeding ratios, lower neutron fluences, improvements concerning the power peaking factor and its variation during burnup, lower control and shutdown reactivity requirements and, especially, a reduced sodium void reactivity, with a potentially favorable influence on the safety behavior of the core. Disadvantages are a higher fissile enrichment, a slightly higher fissile fuel inventory, and slightly larger core dimensions. Problems still to be investigated include the mechanical and thermohydraulic behavior of the internal fertile regions during burnup (bowing, fixed or variable orificing of the fuel elements, thermal stripes in the upper coolant plenum above the core, etc.), and a smaller Doppler reactivity effect. Often the performance of such cores is burdened by a so-called looser neutronic coupling of spatially separated core regions.

5.8 LMFBR Cores with Advanced Oxide and Carbide Fuels

Section 5.5 indicates the design data for SUPERPHENIX, the first commercial size LMFBR plant. The technical data of the fuel elements and the mixed oxide fuel rods of this reactor correspond to a design on the conservative side. In the future, it is hoped to obtain even better systems data for LMFBR's from improved and extended knowledge and experience as well as from advanced oxide or carbide fuels.

The continued development of mixed oxide fuel is mainly aimed at achieving higher burnups, slightly higher power ratings of the fuel rods, and slightly lower wall thicknesses of the cladding tubes and fuel element cans. The percentage of fuel in the core is to be raised and, at the same time, the coolant and structural steel fractions are to be reduced. In this way, the breeding ratio can be stepped up and the core inventory lowered. Both methods result in a reduction of the doubling time of the fissile material inventory.

Mixed carbide PuC/UC fuel development has been underway for approximately two decades within the major LMFBR development projects. Compared to mixed oxide fuel, the main advantages of carbide fuel are its higher density (especially in heavy atoms), smaller fraction of moderating material and higher thermal conductivity, resulting in a rated power of 800–1100 W per cm of fuel rod length, which is a factor of two higher than attainable in mixed oxide fuel.

This leads to smaller cores with lower fissile inventories and, at the same time, even higher breeding ratios than in advanced mixed oxide fuel (see Section 6.2).

5.9 Gas Cooled Fast Breeder Reactors

Large cores of FBR's with oxide fuel can also be cooled by a gas under high pressure, such as helium. At a pressure of 80–120 bar, helium has sufficiently good thermodynamic properties to cool FBR cores with an average power per core unit volume of 280 kW(th)/l. Due to the good neutronic properties of helium, such FBR cores could have high breeding ratios in the range of 1.4 and only a low positive coolant temperature reactivity coefficient. The concept of helium cooled FBR's is still in the development stage and no practical experience is available from a test or demonstration plant of this type. Interest in this FBR variant has been stimulated by the fact that experience could be utilized from the development of helium cooled thermal reactors (HTGR's).

Especially experience with helium blowers, steam generators and prestressed concrete pressure vessels could be used for this design. Gas cooled FBR's have somewhat higher initial fissile inventories than LMFBR's. Helium cooled FBR's could constitute an interesting technical alternative, much like the different solutions evolving in the field of thermal converter reactors.

Selected Literature

General, FBR plant description

Barthold, W.P., *et al.*: Optimization of Radially Heterogeneous 1000 MWe LMFBR Core Configurations. Palo Alto, Cal.: Electric Power Research Institute, EPRI-NP-1000 (1979).

Carle, R.: SUPERPHENIX: First Commercial Plant of the Fast Breeder Line. J. British Nuclear Energy Society *14*, 183–190 (1975).

Clinch River Breeder Reactor Project – A Special Feature Issue. Nuclear Engineering International *19*, 835–865 (1974).

Design, Construction and Operating Experience of Demonstration LMFBR's, Proc. Int. Symposium, Bologna, Italy, 10–14 April 1978. Vienna: International Atomic Energy Agency. 1978.

Dickson, P.W., Doncals, R.A.: Heterogeneous Core Designs for LMFBR's. In: Advances in Nuclear Science and Technology, Vol. 12, pp. 33–91. New York-London: Plenum Press. 1980.

Fast Flux Test Facility (FFTF) – A Special Survey. Nuclear Engineering International *17*, 613–628 (1972).

Häfele, W., *et al.*: Fast Breeder Reactors. In: Annual Review of Nuclear Science, Vol. 20, pp. 393–434. Palo Alto, Cal.: Annual Reviews Inc. 1970.

Häfele, W., *et al.*: Fusion and Fast Breeder Reactors. Laxenburg, Austria: International Institute for Applied Systems Analysis, RR-77-8 (1977).

International Nuclear Fuel Cycle Evaluation, Fast Breeders. Report of INFCE Working Group 5. Vienna: International Atomic Energy Agency. 1980.

Judd, A.M.: Fast Breeder Reactors – An Engineering Introduction. Oxford: Pergamon Press. 1981.

Kazachkovskij, O.D., *et al.*: The Present Status of the Fast Reactor Programme in the USSR. In: Nuclear Power and Its Fuel Cycle, Proc. Int. Conference, Salzburg, 2–13 May 1977, Vol. 1, pp. 393–414. Vienna: International Atomic Energy Agency. 1977.

Khodarev, E.: Liquid Metal Fast Breeder Reactors. International Atomic Energy Agency Bulletin *20* (6), 29–38 (1978).

Köhler, M., *et al.*: Design Considerations for the Primary System and the Primary Components of SNR-2. In: Optimisation of Sodium-Cooled Fast Reactors, Proc. Int. Conference, London, 28 November–1 December 1977, pp. 249–253. London: British Nuclear Energy Society. 1977.

Leipunskii, A.I., *et al.*: A Nuclear Power Station with the BN-600 Reactor. Soviet Atomic Energy (Atomnaya Energiya) *25*, 1216–1221 (1968).

Meshkov, A.G., *et al.*: Prospects of Development of Fast Nuclear Power Reactors in the USSR. In: Nuclear Energy Maturity, Proc. European Nuclear Conference, Paris, 21–25 April 1975 (Zaleski, P., ed.), Vol. 11, pp. 120–124. Oxford: Pergamon Press. 1976.

Morgenstern, F.H.: Der Stand des Projektes Kernkraftwerk Kalkar (SNR 300). Atomkernenergie/Kerntechnik *36*, 250–252 (1980).

PFR – A Special Feature on the British Prototype Fast Breeder Reactor. Nuclear Engineering International *16*, 629–650 (1971).

PHENIX – A Special Feature on the French Prototype Fast Breeder Reactor. Nuclear Engineering International *16*, 557–580 (1971).

Schröder, R., Wagner, J.: Überlegungen zur Einführung schneller Brutreaktoren im DEBENE-Bereich. Kernforschungszentrum Karlsruhe, KfK-Ext. 25/75-1 (1975).

SNR-300, Liquid Metal Cooled Fast Breeder Reactor Prototype Plant. Nuclear Engineering International *21* (246), 39–58 (1976).

Vendryes, G., et al.: Fast Neutron Reactors – Advancement from Initial Research to the PHENIX Power Plant. Argonne National Laboratory, ANL-Trans-1007 (1975).

Waltar, A.E., Reynolds, A.B.: Fast Breeder Reactors. New York: Pergamon Press. 1981.

Yevick, J.G., Amorosi, A.: Fast Reactor Technology. Cambridge, Mass.: The MIT Press. 1966.

Fast reactor physics and safety

Bunz, H., Scholle, U.: Study of the Containment System for the Planned SNR-2 FBR. Proc. 15th DOE Nuclear Aircleaning Conference, Boston, 7–10 August 1978. Springfield, Va.: National Technical Information Servide, US Department of Commerce, CONF 780819 (1979).

Engineering of Fast Reactors for Safe and Reliable Operations. Proc. Int. Conference, Karlsruhe, 9–13 October 1972. Kernforschungszentrum Karlsruhe. 1973.

Fast Reactor Physics. Proc. Int. Symposium, Aix-en-Provence, 24–28 September 1979. Vienna: International Atomic Energy Agency. 1980.

Fast Reactor Safety, Proc. Int. Meeting, Beverly Hills, Cal., 2–4 April 1974. Washington: US Atomic Energy Commission, CONF-740401 (1974).

Fast Reactor Safety and Related Physics, Proc. Int. Meeting, Chicago, Ill., 5–8 October 1976. Springfield, Va.: National Technical Information Service, US Department of Commerce, CONF-761001 (1976).

Fast Reactor Safety Technology, Proc. Int. Meeting, Seattle, Wash., 19–23 August 1979. LaGrange Park, Ill.: American Nuclear Society. 1979.

Fidler, R.S., Collins, M.J.: A Review of Corrosion and Mass Transport in Liquid Sodium and the Effects on the Mechanical Properties. Atomic Energy Review *13*, 3–50 (1975).

Friedrich, H.J.: SNR-300 Tank External Core Retention Device, Design and Philosophy Behind It. Proc. Second Post-Accident Heat Removal Information Exchange, Albuquerque, N.M., 13–14 November 1975. Sandia Laboratories, SAND-76-9008 (1977).

Graham, J.: Fast Reactor Safety. New York: Academic Press. 1971.

Häfele, W.: Prompt überkritische Leistungsexkursionen in schnellen Reaktoren. Nukleonik *5*, 201–208 (1963).

Hummel, H.H., Okrent, D.: Reactivity Coefficients in Large Fast Power Reactors. Hinsdale, Ill.: American Nuclear Society. 1970.

IAEA International Working Group on Fast Reactors, IAEA Study Group Meeting on Steam Generators for LMFBR's, Bensberg, Germany, 14–17 October, 1974. Vienna: International Atomic Energy Agency. 1974.

Jordan, H., et al.: PARDISEKO III – A Computer Code for Determining the Behavior of Contained Nuclear Aerosols. Kernforschungszentrum Karlsruhe, KfK-2151 (1975).

Kessler, G., et al.: Safety of the LMFBR and Aspects of Its Fuel Cycle. In: Nuclear Power and Its Fuel Cycle, Proc. Int. Conference, Salzburg, 2–13 May 1977, Vol. 5, pp. 661–674. Vienna: International Atomic Energy Agency. 1977.

Kiefhaber, E.: The KFKINR-Set of Group Constants. Kernforschungszentrum Karlsruhe, KfK-1572 (1972).

Nicholson, R.B., Fischer, E.A.: The Doppler Effect in Fast Reactors. Advances in Nuclear Science and Technology, Vol. 4, pp. 109–195. New York-London: Academic Press. 1968.

Tang, Y.S., *et al.*: Thermal Analysis of Liquid Metal Fast Breeder Reactors. LaGrange Park, Ill.: American Nuclear Society. 1978.

van de Vate, J.F.: The Safety of SNR 300 and the Aerosol Model; A Summary Report of the RCN Aerosol Research 1967–1971. RCN Petten, Netherlands, RCN-174 (1972).

Vossebrecker, H., Grönefeld, G.: Brennstoffrückhaltung im Reaktortank von natriumgekühlten Reaktoren nach hypothetischen Störfällen mit Kernschmelzen. Atomkernenergie-Kerntechnik *36*, 288–291 (1980).

Gas cooled fast breeder reactors

Dalle Donne, M., Goetzmann, C.A.: Design and Safety Studies for the Gas-Cooled Fast Reactor in the Federal Republic of Germany. In: Gas-Cooled Reactors with Emphasis on Advanced Systems, Proc. Symposium, Jülich, 13–17 October 1975, Vol. II, pp. 377–396. Vienna: International Atomic Energy Agency. 1976.

Melesse-d'Hospital, G., Simon, R.H.: Status of Gas-Cooled Fast Breeder Reactor Programs. Nuclear Engineering and Design *40*, 5–12 (1977).

6 Nuclear Fuel Cycle Options

6.1 Fuel Cycle Options for Converter Reactors

6.1.1 The Once-Through Fuel Cycle

After unloading from the reactor core, spent fuel contains a mixture of U-238 and unused U-235, newly generated plutonium, higher actinides and fission products. U-235 and plutonium are valuable materials, which can be used again for energy generation, whereas fission products are waste materials, the "nuclear ashes" from "burning" uranium.

In the *once-through* (OT)-cycle (Fig. 6-1), valuable uranium and plutonium are not re-used. After removal from the reactor, the fuel is kept in intermediate

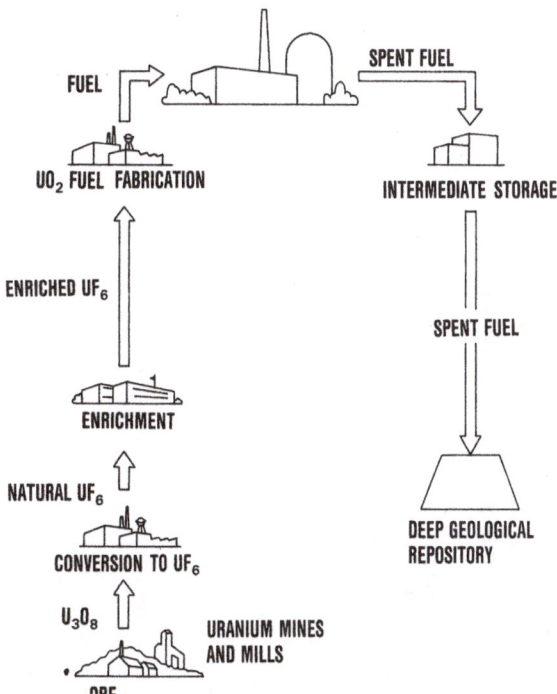

Fig. 6-1. Once-through nuclear fuel cycle for converter reactors with direct spent fuel disposal (NUREG)

Table 6-1. *Fuel cycle design data for converter reactors operating in the once-through fuel cycle (30 years operating time, load factor 0.7) (INFCE)*

		LWR PWR	HWR		HTGR		HTR (pebble bed)		
Total thermal power	MW(th)	3765	3425		3360		3000		
Net electric power	MW(e)	1229	1000		1332		1240		
U-235 enrichment/fuel cycle	%	3.1 (LEU)	natural uranium	1.2 (LEU)	10 (LEU)	20 (MEU)	8.5 (LEU)	20 (MEU)	93 (HEU)
Fraction of core replaced per year		0.33	1	0.3	0.33	0.25	0.23	0.29	0.21
Fuel residence time	d	766	276	790	766	1020	1120	870	1220
Average discharge burnup	MWd(th)/kg	33	7.3	20.9	111	119	100	100	100
Natural uranium requirement	te/GW(e)								
– Initial core		367	131	256	186	180	178	139	168
– Annual eq. reload		139	121	82	118	94	100	95	88
– 30 years cumulative		4224	3716	2651	3446	2800	3096	2960	2720
Separative work requirement	te SWU/GW(e)								
– Initial core		257	–	91	180	212	189	165	218
– Annual eq. reload		111	–	29	129	111	106	112	115
– 30 years cumulative		3318	–	945	3780	3310	3280	3470	3560
Fissile material in spent fuel: Enrichment	%								
– U-235		0.82	0.23	0.1	1.3	5.0	1.6	6	5.7
– Pu-fissile		0.66	0.28	0.34	56	46	56	39	45
Annual eq. discharge	kg/GW(e)								
– U-235		188	276	40.7	58	97	84	115	133
– Pu-fissile		152	331	140	45	19	57	13	0.4

storage facilities for some time (a couple of years up to several decades). This intermediate storage is only an interim solution. In the long run, spent fuel elements must be conditioned such that a permanent storage in deep geological repositories is possible (see Section 7.6.2). In this so-called "direct spent fuel disposal" solution, plutonium and uranium remain preserved in the fuel elements, while the radioactivity of the fission products decays gradually. Uranium and plutonium are no longer available for later power generation in reactors.

Nuclear reactors presently in operation are almost exclusively operated in the OT-cycle up to the point of intermediate storage, because commercial reprocessing of the fuel elements is not yet available on a worldwide basis. This OT-cycle concept implies a high natural uranium consumption.

Table 6-1 shows the main fuel cycle data for converter reactors in the OT-cycle. PWR's are operated with *low enriched uranium* (LEU). HWR's can be operated both with natural and low enriched uranium. Three enrichment stages are under discussion in HTGR's and HTR pebble bed reactors, namely LEU up to about 8–10% enrichment, *medium enriched uranium* (MEU) with 20% enrichment, and *highly enriched uranium* (HEU) with 93% enrichment in U-235.

Among the light water reactors, PWR's have a natural uranium consumption of 4220 te over an operating period of thirty years. BWR's consume almost the same quantity of natural uranium. However, a number of optimization measures could help to reduce the natural uranium consumption of LWR's also to a range of 3700 or even 3000 te. Because of their better neutron economy, HWR's with natural uranium fuel consume slightly less natural uranium (3720 te over 30 years). If the fuel is slightly enriched, their natural uranium consumption can be cut to 2650 te. HTGR's and HTR pebble bed reactors have natural uranium consumption levels between 3450 and 2720 te. The lowest consumption of natural uranium (2720 te) is found in HTR's with HEU fuel. However, that line at the same time requires the highest separative work expenditures for enrichment of the fuel (the figures mentioned above refer to 1 GW(e) and 30 years plant operation).

Spent fuel of converter reactors run on U-235/U-238 fuel always contains plutonium. The largest amount of plutonium is generated in HWR's fueled with natural uranium. However, this HWR spent fuel also has the lowest plutonium concentration. By contrast, HTR reactors with HEU fuel contain very little plutonium in their spent fuel, because the fraction of fertile U-238 in the HEU fuel is only 7% at 93% U-235 enrichment.

6.1.2 Closed Nuclear Fuel Cycles

In a closed fuel cycle the spent fuel, after its radioactivity has decayed in temporary storage facilities, is shipped to a reprocessing plant for chemical reprocessing. After chemical reprocessing, the fissile plutonium or U-233 as well as the residual uranium can be re-used to fabricate new fuel elements and be recycled in converter reactors (Fig. 6-2). A small fraction of the fissile material (about 1%) goes into the radioactive waste during chemical reprocessing and refabrication of the fuel, where it is lost. Recycling improves the utilization of fuel (see Section 2.6) and, consequently, decreases the consumption of natural uranium.

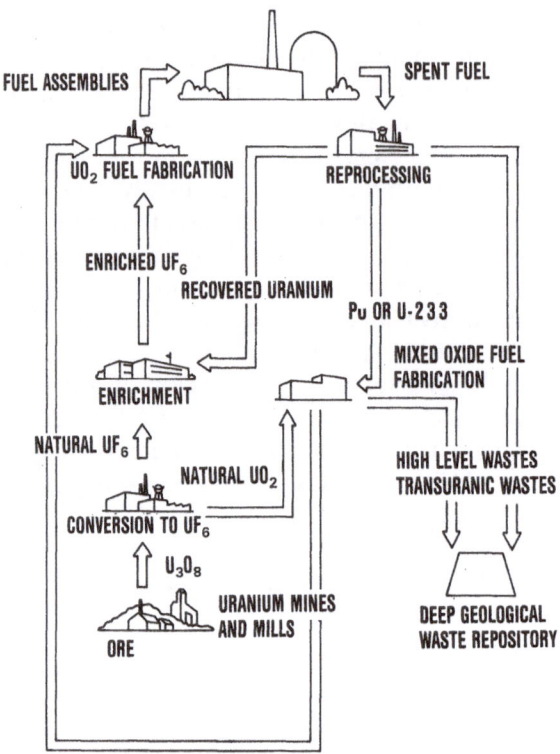

Fig. 6-2. Closed nuclear fuel cycle (recycling of converter reactors) (NUREG)

Recycling is possible both in the U-238/Pu and the Th/U-233 fuel cycles. Fissile plutonium or U-233 from several converter reactors may be collected and used exclusively in special recycle converter reactors. Also, every converter reactor can recycle its own plutonium or U-233 generated in preceding operation cycles. This is then called *self-generated recycling* (SGR).

6.1.2.1 Plutonium Recycling

At the beginning of the first recycle phase of a PWR core, i.e., after unloading spent LEU fuel with a burnup of 33,000 MWd(th)/te, the plutonium fuel has an isotopic composition as indicated in the first column of Table 6-2 (see also Fig. 2-10). After a full recycling period of that same plutonium fuel, i.e., after another burnup of 33,000 MWd(th)/te, the fraction of fissile Pu-239 and Pu-241 isotopes decreases to a composition as indicated in the second column of Table 6-2. Further recycling of the same plutonium would lower the fraction of fissile Pu-239 and Pu-241 isotopes even more. Therefore, the SGR mode is usually applied. In this SGR mode, continuous mixing of recycled plutonium with "fresh" plutonium from spent LEU fuel elements leads to a higher percentage of fissionable plutonium isotopes Pu-239 and Pu-241 (see column 3 of Table 6-2).

In the SGR mode, plutonium is reloaded only in a certain fraction of the fuel elements. These plutonium bearing fuel elements have the same structural

Table 6-2. *Isotopic composition of recycle plutonium in PWR and LMFBR fuels (INFCE, NUREG)*

Isotope		PWR			LMFBR
		Recovered from spent LWR fuel (33,000 MWd(th)/te)	After one recycle period in LWR (33,000 MWd(th)/te)	After 20 recycle periods in the SGR mixing mode	Equi-librium compo-sition
Pu-238	%	1.9	3.5	3.4	1
Pu-239	%	57.9	38.2	41.7	67.3
Pu-240	%	24.8	29.4	29.3	19.2
Pu-241	%	11	17.2	15.2	10.1
Pu-242	%	4.4	11.7	10.4	2.4

design as LEU fuel elements (see Section 4.1.1). Each fuel rod of such a fuel element contains PuO_2/UO_2 mixed oxide (MOX) fuel pellets. The average fissile plutonium enrichment is chosen such that the same discharge burnup as in the LEU fuel elements can be achieved. Radial enrichment zoning within the fuel element is necessary to avoid local power peaks. This is achieved by giving the fuel rods at the periphery of the fuel subassemblies a lower Pu enrichment of 2% Pu_{fiss}, while all remaining fuel rods of the inner part of the fuel element have an enrichment of about 3.2% Pu_{fiss} in quasi-equilibrium recycle periods.

If annual refueling and three years of out-of-core time for reprocessing and refabrication of MOX fuel is assumed, first recycling of self-generated Pu fuel can start only after 4 years (beginning of the 5th cycle). The core fuel management during the first four years (cycles) then remains identical to the PWR OT-cycle. Fig. 6-3, on a cycle by cycle (year by year) basis, shows the percentage of MOX fuel elements per reload, their average enrichment and Pu isotope composition. The fraction of MOX fuel elements starts at about 10% of the core and attains some 30% in later cycles. From the 11th cycle on, the MOX fuel enters its 2nd recycle period (three years for reaching full burnup, three years out-of-core time). The 3rd recycling period starts with cycle 17, etc. The isotopic composition of plutonium changes such that the fraction of non-fissile Pu-240 and Pu-242 isotopes increases, whereas the fraction of fissile Pu-239 and Pu-241 isotopes decreases. From the 4th recycle period on, the Pu composition attains a quasiequilibrium as shown in Table 6-2. The increase in non-fissile plutonium isotopes requires an increase in Pu_{fiss} enrichment in each new recycle period, starting at about 2.9% Pu_{fiss} and attaining some 3.2% Pu_{fiss} in its 4th recycle period.

Compared with the OT-fuel cycle, a Pu-SGR type PWR has a natural uranium consumption of 2730 te/GW(e) over 30 years operating time, if U and Pu are recycled. This corresponds to savings on the order of 35%. Accordingly, only 2490 te SWU/GW(e) over 30 years are needed, which saves 25% of separative work requirements. Data similar to those found for PWR's also result

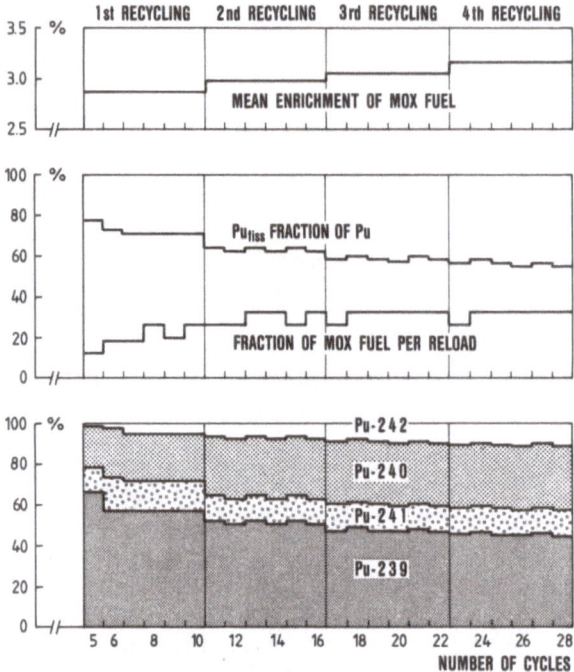

Fig. 6-3. Characteristics of long-term self-generated plutonium recycle in PWR's (H. Stehle)

for BWR's. The Pu-SGR mode can be adopted also for reactor lines other than LWR's, i.e., HWR's, AGR's and HTGR's.

A plutonium recycle LWR fully loaded with PuO_2/UO_2 fuel is called a plutonium burner. In such a plutonium burner, all plutonium bearing fuel elements have the same plutonium enrichment of about 2.8% Pu_{fiss} for the initial core. The initial plutonium loading of the core is roughly 2300 kg of fissile plutonium per GW(e). One third of the fuel is unloaded annually at a burnup of 33,000 MWd(th)/te and can be reloaded after three years of out-of-core time, following reprocessing and refabrication. Since LWR plutonium burners obtain the plutonium for their initial core as well as for the Pu_{fiss} make-up from other LWR's, their natural uranium requirement and their separative work requirement is nil. They need only depleted uranium. However, the whole strategy encompassing LWR's with LEU fuel and an LWR plutonium burner has roughly the same requirements of natural uranium and separative work as an LWR in the Pu-SGR mode.

MOX fuel of LWR's operating in the Pu-SGR mode has been tested extensively and successfully in in-pile tests. In addition, experience with the reprocessing of MOX fuel is already available (see Section 7.2.4).

No LWR plutonium burner reactors have been built or operated so far. Optimization of the neutron physics data of the LWR recycle core can help to further improve the conversion ratio and fuel consumption. This is achieved mainly by reducing the lattice pitch of the plutonium fuel rods in the core.

This reduces neutron moderation, shifting the neutron spectrum towards higher energies. In that case, the conversion ratio could reach levels around 0.9. However, only studies have been conducted so far on such plutonium fueled light water high converter reactors.

6.1.2.2 The Thorium/Uranium-233 Fuel Cycle

Converter reactors operating on U-235 as a fissile fuel and Th-232 as a fertile material generate fissile U-233 which, after removal of the fuel from the reactor core, can be separated by chemical reprocessing (see Section 7.3). The fissile U-233 obtained in this way can be recycled either in the same reactors or in reactor cores with pure U-233/Th-232 fuel. However, in principle, the generation of U-233 must be started with U-235/U-238/Th-232 fuel. Since this fuel also contains fertile U-238, plutonium will be produced besides U-233. The production of plutonium can be restricted by limiting the amount of U-238 contained in the fuel. This applies to highly enriched uranium (HEU) fuel with 93% U-235 enrichment.

In addition, similar to the buildup of plutonium shown in Fig. 2-10, U-233 fuel contains other uranium isotopes building up as a result of neutron capture. The isotopic composition depends on the burnup of the fuel and on the type of reactor. Fig. 6-4 shows the isotopic composition of uranium developing in an LWR core with HEU fuel and thorium. After having attained its maximum burnup within 3 years, the fuel is reprocessed and refabricated. HEU fuel is added to make up for the burnup of fissile U-235. The new fuel is then recycled

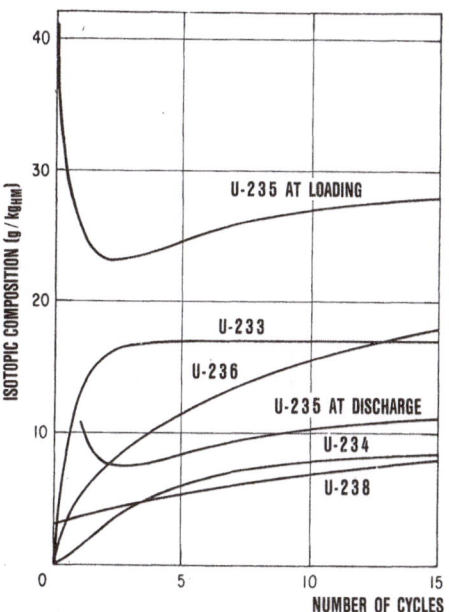

Fig. 6-4. Uranium isotopic composition as a function of continuous recycle for a HEU-235/Thorium fuel cycle (Y.I. Chang)

Table 6-3. *Isotopic composition of recycle uranium in converters (LWR, HWR) and breeder reactors (LMFBR's) (Y.I. Chang, AECL, INFCE)*

Isotope		LWR-HEU/Th quasi-equilibrium cycle[a]	HWR (CANDU) HEU/Th quasi-equilibrium cycle[a]	LMFBR equilibrium cycle
U-233	%	28	35	75.2
U-234	%	14	18.8	21.1
U-235	%	18	15.4	3
U-236	%	29	23.4	0.7
U-238	%	11	7.4	–

[a] Equivalent composition after 15 recycle periods

Table 6-4. *Fuel cycle design data for converter reactors operating in a U-235/Th or U-233/Th recycle mode (30 years operating time, load factor 0.7) (INFCE, P. Kasten, N.L. Shapiro)*

Reactor type		LWR – PWR		
Total thermal power	MW(th)	3765	3765	3800
Net electric power	MW(e)	1229	1229	1270
Fuel cycle		HEU-235/Th	MEU-235/Th	MEU-233/Th
Equilibrium recycle enrichment – recycle feed	%	93(U-235)	20(U-235)	12(U-233)
Fraction of core replaced per year		0.33	0.33	0.33
Average burn-up	MWd(th)/kg	34	33.5	33.4
Fissile requirements	kg/GW(e)	2544	2544	1904
Annual recycled eq. fissile	kg/GW(e)	558	558	478
Annual makeup eq. fissile	kg/GW(e)	337	375	295
Natural uranium requirements	te/GW(e)			
– Initial core		497	493	not adequate
– 30 years cumulative		2015	2211	not adequate
Thorium requirements	te/GW(e)			
– Initial core		73	63	57.9
– 30 years cumulative		902	681	539
Separative work requirements	te SWU/GW(e)			
– Initial core		644	715	–
– 30 years cumulative		2625	2739	–
Fissile material in spent fuel:				
Enrichment	%	63	9.3	7.9
Annual discharge	kg/GW(e)			
– U_{fiss} eq.		541	530	478
– Pu_{fiss}		5	68	65

into the LWR core. As indicated by Fig. 6-4, the U-233 concentration saturates rapidly within 3–4 recycle generations. However, the parasitic U-234 and U-236 isotopes continue to build up even after 15 recycle generations. Table 6-3 (left column) shows the quasi-equilibrium isotopic concentrations for the LWR core (36,000 MWd (th)/te max. fuel burnup) after 15 recycle generations. The middle column in Table 6-3 indicates the quasi-equilibrium isotopic uranium composition after 15 recycle generations in a HWR-CANDU operated in HEU/Th fuel (maximum burnup fuel of 29,300 MWd (th)/te). For comparison, the right column shows the equilibrium composition of plutonium in an FBR fuel cycle.

Table 6-4 shows the main fuel cycle design data of LWR, HWR, HTGR and HTR recycle reactors. The LWR recycle reactor, with HEU/Th and MEU/Th fuel, attains a natural uranium consumption of 2015 and 2210 te of natural uranium, respectively, over thirty years of operation. The separative work expen-

HWR		HTR pebble bed	HTGR	
3425	4029	3000	3360	3360
1000	1260	1240	1332	1332
HEU-235/Th	MEU-233/Th	HEU-235/Th	MEU-235/Th	MEU-233/Th
93 (U-235)	12 (U-233)	93 (U-235)	18 (U-235)	12 (U-233)
0.33	0.5	0.21	0.25	0.33
29.3	14	80	96	48
2740	1859	1910	1260	1260
500	856	270	157	322
129	106	180	371	226
563	not adequate	350	235	not adequate
1489	not adequate	1008	2177	not adequate
117	105.4	48	23.8	31
207	–	261	145	307
704	–	454	297	–
1932	–	1302	2651	–
85	11.2	60	6.9	7.9
505	848	245	152	307
2	22.8	1	25	47

diture is 2625 and 2740 te SWU, respectively. Although the fissile U-233 produced plus the unused U-235 is recycled in the reactor core after chemical separation, an additional 337 and 375 kg of U-235, respectively, must be fed annually either as highly or medium enriched uranium fuels. Table 6-4 also contains data for LWR's operated in a pure Th/U-233 cycle. This U-233 recycle LWR has an initial core inventory of 1904 kg U-233/GW(e) and, in addition to the 478 kg of U_{fiss} recycled (one third of the core inventory), requires an annual 295 kg of U-233 to be obtained by chemical reprocessing of the fuel from other converter reactors run on thorium fuel.

The HWR recycle reactor with highly enriched U-235 fuel has a natural uranium consumption of only 1500 te. Moreover, it only requires an annual makeup of 129 kg of U-235. HWR's with highly enriched U-233/Th fuel, U-233 burners, only need an annual makeup of 106 kg U-233. However, their initial inventory of 1859 kg of U-233 and their annual makeup must come from other reactors which, in turn, use the fissile U-235 of natural uranium. In the HTR pebble bed reactor and the HTGR, improved fuel elements would attain even lower levels of natural uranium consumption, especially in the highly enriched U-235 cycle.

6.1.2.3 Comparison of Various Converter Reactors

Table 6-5 shows the natural uranium consumption and the separative work requirements for different types of nuclear reactors capable of working in the once-through or recycle modes. Improved neutron economy and higher conversion ratios as well as recycling of the fissile material generated reduce the natural

Table 6-5. *Comparison of 30 years' cumulative requirements of natural uranium and separative work for different converter reactors (load factor 0.7)*

	Natural uranium (te/GW(e))	Separative work (te SWU/GW(e))
Once-through cycle		
LWR-LEU	4224	3318
HWR-nat. uranium	3716	0
HWR-LEU	2651	945
HTR-LEU	3096	3280
HTGR-MEU	2800	3310
HTR-HEU	2720	3560
Closed fuel cycle		
LWR-U-238/Pu recycle	2730	2490
LWR-MEU-235/Th	2211	2739
LWR-HEU-235/Th	2015	2625
HWR-HEU-235/Th	1489	1932
HTGR-MEU-235/Th	2177	2651
HTR-HEU-235/Th	1008	1302

uranium consumption. All reactors operating in the recycle mode require technically mature reprocessing in the U-238/Pu fuel cycle or in the Th/U-233 cycle (see Chapter 7).

6.2 Fuel Cycle Options for Breeder Reactors

All breeder reactors must work in a closed fuel cycle. LMFBR's and gas cooled fast breeders as described in Chapter 5 are presently designed to operate in the U-238/Pu fuel cycle, since the highest breeding ratios are attained with a fast neutron spectrum (see Section 2.5). However, in principle it is also possible to design FBR's with U-233/Th fuel, which still attain breeding ratios above 1.

Initially, thermal and fast breeder reactors can be started with plutonium or U-233 available from chemical reprocessing of spent fuel from thermal converter reactors. Later, when a fission breeder reactor economy will have developed, sufficient plutonium or U-233 would be generated also by the FBR's themselves to start additional FBR plants. However, this is not the only way to start breeder reactors. If plutonium or U-233 were not available in sufficient quantities from thermal reactor fuel reprocessing, breeders could also produce their own initial cores of plutonium or U-233 fuel by starting with U-235/U-238 or U-235/Th fuel.

6.2.1 The Uranium/Plutonium Fuel Cycle

Table 5-4 listed the core design and fuel cycle data of SUPERPHENIX as an example of a present commercial size LMFBR plant. Table 6-6 indicates the core design and fuel cycle data of advanced mixed oxide and mixed carbide fuel LMFBR cores. Such advanced LMFBR cores are expected to be applied in commercial size LMFBR plants from about the year 2000 on. Compared to SUPERPHENIX, with a fissile core inventory of 3.7 te Pu_{fiss}/GW(e), these advanced mixed oxide and carbide cores have fissile core inventories of only 3.2 and 2.6 te Pu_{fiss}/GW(e), respectively. At the same time, the breeding ratios increase from 1.18 in the case of SUPERPHENIX to 1.32 and 1.48 for the advanced mixed oxide and carbide cores, respectively.

The fuel remains in the LMFBR core for two years of power generation. Then it will be unloaded and, after a certain cooling period, reprocessed, refabricated and reloaded into the core. This ex-core time of the fuel amounts to some two years, as described in Section 7.4.1. It is expected that the ex-core time can be shortened to about one year in the future. This will lead to relatively low total inventories of the FBR fuel cycle, which will be important for an expanding nuclear energy economy based on FBR's.

For the advanced mixed oxide core, 1.59 te Pu_{fiss} will be loaded per GW(e)·a and $(1.54+0.35)$ te Pu_{fiss} will be unloaded. The net gain of Pu_{fiss} per GW(e)·a thus will be 300 kg. Accounting for losses of $<1\%$ of Pu_{fiss} during reprocessing and refabrication amounts to about 280 kg of excess plutonium per GW(e)·a. For a mixed carbide core the net gain would be about 380 kg Pu_{fiss}/GW(e)·a after accounting for losses in the fuel cycle. The isotopic composition of plutoni-

Table 6-6. *Characteristic core design data of LMFBR cores with advanced mixed oxide and carbide fuels (Reference plant with 1000 MW(e) power and 0.75 load factor) (INFCE)*

		Mixed oxide core	Mixed carbide core
Initial core loading			
Fissile plutonium[a]	te	3.2	2.6
Total U/Pu	te	26.6	25.6
Total breeding ratio		1.32	1.48
Initial blanket loading			
Total U	te	64.8	70.1
Equilibrium loading			
Fissile Pu core[a]	te/a	1.59	1.32
Total U/Pu core	te/a	13.3	12.8
Total U blankets	te/a	18.9	20.5
Equilibrium discharge			
Fissile Pu core[a]	te/a	1.54	1.32
Fissile Pu blankets[a]	te/a	0.35	0.40
Total U/Pu core	te/a	12.6	12.1
Total U/Pu blankets	te/a	18.8	20.4

[a] Fissile plutonium includes U-235 in uranium.

um after many recycling periods (equilibrium FBR-cycle) differs considerably from that in thermal reactor recycling (see Table 6-2). The fast neutron spectrum is responsible for a much higher percentage in fissile plutonium isotopes Pu-239 and Pu-241.

6.2.2 The Thorium/Uranium-233 Fuel Cycle

Besides the abundant reserves of fertile U-238 available from U-235 enrichment plants at 0.2% U-235, the reserves of thorium represent a similarly high energy potential when used as a fertile fuel in breeder reactors. Thorium and U-233 could be used in FBR's as mixed oxide or mixed carbide fuels. However, the scientific and technical basis explaining the behavior of these fuels at high burnups in FBR cores is limited.

The isotopic composition of uranium in an FBR core after many recycling periods (equilibrium FBR cycle) also differs considerably from that of thermal reactor recycling (see Table 6-3). The fast neutron spectrum creates an isotopic composition with a higher percentage of the fissile isotopes U-233 and U-235. Starting from Pu/U-238 fueled LMFBR's, thorium may first be used in the radial blanket instead of U-238. This would breed U-233 from Th-232 following neutron capture. In addition, the present Pu/U-238 fuel could then be replaced in the FBR core either by U-233/Th fuel or by U-233/U-238/Th fuel. In such

Table 6-7. *Core characteristics of LMFBR's with alternative fuel cycles (Reference plant with 1000 MW (e) power output and 0.75 load factor) (INFCE)*

		Mixed oxide fuel				Mixed carbide fuel
Core fuel		Pu/U-238	Pu/U-238	U-233/Th	U-233/Th	Pu/U-238
Ax. blanket fuel		U	U	Th	Th	U
Rad. blanket fuel		U	Th	Th	Th	U
Total breeding ratio		1.32	1.31	1.10	1.24	1.48
Initial core loading						
U-233	te	–	–	3.145	2.760	–
U-235	te	0.174	0.084	0.127	0.111	0.184
Fissile Pu	te	3.142	3.152	–	–	2.602
Equilibrium loading						
U-233	te/a	–	0.130	1.586	1.078	–
U-235	te/a	0.045	0.030	0.075	0.067	0.047
Fissile Pu	te/a	1.946	1.703	–	0.461	1.680
Equilibrium discharge						
U-233	te/a	–	–	1.573	1.380	–
U-235	te/a	0.060	0.042	0.063	0.056	0.063
Fissile Pu	te/a	1.571	1.576	–	–	1.301

FBR cores, the U-233/Th fuel would need approximately 12% of U-233 in depleted uranium ("denatured fuel cycle"). Table 6-7 shows characteristic data of advanced oxide and carbide fuel cores. All variants indicated in Table 6-7 have the same design parameters for the fuel elements as the reference cores in Table 6-6.

From Table 6-7 it can be noted that the breeding ratio is not affected very much when thorium is utilized in the radial blanket instead of U-238. However, when thorium is introduced in the core, the breeding ratio is reduced. The main reason for this lower breeding ratio is the smaller fast fission effect of Th-232 compared to U-238. When U-233 is substituted for Pu in the FBR core, the smaller η-value of U-233 (see Section 2.5) is mainly responsible for the reduction in breeding performance. The breeding ratio of the U-233/U-238/ Th fueled core would be about halfway between the Pu/U-238 reference core and the U-233/Th fueled core. This can be explained by the higher fast fission contribution of U-238 and the improved η-value arising from a buildup of Pu-239 (conversion of U-238) during reactor operation. The buildup of Pu-239 can attain up to 30% of the total fissile mass. As a conclusion, it can be stated that FBR's operating in the pure Th/U-233 fuel cycle would have breeding ratios approximately 20% lower than the reference FBR core with mixed oxide PuO_2/UO_2 fuel operating in the U-238/Pu fuel cycle.

6.3 Natural Uranium Consumption in Various Reactor Scenarios

Developing the nuclear energy potential of the U-238 and Th-232 fertile materials requires first that the plutonium or U-233 fissile materials are generated by converter reactors using fuel containing U-235. This plutonium or U-233 may then be used to start FBR's. When a sufficient number of FBR's are built and operated it is also possible to produce in their blankets plutonium from U-238 or U-233 from Th-232. This development strategy is followed because many converter reactors producing, e.g., fissile plutonium are operating already. If this were not the case, FBR's operating in the U-238/Pu fuel cycle could also be started with U-235 enriched UO$_2$ fuel and breed their own plutonium. In any case, this implies a relatively high natural uranium consumption until a condition will have been reached in which FBR's in a symbiosis with converter reactors can use the U-238 or Th-232 fertile materials. FBR's, which produce a surplus of plutonium or U-233 can make this fuel available for the construction of new FBR's or they can be operated in a symbiosis with converter reactors, which also contain plutonium or U-233 as fissile materials.

Fig. 6-5 shows the annual and cumulative natural uranium consumption for various reactor scenarios in the U-238/Pu fuel cycle. For the "INFCE low" nuclear energy requirement projection shown in Fig. 1-1, the following combinations are compared:
- LWR and HWR in the OT-cycle,
- LWR with U/Pu recycling,
- LWR combined with LMFBR (mixed oxide or mixed carbide fuels).

In each of these scenarios, a certain mixture of LWR's, HWR's and gas cooled reactors is assumed for the pre-2000 period. For the post-2000 period, each strategy is characterized by the rapid dominance of the reactor type favored. While the annual natural uranium consumption rises steadily for LWR-OT, HWR-OT and LWR with U/Pu recycling, it remains nearly constant at the end of the time period to be considered for HWR-OT with low enriched (LEU). With the introduction of LMFBR's, the annual natural uranium consumption decreases substantially from 2010 on and the cumulative consumption remains limited to some 2.5–4.5 million te from 2030–2050 on. The upper curves of the LWR-LMFBR system apply to present design data of FBR's with oxide fuel; the lower curves hold for advanced LMFBR's with oxide and carbide fuels. It is assumed that from the year 2000 onward, LMFBR's will be operated on advanced oxide or carbide fuels and that the design data of SUPERPHENIX apply prior to the year 2000. Only FBR's can reduce to zero the annual natural uranium consumption, thus limiting the cumulative consumption to some million te in the long run.

In Fig. 6-6, the annual and cumulative natural uranium consumption for reactor scenarios in the Th/U-233 fuel cycle is shown. LWR's, HWR's and HTGR's can initially be started on enriched U-235/Th fuel. Depending on the availability of U-233, HWR's or HTGR's can later be added with MEU-233/Th or HEU-233/Th fuel and thus dominate the scenario. As can be seen from Fig. 6-6, HWR's and HTGR's, both with MEU and HEU fuels, show almost the same annual and cumulative natural uranium consumption levels.

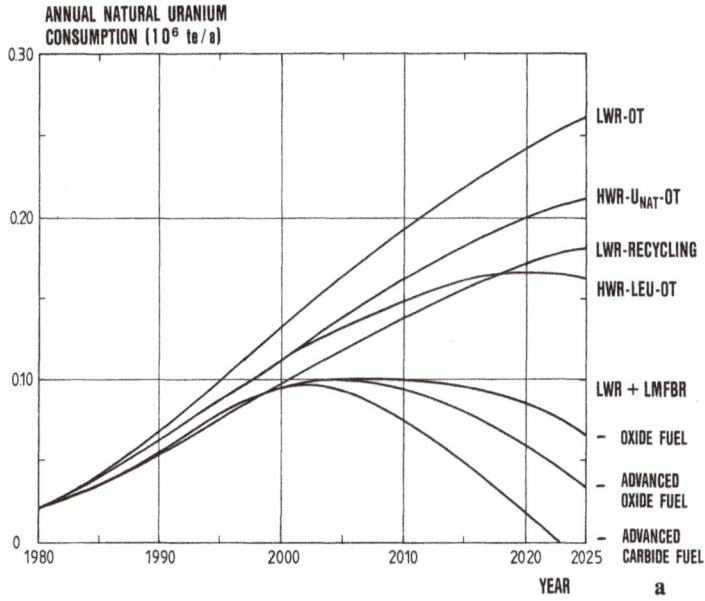

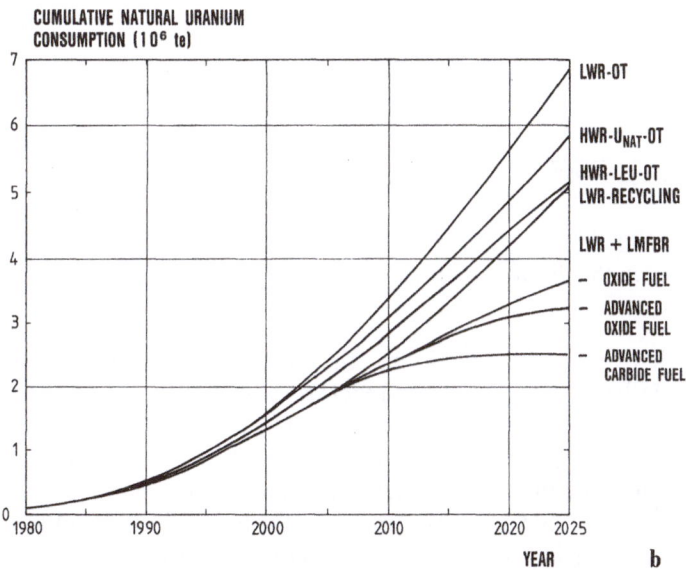

Fig. 6-5a and b. Annual and cumulative natural uranium consumption for reactor scenarios operating in the uranium/plutonium fuel cycle (INFCE)

Similar studies of scenarios with FBR's operating with U-233/Th fuel in symbiosis with converter reactors show that these symbiotic systems need more natural uranium, as more thorium is used in the blankets or even in the core. The highest natural uranium consumption is required for the case where the FBR contains thorium in both the core and the blankets.

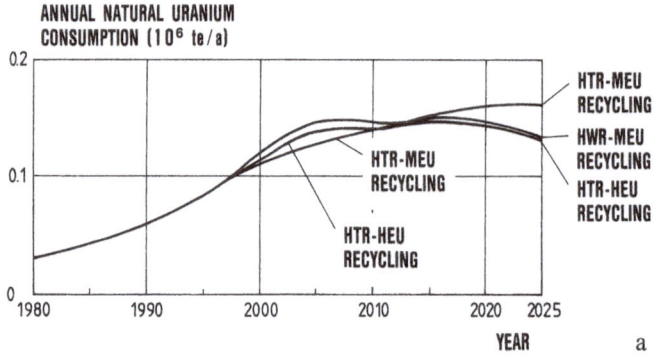

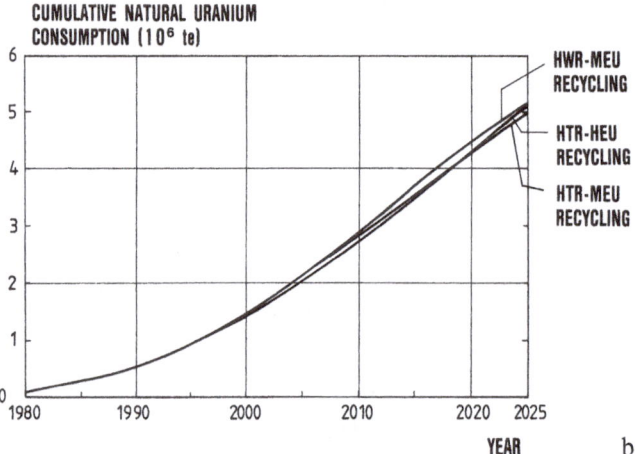

Fig. 6-6a and b. Annual and cumulative natural uranium consumption for single-type reactor scenarios operating in the thorium/uranium fuel cycle (INFCE)

From these results it can be concluded that the U-238/Pu fuel cycle using the combination of LMFBR's with converter reactors, e.g., LWR's or HWR's, will result in the lowest natural uranium consumption. The nuclear energy potential of U-238 tapped in this way in itself will be sufficient for several thousand years of energy generation on the basis of energy consumption estimates for the year 2000. In the Th/U-233 fuel cycle, more natural uranium would be consumed in all reactor scenarios. However, in principle also the exploitation of thorium as a fertile material is feasible, depending on when the Th/U-233 fuel cycle will be available commercially and technically.

Selected Literature

Chang, Y.I., *et al.*: Alternative Fuel Cycle Options. Argonne National Laboratory, ANL-77-79 (1977).

Conolly, Th.J., *et al.*: World Nuclear Energy Paths. New York-London: The Rockefeller Foundation/The Royal Institute of International Affairs. 1979.

Data Base for a CANDU PHWR Operating on the Thorium Cycle. Atomic Energy of Canada Ltd., AECL-6595 (1979).

Edlund, M.C.: High Conversion Ratio Plutonium Recycle in Pressurized Water Reactors. In: Proceedings of the Wingspread Conference on Advanced Converters and Near Breeders, Racine, Wis., 14–16 May 1975. Annals of Nuclear Energy, Vol. 2, pp. 801–807. Oxford: Pergamon Press. 1975.

Final Generic Environmental Statement on the Use of Recycle Plutonium in Mixed Oxide Fuel in Light Water Cooled Reactors (GESMO). Washington: US Nuclear Regulatory Commission, NUREG-002 (1976).

International Nuclear Fuel Cycle Evaluation, Fuel and Heavy Water Availability. Report of INFCE Working Group 1. Vienna: International Atomic Energy Agency. 1980.

International Nuclear Fuel Cycle Evaluation, Reprocessing, Plutonium Handling, Recycle. Report of INFCE Working Group 4. Vienna: International Atomic Energy Agency. 1980.

International Nuclear Fuel Cycle Evaluation, Fast Breeder Reactors. Report of INFCE Working Group 5. Vienna: International Atomic Energy Agency. 1980.

International Nuclear Fuel Cycle Evaluation, Advanced Fuel Cycle and Reactor Concepts. Report of INFCE Working Group 8. Vienna: International Atomic Energy Agency. 1980.

Kasten, P., et al.: Assessment of Thorium Fuel Cycles in Power Reactors. Oak Ridge National Laboratory, ORNL/TM-5565 (1977).

Report to the American Physical Society by the Study Group on Nuclear Fuel Cycles and Waste Management. (Pines, D., ed.). Reviews of Modern Physics 50 (1), Part II (1978).

Shapiro, N.L., et al.: Assessment of Thorium Fuel Cycles in PWR's. Palo Alto, Cal.: Electric Power Research Institute, EPRI-NP-359 (1977).

Stehle, H., et al.: Experience with Plutonium Recycle Fuel for Large Light Water Reactors in the Federal Republic of Germany. In: Nuclear Power and Its Fuel Cycle, Proc. Int. Conference, Salzburg, 2–13 May 1977, Vol. 3, pp. 305–322. Vienna: International Atomic Energy Agency. 1977.

7 Technical Aspects of Nuclear Fuel Cycles

7.1 Discharge and Storage of Spent Fuel Elements

After discharge from the reactor core, the fuel elements are stored on the reactor site for a period of at least one year to allow for radioactivity decay and cooling. The reactor plant proper usually has a fuel storage capacity of at least three years' discharge volume in addition to a full standby core inventory. The use of compact storage racks with neutron absorbers allows this storage capacity even to be extended to nine years' discharge volume. Spent fuel elements are then transported in spent fuel casks either to intermediate storage facilities or to storage pools at reprocessing plants. Table 7-1 shows the fuel characteristics of discharged fuel elements for different converter reactors and LMFBR's. The total activity, expressed in Ci/kg, for various times after discharge is very similar for fuel of converter and breeder reactors.

7.1.1 Shipping Spent Fuel Elements

Spent fuel elements are shipped in special fuel casks, which weigh between 30 and some 100 te and have load capacities for up to 6 te of spent fuel (Fig. 7-1). Fuel casks weighing up to 35 te are usually transported by special trucks on the road. Heavier fuel casks are transported on special rail cars. Also barge shipments on both inland waterways and oceans are made. The spent fuel elements are cooled within the casks either by air (dry casks) or water (wet casks).

The casks contain the necessary shielding with steel, lead and water or borated water. They are cooled by natural airflow over fins on the outer surface or by forced air circulation. Spent fuel casks are designed to withstand severe accident conditions during shipment. Releases of radioactivity under such conditions must be rendered impossible. Therefore, the casks must be able to withstand such impacts as thermal tests (fire), drop tests under gravity, penetration tests, and water immersion tests before being cleared for actual shipment. Special international shipping regulations have been elaborated.

7.1.2 Interim Storage of Spent Fuel Elements

Spent fuel elements can be stored for interim periods in water pools (wet storage), air cooled vaults (dry storage) or in special containers. For wet storage in intermediate storage pools or storage pools of reprocessing plants, the spent

Table 7-1. *Fuel element characteristics for different nuclear reactors after full burnup and discharge from the reactor core (INFCE)*

Characteristics of fuel element[a]		LWR		HWR	Gas Cooled Reactor			FBR
		PWR 1000 MW(e)	BWR 1000 MW(e)	CANDU 540 MW(e)	AGR 660 MW(e)	THTR 300 300 MW(e)	HTGR 342 MW(e)	LMFBR 1000 MW(e)
Total length	mm	3200–4827	4470	495	1049		793	5400
Cross section								
Side	mm	197–230	138–152			60 (sphere)		166
Diameter	mm			81.4–102.5	238			
Cladding material		Zircaloy 4	Zircaloy 2	Zircaloy 4	Stainless steel	Graphite	Graphite/SiC	Stainless steel
Total weight per element	kg	480–840	250–307	16.6–24.7	83.5	0.205	128.6	470
Heavy metal weight per element (initial)	kg	122–548	172–194	13.4–19.8	42.7		12.6	155
Fuel		UO_2	UO_2	UO_2	UO_2	$(U,Th)O_2$	$(U,Th)O_2$	UO_2/PuO_2
Design burnup	MWd(th)/kg	26–40	27.5–30	6.5–8.1	10–25	100	110	70–100
Total activity after:	Ci/kg							
150 days		4.6×10^3	3.8×10^3	7.9×10^2	1.2×10^3	1.4×10^4	7.2×10^3	6.4×10^3
1 year		2.3×10^3	1.9×10^3	8.4×10^1	6.1×10^2	5.5×10^3	3.7×10^3	3.8×10^3
10 years		3.2×10^2	2.9×10^2		10^2	1.1×10^3	1.1×10^3	0.7×10^3
Decay heat after:	W/kg							
150 days		24.3	18.7	3.15	4.9	60	28	27
1 year		10.4	8.2	0.22	2.4	26	15	14
10 years		2.3	2.2		0.3	4.2	3.8	1.3
Annual discharge:								
Number of fuel elements		41–64	170–210	4863	820	1.7×10^5	240	154
Uranium	te	26.3–32.9	35.8–38	90.9	35	8.9×10^2	0.125	25
Plutonium	kg	258–316	260–313	345	164	1.5	2.81	2100

[a] Range of existing reactors

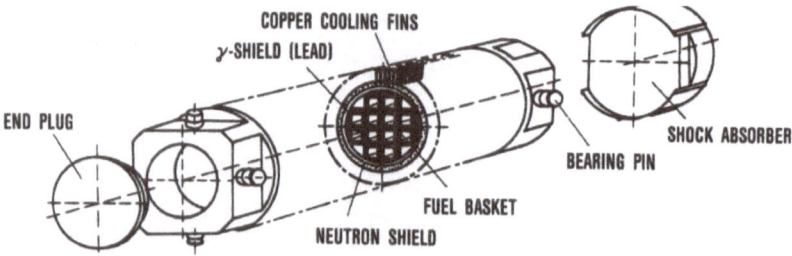

Fig. 7-1. Spent fuel shipping cask for LWR fuel elements (KFA Jülich)

fuel elements are arranged in racks or baskets kept in water pools. The water serves as a heat transfer medium for the heat generated in the fuel elements and provides the necessary shielding of the fuel elements. It is maintained at a sufficiently high level to provide shielding during all fuel handling operations. The walls and floors of storage pools are made of reinforced concrete lined with stainless steel. Water pools with a capacity of up to 1000 te of spent fuel are technically feasible. An intermediate storage facility may be equipped with several such water pools.

Fig. 7-2 shows an intermediate storage facility for LWR spent fuel elements. It has a storage capacity of 1500 te of uranium or 5400 LWR fuel elements and consists of four water pools for intermediate storage, two pools for receiving and two pools for discharging spent fuel elements. All water pools are equipped with heat exchange systems to keep the temperature at about 40 °C. The water is purified by ion exchange to ensure the specified water quality and good visibility. Fission and corrosion products are eliminated by the purification system.

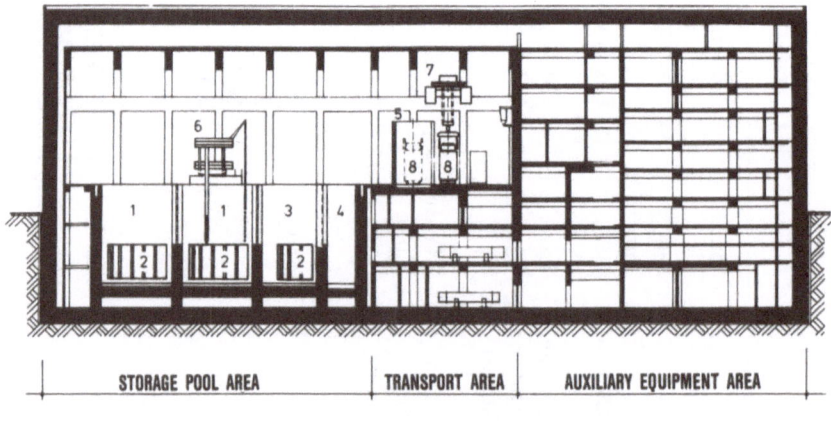

1 STORAGE POOL	5 DECONTAMINATION BOX
2 STORAGE RACKS	6 FUEL ELEMENT HANDLING MACHINE
3 RECEIVING POOL	7 CRANE
4 DISCHARGE POOL	8 STORAGE CASK

Fig. 7-2. Spent fuel intermediate storage facility (DWK)

The intermediate storage facilities are also equipped with lifting, handling and transfer devices to handle the spent fuel elements and heavy fuel transport casks. Criticality questions have to be taken into consideration in the design and construction of fuel element storage racks or baskets for enriched fuel elements in storage pools. LWR spent fuel elements can be stored, if needed, in water pools for much more than 10 years. During this time period, the fuel elements will not experience any water corrosion on their outer surfaces.

Dry storage of LWR spent fuel elements is also feasible in air cooled vaults made of cast iron. These cast iron spent fuel vaults take up to 4 PWR or 16 BWR fuel elements. They are equipped with outside cooling fins and can be stored in large intermediate storage buildings.

Dry storage is also used for HWR and HTGR graphite fuel elements. Spherical graphite fuel elements of HTR's can be stored under dry conditions in gastight cans.

After some years of intermediate storage, fuel elements may be conditioned for permanent storage without reprocessing (see Section 7.6.2).

LMFBR fuel elements are kept first in sodium cooled storage tanks on the reactor site. For intermediate storage they are filled in cans, cooled either with sodium and then stored under water or only cooled by air or an inert gas (nitrogen). Before reprocessing, the sodium is removed from the fuel element surface by melting or steam cleaning in a hot inert gas atmosphere.

7.2 The Uranium-238/Plutonium Fuel Cycle

Natural uranium can be utilized more efficiently in a closed fuel cycle with reprocessing and recycling of fissile material. This applies to fuel used in LWR's, HWR's or HTR/HTGR's. For near-breeders and FBR's, the closed fuel cycle is imperative. Technical aspects of reprocessing and recycling (refabrication) in the uranium/plutonium fuel cycle will be described in the following sections.

7.2.1 Reprocessing Spent UO$_2$ Fuel Elements

Spent fuel elements with UO$_2$ fuel and stainless steel or zircaloy claddings are transported to the reprocessing plant and stored there prior to chemical reprocessing. The steps of disassembly of such fuel elements, dissolution of the fuel as well as chemical separation are the same in principle for all fuel elements of converter reactors operated on UO$_2$ fuel (PWR, BWR, HWR, AGR). It is therefore sufficient to describe, as representative aspects, the technical steps of fuel disassembly and dissolution in the so-called head end of the reprocessing plant for spent UO$_2$ fuel from an LWR.

7.2.1.1 LWR Fuel Element Disassembly and Spent Fuel Dissolution

In a reprocessing plant (Fig. 7-3 shows the head end of such a plant), the storage pools are arranged close to the fuel element disassembly cells. The fuel elements are moved by means of a crane from the storage pool into the

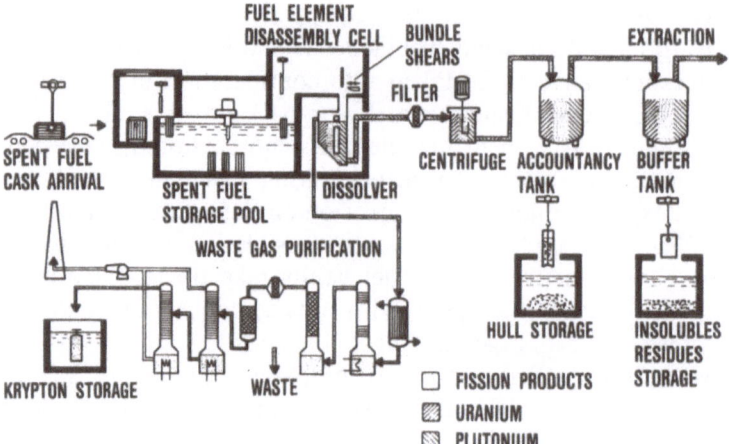

Fig. 7-3. Head end and waste gas purification system of a reprocessing plant (KFA Jülich)

disassembly cell above it. In this cell, LWR fuel elements are cut up by large bundle shears. After the end caps have been removed from the fuel elements, the fuel rod bundles are chopped into pieces approximately 5 cm long. The bundle shear is operated remotely and is designed so that it can also be repaired by remotely operated tools. The fuel element and fuel rod sections drop through a chute directly into a dissolver basket located in the dissolver cell underneath. The basket is filled with boiling nitric acid, which leaches the fuel out of the chopped fuel rod hulls. After leaching of the fuel, the remaining hulls and fuel element sections are dumped from the basket into a container, and the container is moved into the hull storage facility.

The fuel solution still contains small solid parts, such as zirconium or steel chips from chopping. Moreover, it includes undissolved particles of fission and corrosion products, such as ruthenium, palladium, rhodium, molybdenum, technetium and zirconium. The undissolved fraction of plutonium, which may be contained in these undissolved particles, is about 1%. The undissolved solid particles are removed through coarse filters or by centrifuges. Solid particles separated in the dissolution of LWR fuel elements may contain up to 0.3 kg/te fuel of zirconium chips and some 3.3 kg/te fuel of undissolved corrosion and fission products.

7.2.1.2 Gas Cleaning and Retention of Gaseous Fission Products

During the processes of chopping and dissolution of the fuel, gaseous and volatile fission products are released. They must be removed together with water vapor, nitrous gases (NO, NO_2, N_2O) and the nitrogen which may have been applied as a scavenging gas in fuel element chopping. This mixture of volatile fission products, vapors and gases must be treated in the waste gas cleaning system. Gaseous and volatile fission products are made up of the following components:

Tritium produced by ternary fission and by (n,T)-reactions in light atomic nuclei. Some 40% of the tritium generated remain in the metal structure of the zircaloy cladding tubes. The balance is released as tritiated water, HTO, during dissolution and may enter the gaseous effluent section together with water vapor. Less than 1% of the tritium is found there as gaseous tritiated hydrogen, HT.

Carbon, C-14, is produced by an (n,α)-reaction from O-17 and by the (n,p)-reaction of N-14. In the gaseous effluent it appears as $^{14}CO_2$.

Krypton is generated as a gaseous fission product. Some 7% of the krypton fission products produced consist of Kr-85 isotopes.

Xenon is another gaseous fission product. However, only traces of the Xe-133 isotope produced must be considered, because it has a relatively short halflife of 5.27 d.

All the other fission product noble gases generated are either stable or have very short halflives.

I-129 and traces of I-131 are partially volatile isotopes initially found in dissolved fuel. They may be carried into the gas stream through boiling and as a result of passing an inert gas through the dissolved fuel solution, which entraps the iodine in the gas stream.

Ru-106 may volatilize as ruthenium tetroxide evaporating from strong nitric acid solutions, but only some 10^{-4} fractions of Ru-106 enter into the gaseous effluent stream. In a similar way, small traces of such β-emitters as strontium or α-emitters as uranium and plutonium can penetrate into the gaseous effluent as aerosols. However, only some 10^{-4} to 10^{-6} fractions of the fuel inventory are carried into the gas stream as aerosols.

These gaseous effluents are first passed through a condenser. Afterwards, the nitrogen oxides are oxidized and washed out. This already removes 99% of the aerosols. The remaining aerosol fractions only amount to 10^{-6} to 10^{-8} times the inventory. Scrubbers and high-efficiency particulate aerosol (HEPA) filters are used next to remove the aerosols. Iodine is retained very efficiently in silver impregnated ($AgNO_3$) filter materials. Tritium as HTO contained in water vapor and $^{14}CO_2$ are retained in molecular sieves. Tritium present as HT hydrogen is converted into HTO water and is recycled. The removal of Kr-85 can be achieved by means of low temperature rectification. In the same process, the xenon noble gas can also be removed. The separated krypton can be stored in compressed gas cyclinders. Alternatives may be the entrapment in zeolites (crystallized silicates) and ion implantation in metals.

7.2.1.3 Chemical Separation of Uranium and Plutonium

Although a number of chemical separation techniques have been proposed and developed in the past few decades, the most efficient process to date has remained the PUREX process (plutonium and uranium recovery by extraction). The PUREX process uses tri-n-butyl phosphate (TBP), which may be diluted, for instance, by kerosene or n-paraffin (hydrocarbon) solvents as organic solvents to extract uranium and plutonium. TBP is stable in nitric acid and can selectively extract tetravalent and hexavalent uranium and plutonium nitrate

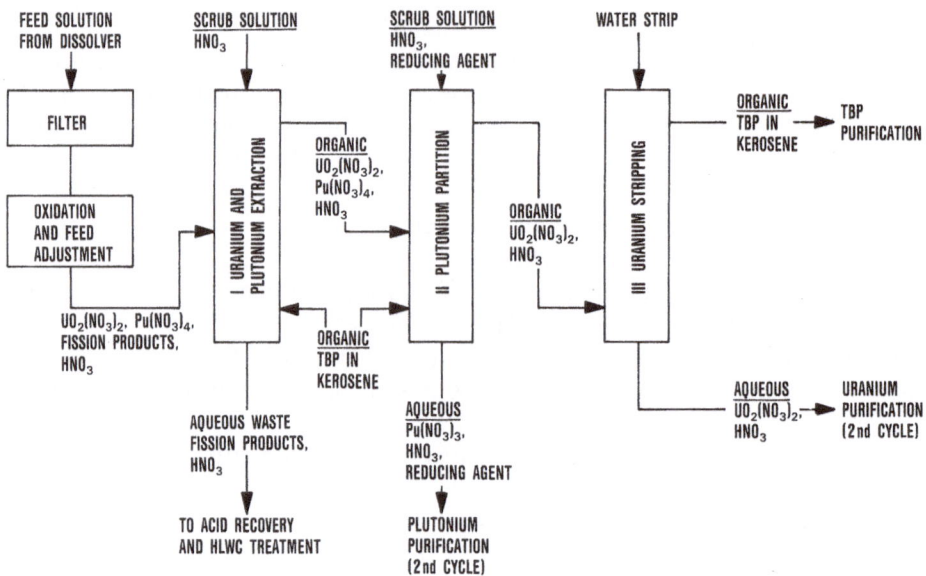

Fig. 7-4. Simplified PUREX process flowsheet (M. Benedict)

complexes. However, this selective extraction capability of TBP does not apply to trivalent plutonium nitrate complexes.

For extraction, the fuel solution acidified with nitric acid and containing uranium, plutonium, higher actinides and fission products is moved from the middle of column I (Fig. 7-4) in a liquid-liquid countercurrent extraction flow past the specifically lighter organic solvent (TBP in kerosene) rising from the bottom. In that process, the organic solvent extracts uranium and plutonium, while the fission products and actinides remain in the aqueous solution. The solution with nitric acid leaves the column at the bottom as *high level* aqueous *waste* (HLW). It contains the fission products and higher actinides. The aqueous waste is evaporated to recover the nitric acid. The remaining concentrate is further treated as *high level waste concentrate* (HLWC).

The rising organic solvent contains uranium and plutonium and small traces of fission products, which are removed by a nitric acid solution injected at the top of the column. The organic solvent leaves the column at the top and is introduced into column II, where the tetravalent and hexavalent plutonium is reduced to trivalent plutonium by means of a reducing agent stream, e.g., U(IV) nitrate with hydrazine nitrate, hydroxylamine nitrate or, formerly, Fe(II) sulfamate. (The most elegant method developed recently, uses electrolytic reduction within the extraction apparatus). This trivalent plutonium is sparingly soluble in organic TBP-kerosene and, as a consequence, is re-extracted into the aqueous phase, while hexavalent uranium remains in the organic TBP-kerosene phase. Small amounts of re-extracted uranium are extracted again by organic TBP-kerosene introduced at the bottom of the second column. The aqueous plutonium product stream leaves the second column at the bottom, while the organic uranium product stream leaves at the top and enters the bottom of

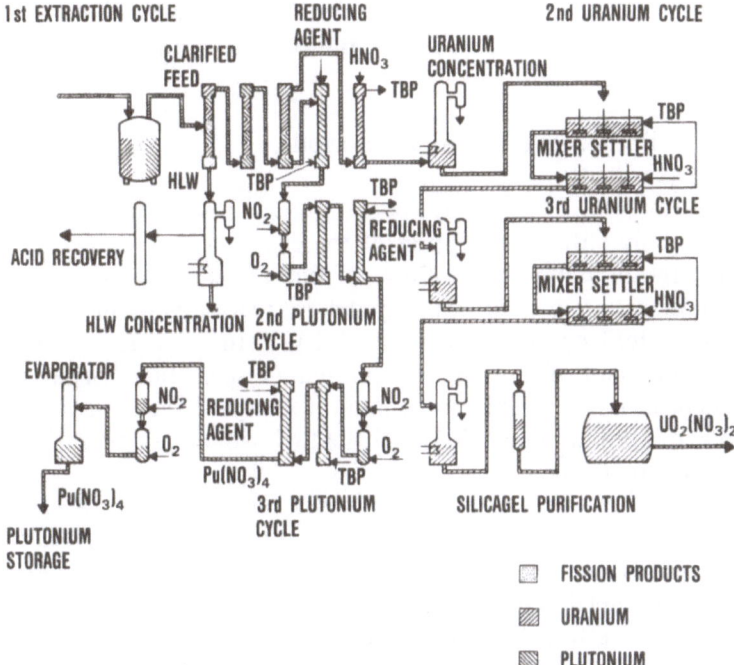

Fig. 7-5. Simplified flowsheet of the uranium/plutonium extraction cycles of a reprocessing plant (KFA Jülich)

the third column, where it is met by a countercurrent stream of diluted nitric acid as an aqueous re-extraction solution flowing from the top. The uranium product stream with nitric acid then leaves column III at the bottom, while the organic solvent leaves at the top. After removal of organic decomposition products and fission products by washing, the organic solvent can be recycled into the system.

For sufficient decontamination of uranium and plutonium, the uranium and plutonium product streams are required to pass through two further decontamination cycles as shown in Fig. 7-5. The final products, after concentration and purification, are plutonium nitrate, $Pu(NO_3)_4$, and uranyl nitrate, $UO_2(NO_3)_2$. The resulting waste streams must be treated separately, as will be described in Section 7.5.

The extraction apparatus shown in Fig. 7-5 can be used in three different technical designs, i.e., as pulsed perforated plate columns, mixer-settlers, or centrifugal contactors. Centrifugal contactors have very short contact times for the aqueous and the organic phases, largely protecting the organic solvent from degradation by radiation. This makes them particularly suitable for fuel with short cooling time and high burnup, such as LMFBR fuel. Pulsed columns also have relatively short contact times of the organic solvent, permitting the installation of heterogeneous neutron lattices for criticality control. Mixer-settlers are very reliable, flexible and simple systems with longer contact times. They have been used most successfully in reprocessing fuel elements with low burnups and in the second and third uranium purification cycles. However, in the first decontamination cycles and in the plutonium purification cycles, preferably pulsed columns are used.

7.2.1.4 Mass Flows of Radioactive Material in a Model LWR Fuel Reprocessing Plant

After this short and simplified description of the chemistry of reprocessing LWR fuel elements, the individual mass flows of fuel and nuclear waste will now be considered in a commercial scale model reprocessing plant with a throughput of about 4 te/d or 1000–1200 te/a of LWR fuel, as described in Fig. 7-6. A reprocessing plant of this capacity can handle spent fuel discharged from some 30–35 GW(e) LWR's. The model plant is designed for a maximum burnup of LWR fuel of approx. 40,000 MWd(th)/te fuel with a maximum fissile material enrichment of some 4%.

The fuel is assumed to have been kept in intermediate storage for an average of three years before being reprocessed. (Extended interim storage as envisaged in a number of countries would reduce the levels of radioactivity described in Fig. 7-6). During spent fuel dissolution, a model plant of this type will produce some $1400 \ m^3/d$ of gaseous effluent, some $2 \ m^3/d$ of hulls and structural materials, and roughly $0.2 \ m^3/d$ of sludge of insoluble fuel residues. In the gaseous effluent, Kr-85 has the highest radioactivity with 10^4 Ci/te of fuel, while I-129 and C-14 with 0.04 Ci/te and 0.7 Ci/te of fuel, respectively, only make minor contributions. The hulls and structural components of the fuel elements are radioactive after prolonged exposure to neutrons in the reactor core. In addition, the hulls contain small amounts of undissolved uranium and plutonium. The radioactivity of the hulls and structural material amounts to $5 \times 10^4 \ Ci/m^3$. The insoluble residues also contain plutonium and have a bulk radioactivity of roughly $10^6 \ Ci/m^3$.

In the first extraction cycle, the fission products, higher actinides and small quantities of unextracted U/Pu produced amount to $2 \ m^3/d$ of HLWC with a radioactivity of $1.2 \times 10^6 \ Ci/m^3$. In the second and third uranium and plutonium decontamination cycles, some $0.4 \ m^3/d$ of organic solvent is produced as organic *medium level waste* (MLW), which contains small traces of U/Pu. It has a radioactivity of approx. $1 \ Ci/m^3$. Moreover, about $6 \ m^3/d$ of aqueous medium level waste of $100 \ Ci/m^3$ are produced, which also contains small traces of U/Pu. All these quantities of HLWC and MLW are further treated by special waste conditioning and storage techniques, which will be described in Section 7.5. Some $160 \ m^3/d$ of liquid effluents with a very low radioactivity of $10^{-8} \ Ci/m^3$ can be discharged directly without any further treatment. Tritium enriched in water is being recycled and conditioned as described in Section 7.5.

The model reprocessing plant generates $0.16 \ m^3/d$ of plutonium nitrate solution with 40 kg of plutonium and $8.8 \ m^3/d$ of uranyl nitrate with 3.96 te of uranium.

Commercial reprocessing plants with capacities of 1000–1200 te/a of LWR fuel must also have the corresponding storage capacities. For spent fuel elements, the storage capacity should be 1000–3000 te U, for HLWC,

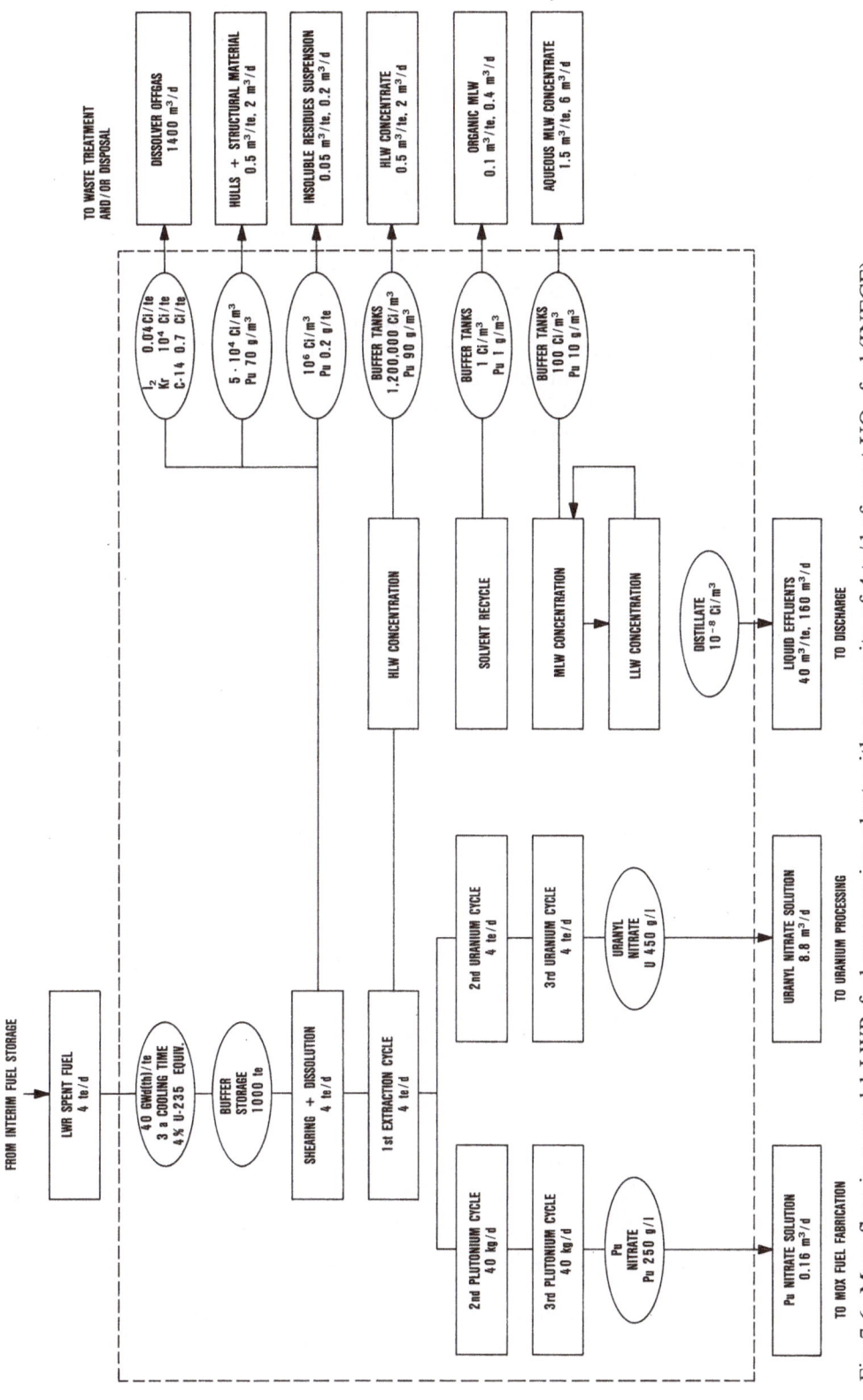

Fig. 7-6. Mass flow in a model LWR fuel reprocessing plant with a capacity of 4 te/d of spent UO₂ fuel (INFCE)

$1000-2000 \text{ m}^3$, for aqueous MLW, $1500-10,000 \text{ m}^3$, and for organic MLW, approx. $200-500 \text{ m}^3$. For plutonium nitrate, buffer storage capacities of $1-2 \text{ m}^3$ are provided.

7.2.1.5 Radioactive Inventories of Spent Fuel and Waste

Besides the mass flows of fuel, fission products, actinides and waste it is mainly their inventories of radioactivity, their heat generation and their potential of radiotoxicity, which constitute important parameters on which to base engineered safety measures within the reprocessing plant as well as in waste treatment and storage. Customarily, these data are based on 1 te of *heavy metal* (HM) fuel. In that case, roughly 1.14 te of UO_2 or UO_2/PuO_2 correspond to 1 te_{HM}. When loaded into the core, 1 te_{HM} of fresh LWR fuel in an equilibrium cycle with 3.2% U-235 enrichment contains 32 kg of U-235 and 968 kg of U-238. When unloaded from the LWR core after a burnup of 36,000 MWd(th)/te, 1 te_{HM} of spent fuel still contains 6.8 kg of U-235 and 942.0 kg of U-238, but 4.1 kg of U-236, some 9.5 kg of different plutonium isotopes, 36.9 kg of fission products, 0.5 kg of Np-237, 0.15 kg of americium, and 0.04 kg of curium.

Fig. 7-7 indicates the radioactivity in Ci/te_{HM} of spent LWR fuel with a burnup of 36,000 MWd(th)/te_{HM} as a function of time after discharge from the core. The bulk of radioactivity is constituted by the fission products. Over a three-year period of storage for cooling at the reactor site, transport and interim storage, the radioactivity of fission products and actinides decreases by some two orders of magnitude. In the reprocessing step, the fission products and all actinides, such as neptunium, americium and curium, and approx. 1% Pu- and U-losses

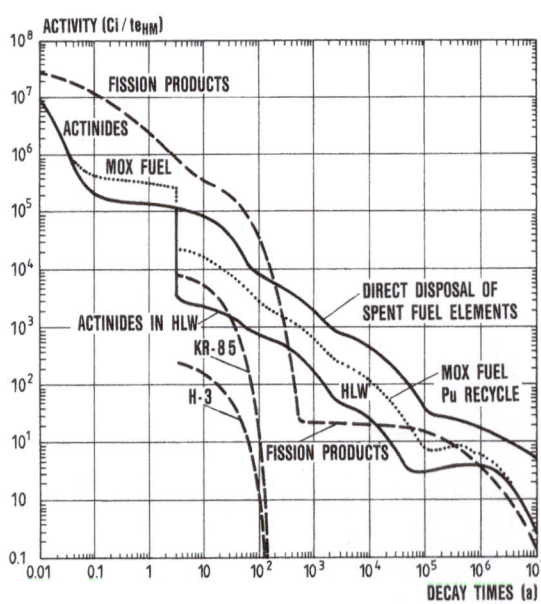

Fig. 7-7. Activity of fission products and actinides in high-level waste and LWR fuel; reprocessing after 3 years of cooling time (H.O. Haug)

are separated and go into the high level waste, while 99% of the plutonium and uranium is recovered. The bulk of the fission products decay rapidly within the first 500 years. After some 200 years, the radioactivity of the actinides plays the dominant role in high level waste until around 11,000 a, when fission products like Tc-99 and others take over for a long time.

Fig. 7-7 also presents a comparison of low enriched UO_2 fuel with PuO_2/UO_2 mixed oxide (MOX) fuel for LWR's and compares the activity of HLW (reprocessing case) with the alternative of direct disposal of spent LWR fuel elements.

7.2.2 Recycling of Plutonium and Uranium

To make MOX fuels, e.g., for LWR cores operated in the Pu-recycle mode or for FBR cores, the plutonium nitrate and uranyl nitrate solutions produced in the reprocessing plant must be converted into uranium and plutonium oxides. This may be done either at the end of the reprocessing plant or at the head of the refabrication plant.

In a fuel cycle center, the plants for reprocessing, conversion and MOX fabrication are co-located on one site to avoid movements of plutonium nitrate or plutonium oxide. In this case, conversion can be linked straight to the head end of the MOX fuel fabrication plant. Also, plutonium nitrate and some of the uranyl nitrate can be directly mixed and co-converted into PuO_2/UO_2. The PuO_2/UO_2 may then be stored in a buffer store before being turned into fuel elements.

7.2.2.1 Converting of Plutonium Nitrate into Plutonium Oxide

In the conversion process, plutonium nitrate is mixed with oxalic acid to form a plutonium oxalate, $Pu(C_2O_4)_2$, precipitate. The suspension is pumped into a filtering unit where the precipitate is separated as a cake. The cake is then calcined at temperatures > 300 °C. The resulting plutonium oxide, PuO_2, is milled, screened and finally stored in a plutonium oxide buffer store for subsequent mixed oxide fuel fabrication.

Alternatively, mixed PuO_2/UO_2 powder can be produced following co-precipitation of ammonium diuranate and plutonium hydroxide after the addition of ammonia to a mixture of plutonium/uranium nitrate solution.

7.2.2.2 Converting Uranyl Nitrate into Uranium Oxide

In uranyl nitrate conversion, ammonia and carbon dioxide is fed to the uranyl nitrate solution. Ammonium uranyl carbonate is generated, which is precipitated and can then be decomposed thermally in a fluidized bed furnace. After splitting by thermal decomposition and separation of ammonia and carbon oxide, the uranium trioxide (UO_3) product is obtained. This uranium trioxide can be reduced to uranium dioxide, UO_2, by the simultaneous addition of hydrogen at 500–600 °C. The uranium dioxide powder is homogenized and stored for subsequent fuel element fabrication.

7.2.2.3 Mixed Oxide Fuel Fabrication

Plutonium recycling in thermal reactors, e.g., LWR's, requires the fabrication of MOX pellets. If the reprocessing plant and the MOX fuel fabrication plant are not co-located on one site, and if plutonium nitrate and uranyl nitrate have already been converted into oxides, PuO_2 and UO_2 will have to be transported to the MOX fuel fabrication plant. After storage in a buffer store, plutonium oxide and uranium oxide are mixed to the required Pu_{fiss} enrichment. The mixed oxide powder is then precompacted and granulated into a freely flowing powder. This is turned into pellets, which are first sintered at temperatures of 1000 and 1700 °C and then ground to the required dimensions. The pellets are heated to remove their moisture and gas contents. Finally, they are loaded into zircaloy or steel tubes, to which end caps are welded, and assembled into fuel assemblies.

During storage the Pu-241 isotope decays to Am-241, which emits 60 keV γ-radiation. This requires either chemical separation of Am-241 prior to fabrication or appropriate shielding against γ-radiation during the fabrication process, if PuO_2 had been stored for more than two years. Also, a certain fraction of the mixed oxide pellets are imperfectly made and are rejected during inspection procedures. Such material and grinder fines, which are designated clean rejected oxide, can be recycled directly into the manufacturing process. However, a small fraction of pellets and powder are contaminated by corrosion products, etc. These constitute the so-called dirty scrap requiring chemical purification. Dirty scrap is processed through a series of roast, scrub, wash-leach and filter systems. Dirty mixed oxide scrap is dissolved in a nitric acid/fluoride solution and, along with filtrates from wash-leach processes, is treated chemically as in a PUREX reprocessing step to recover the uranium and plutonium. Other process residues include such contents as metal scrap, plastics, rubber, cellulose, cleaning materials and organic substances. These wastes are treated as low level waste (see Section 7.5).

As MOX fuel must be reprocessed after having attained its design burnup, the fuel fabrication process must guarantee high solubility ($>99\%$) of the MOX fuel in nitric acid. Such high solubility is required to minimize the plutonium loss in the residues during chemical reprocessing. This requires special procedures (e.g., milling) in the MOX fabrication process described above.

If the plants for chemical reprocessing and MOX fuel refabrication are co-located on one site, the process of co-conversion can be adopted. In that case, two other refabrication processes can be applied. The sol-gel process, described in more detail in Section 7.3.3, allows the direct fabrication of MOX particles, which can be pressed and sintered into fuel pellets. The AUPuC (*ammonium* (*U,Pu*) *carbonate*) refabrication process also allows the fabrication of relatively coarse grain MOX powder. This crystal powder is fabricated essentially free of Am-241 and then pressed and sintered into MOX fuel pellets. Both the sol-gel and the AUPuC fabrication processes avoid the generation of plutonium dust within the glove boxes of the fuel fabrication line and therefore lead to relatively lower radiation doses to the staff working in MOX fuel refabrication plants.

Table 7-2. *Main design characteristics of a MOX fuel fabrication plant (INFCE)*

Fuel fabrication capacity	te$_{HM}$/a	300
Number of fabrication lines		6–7
Enrichment of MOX fuel	%	4.5 (Pu total)
Clean rejected oxide reprocessed	te/a	12–24
Dirty rejected oxide reprocessed	te/a	4–6
Pu-contaminated material treated	m^3/a	1000
Pu-contaminated material disposal	m^3/a	200–250
Pu-contaminated liquid treated	m^3/a	300
Pu-contaminated liquid disposal	m^3/a	20–25
Other active liquid wastes	m^3/a	800–1000
Inactive liquid wastes	m^3/a	20,000

Table 7-2 lists the main design characteristics of a MOX fuel fabrication plant with an annual fabrication capacity of 300 te$_{HM}$/a. This capacity roughly corresponds to the plutonium mass flow produced by the 4 te/d or 1000–1200 te/a reprocessing plant described in Section 7.2.1.

7.2.3 Status of Uranium Fuel Reprocessing Technology

The largest body of experience is presently available in reprocessing spent metallic fuel of gas graphite reactors. A reprocessing plant with a capacity of 2000–2500 te$_{HM}$/a operated at Windscale, UK, by BNFL has reprocessed more than 20,000 te$_{HM}$ of metallic fuel. Similar reprocessing plants of 1000 te$_{HM}$/a capacity were and are being operated at Marcoule and La Hague, France.

The basic technology of the PUREX process, of plutonium storage, handling and transport is well established. Some countries have built, are building, or plan to build reprocessing plants for spent LWR oxide fuel on an industrial scale. Some hundred tonnes of uranium oxide fuel have been reprocessed in such countries as France, the United Kingdom and the United States. Slightly smaller quantities have been reprocessed on a pilot scale in Belgium, the Federal Republic of Germany and Japan. Table 7-3 indicates the annual capacities of reprocessing plants already in operation or planned in the Western world. The development of spent fuel reprocessing as a commercial industry was handicapped, especially in the USA, during the late 1970's by the non-proliferation policy and safeguards discussions about large reprocessing plants (see Section 8.3).

7.2.4 Status of Experience in Mixed Oxide Fuel Fabrication and Reprocessing

A considerable amount of MOX fuel has been fabricated already in small scale refabrication plants in countries including the USA, UK, France, Belgium, Japan and the Federal Republic of Germany. MOX fuel elements have been irradiated successfully in LWR's and other types of reactors up to burnups

Table 7-3. *Reprocessing plant capacity for spent LWR fuel in the Western World (INFCE)*

Country	Plant	Reprocessing capacity (te_{HM}/a)	Storage capacity of spent fuel (te_{HM})
France	Cap de la Hague	UP2: 250–800 UP3: 800	2400–8000
UK	Sellafield	Thorp 1: 650 Thorp 2: 650	1500–6000
Belgium	Mol EUROCHEMIC[a]	75	
Federal Republic of Germany	WAK	35	
	DWK-Hessen[b]	350	1500 (Ahaus) 1500 (Gorleben)
Japan	Tokai Mura[b]	200 1500	2000
USA	NFS (West Valley)[c]	300–750[b]	
	AGNS (Barnwell)[d]	1500	
	EXXON Nuclear (Oak Ridge)[b]	1500	7000

[a] Shut down.   [c] Shut down since 1972.
[b] Planned capacity.   [d] Operation deferred since 1977.

of 33,000 MWd(th)/te_{HM}. Some of these MOX fuel elements have also been successfully reprocessed.

Spent MOX fuel differs from low enriched spent UO_2 fuel in its slightly different content of fission products and in its content of higher actinides. Plutonium fuel builds up a higher percentage of americium and curium as a consequence of neutron capture in the thermal neutron spectrum of the reactor core.

The higher plutonium content only requires a slight adjustment of the flowsheet of the reprocessing plant described in Section 7.2.1. If fabrication specifications for the MOX fuel are observed carefully (high solubility and homogeneity), there is no great difference between the dissolution capability of spent MOX fuel and that of spent low enriched UO_2 LWR fuel. The increased fractions of americium and curium isotopes become part of the HLW, where they will cause a somewhat higher decay heat generation. This must be taken into account in waste conditioning.

7.2.5 Safety Aspects

7.2.5.1 Safety Design Measures in Reprocessing Plants

Unlike nuclear reactors, reprocessing plants are characterized by the following differences in hazard potential:
– The nuclear fuel is not arranged in a neutronically critical geometry ($k_{eff} < 1$).

- The fuel is not used for power generation and, correspondingly, is only at a low temperature and low pressure.
- The radioactivity of the spent fuel decays by a factor of 65 within one year after unloading from the reactor core, as a result of the decay of fission products and actinides.

Accordingly, the radioactive inventory per te_{HM} of LWR fuel in a reprocessing plant is much smaller than in a nuclear reactor. These are the reasons why large reprocessing plants can have capacities which allow the spent fuel from many GW(e) of power reactors to be reprocessed.

In reprocessing plants, as in nuclear reactors, the safety design principles of diversity and redundancy are applied in supplying electricity and cooling water to ensure the reliability of heat removal. Appropriate reserves for cooling water are taken into account in plant design. The multiple barrier principle between radioactive substances and the environment is observed. The radioactive materials are enclosed in leaktight systems of stainless steel pipes, vessels and other equipment which, in turn, are enclosed in leaktight cells with high-density concrete walls up to 2 m thick. In case of a leak in a pipe or vessel, stainless steel catch pans prevent radioactive liquid from penetrating the floor of the containment cells. The inner containment is surrounded by an outer protective shield. The inner and the outer containments are operated at pressures lower than atmospheric. Rooms with the highest radioactivity levels are kept at the lowest pressures. Contaminated air is filtered by at least two redundant filter systems and treated to reduce any radioactivity to acceptable levels. After filtering, the air is released into the environment through a stack (for details, see Section 8.1). Releases to the atmosphere or to rivers, lakes or the sea are continuously monitored.

As in nuclear reactor plants, the containment of a reprocessing plant must be designed to withstand earthquakes, floods, tornados, airplane crash impacts, shockwaves caused by explosions, fires and sabotage. Engineered safety features also include measures to prevent criticality in dissolvers, extraction columns and buffer tanks. This can be achieved by limiting the geometries of the equipment, adding such neutron poisons as boron, hafnium and gadolinium, and by strictly limiting the fissile enrichment of spent fuel elements to be reprocessed, respectively. Moreover, the fissile enrichment of solutions is continuously monitored by measurements at critical points of a facility.

The design base accidents of the plant to be considered are the explosion of an evaporator for high level waste concentration and a criticality accident in one of the components carrying fissile material. Kerosene, TBP and nitric acid have a potential for exothermic reactions only if organic products were able to reach the evaporator and only if temperatures above 140 °C were attained. Such conditions are avoided during operation by keeping the temperature of the process steam for evaporation at 130 °C.

Despite such safety design measures, design base accidents are analyzed and the design of the inner cells must limit the consequences of such accidents. If the evaporator or another container were destroyed by an explosion, the waste solution would leak into the catch pan on the cell floor and could be pumped back into another container held in reserve. Radioactive aerosols would be retained in the filter system. The radioactive impact upon the environment would be limited. Similar design measures are taken to limit the consequences of a criticality accident. For details of risk analyses of reprocessing plants, see Section 8.2.

7.2.5.2 Safety Considerations for Mixed Oxide Fuel Fabrication Plants

The multiple barrier system is also applied to limit the release of plutonium from MOX fabrication plants. Primary confinement is provided by shielded glove boxes or hot cells containing the plutonium pellet fabrication equipment. This primary confinement is surrounded by an operating and maintenance building accessible only through locks. It also constitutes a firewall and a shielding protection. An outer shell protects these two inner confinements against natural and external events (earthquakes, airplane crashes, etc.). Pressure differentials are maintained within the different confinements such that the lowest pressure applies to the rooms with the highest plutonium concentrations. Air is exhausted only through a number of HEPA filters connected in series. The safety design basis of a mixed oxide fuel fabrication plant is determined by the potential of criticality accidents, fires and explosions in the manufacturing equipment.

7.3 The Thorium/uranium-233 Fuel Cycle

Thorium may be used as a fertile material in converter or breeder reactors, as described in Chapter 6. In that case, U-233 must be recovered from spent fuel elements by chemical reprocessing. After storage on the reactor site or in interim storage facilities, spent fuel elements with U-233/Th fuel are transported to the reprocessing plant.

7.3.1 Fuel Element Disassembly

The design of U-233/Th fuel elements to be used in LWR's or HWR's roughly corresponds to that of today's standard fuel elements with low enriched uranium. Consequently, the head end of the U-233/Th reprocessing plant required for chopping the fuel element into short pieces and subsequent dissolution may be the same as described in Section 7.2.1.

The fuel of HTGR's or HTR's consists of small fissile particles coated with pyrolytic carbon and silicon carbide; fertile particles made of ThO_2 are coated with carbon. These particles are imbedded in a graphite matrix of the fuel element (prismatic block or sphere). Prior to dissolution of the fuel, the graphite must be separated from the fuel. This is done by crushing the blocks or spheres and burning the graphite in a fluidized bed. Fissile particles coated with SiC must be crushed and the residual inner pyrolytic carbon layer must be burnt. C-14 produced by the neutron activation of carbon and by (n,p)-reactions with residual nitrogen must be specially treated in the waste gas cleaning system as ^{14}CO and $^{14}CO_2$. The ash produced is made up of the fissile particles containing U-235/U-238, plutonium and fission products and of the fertile particles containing $^{233}UO_2/ThO_2$ and fission products. For further treatment, the fertile and fissile particles are separated.

The $^{233}UO_2/ThO_2$ fuel must be treated by the THOREX (*tho*rium *o*xide *r*ecovery by *ex*traction) reprocessing technology. The fissile fuel, which contains U-235/U-238, plutonium and fission products, may be reprocessed by the PUREX technology. If medium enriched uranium were mixed with thorium, this fuel would contain thorium, uranium isotopes, plutonium, fission products and higher actinides after irradiation in the reactor core. In this case, a combined PUREX-THOREX-reprocessing technology would have to be applied.

7.3.2 THOREX Process

The U-233/Th fuel is dissolved in very highly concentrated 13 M nitric acid, 0.05 M hydrofluoric acid and 0.1 M aluminum nitrate held at boiling temperature. The residual solids are removed from the solution by centrifuging. The solution with $Th(NO_3)_4$ and $UO_2(NO_3)_2$ then enters the first extraction column (Fig. 7-8) and is moved in a countercurrent flow against TBP dissolved in a hydrocarbon solvent. TBP selectively dissolves thorium nitrate and uranyl nitrate while moving upward in the column. The fission products, protactinium and aluminum nitrate leave the column at the bottom together with the scrub solution, which is added at the top of the column. Careful adjustment of these chemical processes is necessary to separate fission products, especially Zr-95, from thorium. In the tetravalent state, thorium is chemically very similar to

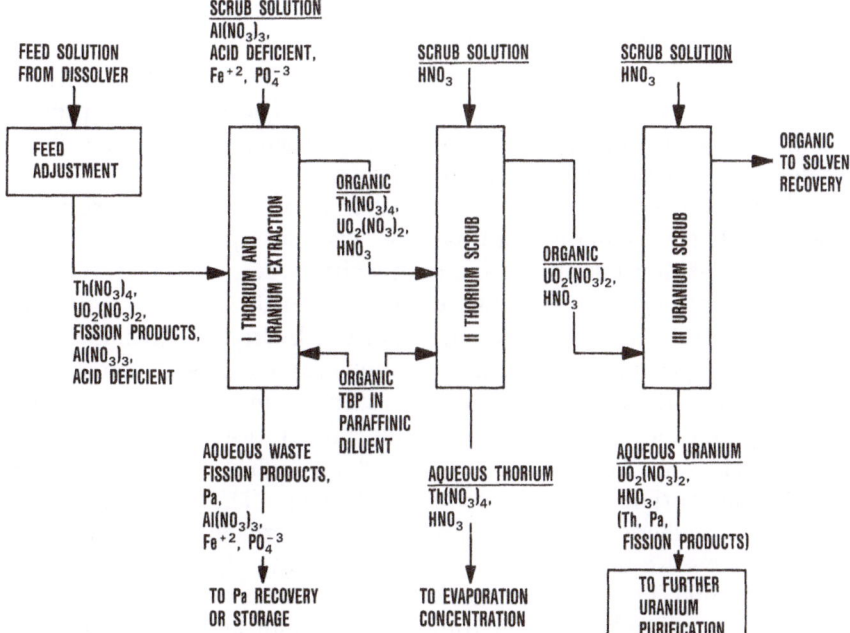

Fig. 7-8. Simplified THOREX process flowsheet (M. Benedict)

zirconium. Pa-233 with a halflife of 27.0 d is the precursor of U-233 and must be chemically recovered from the HLW. Alternatively, the fertile fuel can be cooled until Pa-233 has decayed into U-233. In the second column, $Th(NO_3)_4$ is recovered from the TBP by being moved in a countercurrent flow against diluted nitric acid. $Th(NO_3)_4$ and nitric acid leave the second column at the bottom. The organic solution together with $UO_2(NO_3)_2$ flows into the third column, where uranium is re-extracted. Uranium is then purified further in additional solvent extraction separation steps. Small traces of plutonium and neptunium may be separated by additional extraction chromatography. In case more plutonium is built up, e.g., in medium enriched U-235/U-238/Th fuel (see Chapter 6 for MEU-Th reactor cores), the separation from plutonium and neptunium by extraction chromatrography is not sufficient. In these cases the plutonium must be co-extracted with uranium and thorium. This may also be achieved by solvent extraction in contact with TBP. However, this PUREX-THOREX process is more complicated than the THOREX process.

The THOREX process: Status and experience. The technological basis of the THOREX process is well understood, but much less experience has been accumulated in its implementation than with the PUREX process. Small THOREX pilot facilities were operated successfully at Oak Ridge, Hanford and Savannah River in the USA. Within the US LWBR and HTGR programs, approximately 870 te of thorium (mostly ThO_2) was reprocessed. In the Federal Republic of Germany and in Italy, some experience with small pilot facilities is also available. The availability of commercial scale reprocessing plants for thorium fuel would certainly need future development efforts.

7.3.3 Uranium-233/Thorium Fuel Fabrication

$^{233}UO_2/ThO_2$ pellet type fuels for LWR's or CANDU's can be fabricated by remote fabrication lines in shielded concrete cells. The fabrication process follows similar steps of powder mixing, pressing, sintering and grinding of the pellets, as described for the MOX fuel fabrication in Section 7.2.2.

Wet chemical processes are applied to produce thorium-uranium fuel particles for HTGR fuel. These techniques are used because, in reprocessing, thorium and uranium occur as thorium nitrate and uranyl nitrate, respectively. If thorium nitrate is brought into contact with ammonium hydroxide, thorium hydroxide and ammonium nitrate will be produced. Thorium hydroxide precipitates as an amorphous, gel type structure. Similarly, uranyl nitrate reacts to form ammonium diuranate, but special additives are necessary to modify the precipitate into a gel. The first step in fabrication is the preparation of a solution of the right viscosity (sol). The sol is then passed through a vibrating jet and dispersed into appropriately shaped spherical droplets. The sol droplets fall into a gelation bath where the gelation process into microspheres is completed (gel). The gel spheres are then washed and dried. Then the kernels are sintered at approximately 1400 °C. In this way, either ThO_2 kernels or U-233/Th mixed oxide kernels are produced. The preparation of uranium kernels requires one additional calcination step in a fluidized bed to decompose organic compounds of the broth. Carbides are produced by carbothermic reduction at temperatures up to 2500 °C.

The outer layer of the kernels of pyrolytic carbon or silicon carbide is applied in an electrically heated fluidized bed. The kernels are then assembled into graphite fuel elements. Assembly is performed remotely in shielded cells. This remote fabrication is necessary because of the U-232 contamination of uranium and the Th-228 contamination of thorium. U-232 is formed by an (n,2n)-reaction with Th-232 and a subsequent (n,γ)-reaction with Pa-231 as well as by an (n,2n)-reaction with U-233. It is also formed by the decay chains initiated by the (n,γ)-reaction with U-235 and the (n,γ)-reaction with Th-230. The latter isotope is a byproduct of thorium mining. U-232 decays to Th-228, which decays to Bi-212 and Tl-208 through a chain of shortlived isotopes. Both end products are emitters of very high-energy γ-radiation. Recycled uranium in the Th/U-233 fuel cycle will contain several 100 ppm of U-232. Soon after reprocessing, this trace amount of U-232 continuously builds up Th-228 and its γ-emitting daughters, Bi-212 and Tl-208. This high-energy γ-radiation together with neutrons produced from (α,n)-reactions (α-decay from uranium and thorium isotopes) with light elements, such as oxygen or carbon, require shielding during refabrication.

Thorium separated from reprocessing is not recycled directly. Th-228 appearing in the separated thorium results in appreciable radioactivity. Therefore it has to be stored for some ten years before it may be recycled.

Most of the experience in $^{233}UO_2/ThO_2$ pellet fuel fabrication has been gained in the US LWBR project. Actual experience, however, is limited to low U-232 (about 10 ppm) feed material. Experience with fresh $^{235}UO_2/ThO_2$ fuel for HTGR's is available in pilot plants, but not yet for HTGR fuel refabrication with several 100 ppm U-232 content.

7.4 The Uranium/Plutonium Fuel Cycle of Fast Breeder Reactors

Like all recycling converter reactors and near-breeder reactors, LMFBR's must work in a closed fuel cycle. Their systems inventory, consisting of the core fuel inventory and the fuel inventory passed through reprocessing and refabrication, should be as small as possible for strategic reasons. On the one hand, this leads to the requirement of a small in-core fuel inventory and high burnup (long utilization of the fuel for energy generation). On the other hand, it means short ex-core times of the fuel inventory in the fuel cycle.

7.4.1 Ex-Core Time Periods of LMFBR Spent Fuel

At present, it is generally assumed that an ex-core time of two years is feasible for the LMFBR fuel cycle. For future advanced FBR fuel cycles, however, ex-core times of roughly one year are envisaged. Fig. 7-9 shows the LMFBR fuel cycle, indicating the different time spans for the fuel outside the reactor plant. A model fuel cycle is assumed for this diagram, which corresponds to a fuel reprocessing capacity for roughly 10 GW(e) LMFBR's.

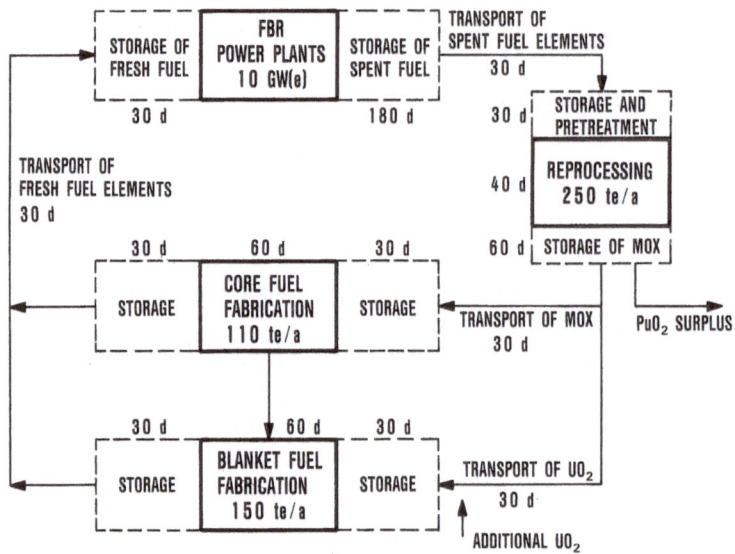

Fig. 7-9. Ex-core time of a fast breeder reactor spent fuel cycle (INFCE)

After unloading from the reactor core, the core elements and the radial blanket elements are first stored on the reactor site for some 180 days. Then they are transported to the reprocessing plant in shipping casks, which can contain six to twelve fuel elements each. Shipping the fuel elements takes about thirty days. Another thirty days are assumed for intermediate storage and pre-treatment of the fuel elements prior to cutting and dissolution. Assuming a reprocessing plant with annual reprocessing capacities of 250 te/a (165 te of core and axial blanket fuel mixed with 85 te of radial blanket fuel), the total time required for all steps, from chopping the fuel pins to conversion to PuO_2 and UO_2 powder, is estimated to be forty days. Sixty days are assumed for intermediate storage of the oxide powder, another thirty days for transport to the fuel refabrication plant.

The associated PuO_2/UO_2 fuel refabrication plant will have an annual capacity of about 110 te of mixed PuO_2/UO_2 fuel for the core and an annual capacity of 150 te for UO_2 blanket fuel (65 te for the axial blankets and 85 te for the radial blanket). The UO_2/PuO_2 powder will be stored for about thirty days and then transferred in batches to the fabrication lines. The fabrication process takes about sixty days, and another thirty days are required for fuel element storage prior to shipment to the FBR power plant. Shipment requires some thirty days; another thirty days are assumed for storage on the reactor site before the fuel is loaded into the core for power generation.

These time periods add up to 550 days. Assuming another 180 days for unforeseen delays, which may arise from imperfect synchronization between the various fuel cycle operations, the total ex-core or fuel cycle time adds up to 730 days or two years. However, it is obvious that co-location of reprocessing and refabrication plants, good synchronization of the different fuel cycle activities, and reductions in storage time can help to achieve ex-core times of about one year.

7.4.2 Mass Flow in a Model LMFBR Fuel Cycle

A model fuel cycle for reprocessing the PuO_2/UO_2 fuel discharged from 10 GW(e) LMFBR's roughly corresponds to a capacity of 1 te_{HM}/d or, at 250 full-load days, an annual capacity of 250 te_{HM}. Such a fuel cycle includes reprocessing and refabrication plants on an industrial scale. Fig. 7-10 indicates the mass flows of the most important materials in this model LMFBR fuel cycle.

From the 10 GW(e) LMFBR's, at a load factor of 0.7, an annual 166 te_{HM} of uranium and plutonium in core fuel elements and 85 te_{HM} of uranium and plutonium in radial blanket elements are discharged and shipped to the reprocessing plant. These spent fuel and blanket elements in addition contain 6.45 te of fission products. Of this fission product volume, some 5.8 te are contained in the core fuel elements and in the axial blankets, and some 0.65 te in the radial blanket elements. The distribution into different fission product isotopes differs slightly from that encountered in LWR's, because of the fast neutron spectrum and the fissile plutonium. In addition to these quantities of fuel and fission products, there are approximately 200 kg of higher actinides (Np-237, Am-241, Am-242, Am-243, Cm-242 and Cm-244).

In the reprocessing plant, the fission products and the actinides are separated and go into the HLWC concentrate. Some 227 te_{HM} of uranium and 21.7 te of plutonium are recovered, of which 1.59 te of plutonium can be diverted as a breeding gain to start new LMFBR's or to feed converter reactors. Roughly 1%, i.e., 200 kg of plutonium and some 2000 kg of uranium, initially remain as high level solid or liquid wastes accumulating in the reprocessing and refabrication plant. However, in the waste treatment step, most of the plutonium

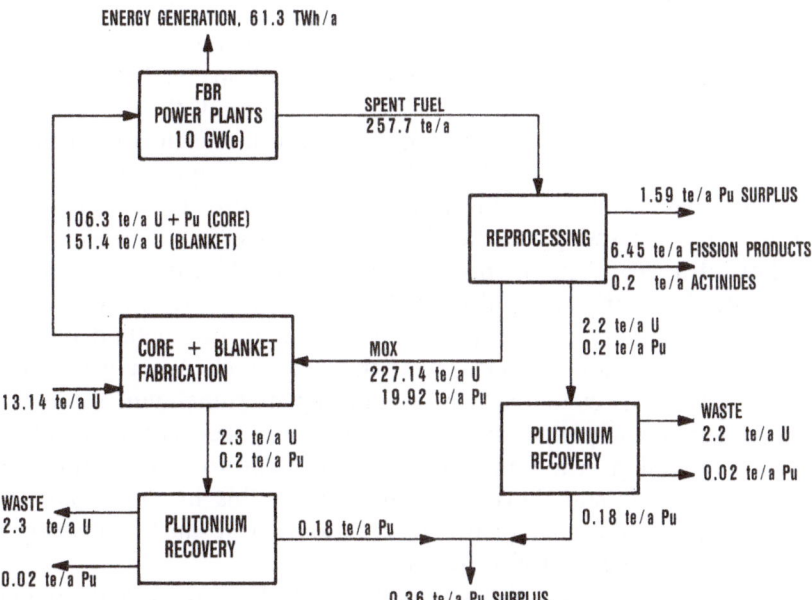

Fig. 7-10. Mass flow within a model fast breeder reactor fuel cycle (INFCE)

is recovered so that ultimately only some 20 kg/a of plutonium will be lost to the HLW and MLW.

In refabrication, the spent U-238 must be made up for by depleted uranium from enrichment. The quantity of uranium to be replaced is a result of the fractions accumulating as a result of fission, conversion and losses during reprocessing and refabrication. In this case, it is approx. 13 te$_{HM}$ of depleted uranium per annum.

7.4.3 Radioactive Inventories of Spent LMFBR Fuel

When comparing the radioactive inventories of spent LMFBR fuel elements with those for LWR's and Pu recycle LWR's, it is seen that the activities of fission products are roughly the same. Although the fission product contents in the core fuel rods are slightly higher in LMFBR's than in LWR's, as a result of the higher power densities and higher burnups, this is almost completely offset by the mixture of highly loaded core fuel and low-loaded blanket fuel. As a result of the higher plutonium enrichment, the activity of plutonium isotopes is highest in the LMFBR fuel. When comparing the α-activity of plutonium of LWR:LWR Pu-recycle:LMFBR, the ratio is 1:2:5. For the α-activity of americium, there is an even more pronounced 1:4:15 ratio for the LMFBR. However, the bulk α-activity is due to the curium isotopes, and a ratio of 1:9:2.5 is obtained for curium. This means that the α-activity is higher in the LWR Pu-recycle than in the LMFBR spent fuel.

7.4.4 LMFBR Fuel Reprocessing

Also for LMFBR spent fuel, the PUREX process is used as described in Section 7.2.1. However, technical modifications are required to take into account the specific characteristics of LMFBR fuel elements enriched in plutonium.

Unlike LWR fuel assemblies, the LMFBR fuel elements are first dismantled. The present state of the art does not permit direct chopping of the fuel element assemblies. The end pieces of the fuel elements are cut off, and the fuel element wrapper is removed mechanically. The fuel rods are then separately cut into pieces 2.5 cm long by means of a shear. In this step, core fuel and blanket fuel are mixed. The fuel rod pieces fall into the dissolver, where the fuel is dissolved in hot nitric acid. The dissolver geometry must be carefully adapted to the higher plutonium enrichment of LMFBR fuel to avoid criticality.

The dissolution capability of the fuel is influenced by the method of fuel fabrication and by the irradiation history of the spent fuel. A small fraction of insoluble particles will remain, which is made up of ruthenium, rhodium, tellurium, molybdenum and palladium and of undissolved fuel. The noble metal fission products, ruthenium, rhodium and palladium, tend to form alloys with part of the plutonium at high burnup of the fuel. If this insoluble particle fraction contains more than 0.5% of the total plutonium, it must be dissolved in a separate step by adding hydrofluoric acid.

The fuel solution coming from the dissolver is first clarified by centrifuging or filtration, as in LWR fuel reprocessing. Then the PUREX countercurrent solvent extraction process is applied. However, unlike LWR reprocessing, the contact times of the solvent and the nitric acid solution must be shorter to limit radiolysis of the solvent. For this purpose, pulsed columns or centrifugal contactors are used.

In the subsequent process of partitioning plutonium and uranium, a larger volume of the chemical reducing agent must be used because of the higher plutonium enrichment. This generates larger volumes of process solutions and waste streams. When decontaminating plutonium and uranium, the higher plutonium concentration again must be taken into account to ensure that the plutonium fraction in the waste is kept as small as possible. The process for conversion of plutonium nitrate and uranyl nitrate into PuO_2 and UO_2 is the same as in an LWR reprocessing plant. Also, plutonium nitrate and uranyl nitrate can be directly mixed, co-converted and co-precipitated into mixed PuO_2/UO_2.

The modifications of the PUREX process described above, the introduction of special technical components, and the smaller dimensions of all tanks (higher plutonium enrichment, criticality) ultimately require the construction of special LMFBR reprocessing plants.

7.4.5 LMFBR Fuel Fabrication

In general, the same fabrication process is applied for LMFBR fuel as for LWR recycle (MOX) fuel (Section 7.2.2). If the reprocessing plant and the LMFBR fuel refabrication plant are not co-located on one site, PuO_2 and UO_2 will be shipped to the refabrication plant. After storage in a buffer store, the MOX fabrication process begins with mechanical blending of UO_2 and PuO_2 powders to establish the desired enrichment. Afterwards, the fabrication process proceeds with pressing, sintering, grinding and drying of sintered pellets. This is followed by assembling core pellets into stacks, adding axial blanket pellets, inserting the pellets into cladding tubes and introducing an inert atmosphere. Finally, the fuel pins are welded and assembled into fuel elements.

The PuO_2/UO_2 fuel must be highly soluble in nitric acid. Even after a burnup of about 100,000 MWd(th)/te, the LMFBR fuel must have a solubility in nitric acid of >99% in order to minimize the volume of insoluble residues and fuel losses during reprocessing. This requires special attention to be paid in the milling and sintering processes.

LMFBR fuel pellets have smaller diameters than LWR pellets. The pellet fabrication and pin loading operations are carried out in glove boxes. To protect the workers against γ-radiation and neutrons originating from various plutonium isotopes and their radioactive daughters (spontaneous fission and (α,n)-reactions with oxygen), shielding must be provided at the fabrication lines. Future plants are expected to be operated remotely to a large extent. Besides the present reference LMFBR fuel fabrication technology described above, also the sol-gel precipitation technique, the vibro-compaction technique, and the AUPuC process are being employed.

The sol-gel process allows the co-conversion of Pu nitrate and uranyl nitrate followed by the fabrication of spherical PuO_2/UO_2 particles, which can be pressed and sintered into fuel pellets. The AUPuC refabrication process following after co-conversion allows the fabrication of a coarse grain PuO_2/UO_2 and Am-241 free crystal powder, which can be pressed and sintered into pellets. Both refabrication processes avoid the formation of Pu dust (see Section 7.2.2.3). They are favored for application in future advanced refabrication plants of FBR fuel cycle centers.

In the fabrication process, different types of contaminated waste are produced. In a refabrication plant of the reference fuel cycle shown in Fig. 7-10, it must be taken into account that roughly 200 kg of Pu (1%) annually would go into the waste. Consequently, special efforts must be made to recover this plutonium. The resultant waste should not contain more than 20 kg of plutonium annually.

7.4.6 Status of LMFBR Fuel Reprocessing and Refabrication

Several small test and pilot plants for LMFBR fuel reprocessing and MOX fuel fabrication in the USA, the UK and France have been operated for almost twenty years. The UK runs a small pilot reprocessing plant of 5 te/a throughput at Dounreay for reprocessing fuel from PFR. Japan and Germany together with Belgium have laboratory scale reprocessing facilities in operation. France between 1969 and 1979 operated a small experimental test facility at La Hague with a capacity of 1 kg_{HM}/d. It reprocessed about 1 te_{HM} of fuel from RAPSO-DIE with fuel burnups between 40,000 and 130,000 MWd(th)/te. The SAP reprocessing pilot plant at Marcoule, France, has reprocessed some 6 te_{HM} of fuel from RAPSODIE and PHENIX between 1975 and 1980. The TOR facility at Marcoule, France, has a higher yearly capacity of up to 5 te_{HM} and can reprocess spent fuel from PHENIX, SNR 300 and some fuel from SUPER-PHENIX. The USA, Belgium, UK, Japan and the Federal Republic of Germany have PuO_2/UO_2 fuel fabrication plants with capacities of 5–10 te_{HM}/a. France operates a MOX fuel fabrication plant of 20 te_{HM}/a throughput at Cadarache.

With SUPERPHENIX and other commercial demonstration LMFBR plants coming into operation between 1983 and 1990, reprocessing and refabrication plants with 50 to 100 te/a throughput will be needed.

7.5 Waste Conditioning

Solid and liquid radioactive wastes must be conditioned so that they can be safely handled and stored, both for intermediate and secular time spans. This is achieved by volume reduction and solidification. The final product is to be stable, both mechanically and chemically; it must dissipate by conduction the heat it generates, and it is also expected to be stable against its characteristic ionizing radiation.

7.5.1 Conditioning Waste From Spent LWR Fuel Reprocessing

7.5.1.1 Solidification and Storage of Liquid High Level Waste

The HLW solution generated in the extraction of uranium and plutonium during reprocessing of spent LWR fuel is concentrated by approximately a

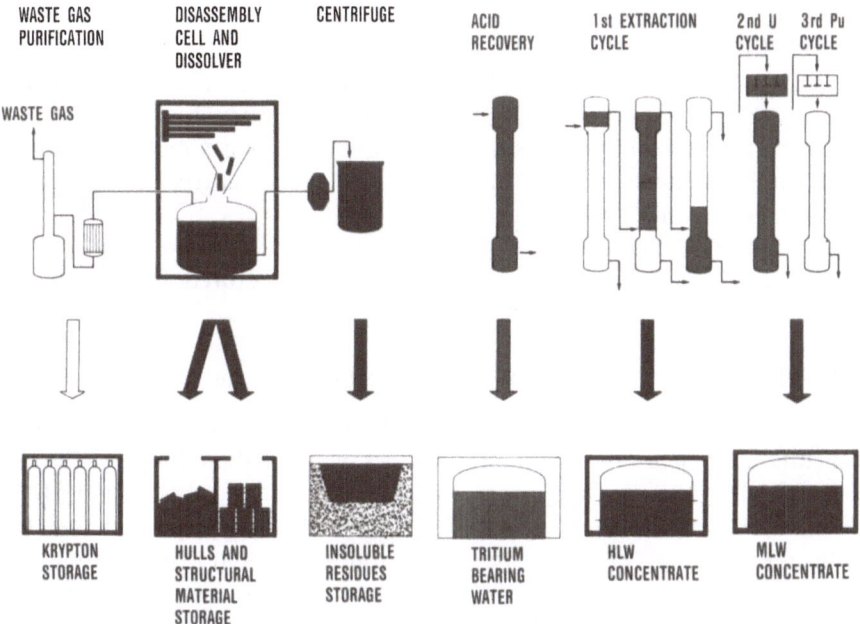

| WASTE GAS PURIFICATION | DISASSEMBLY CELL AND DISSOLVER | CENTRIFUGE | ACID RECOVERY | 1st EXTRACTION CYCLE | 2nd U CYCLE | 3rd Pu CYCLE |

WASTE GAS

| KRYPTON STORAGE | HULLS AND STRUCTURAL MATERIAL STORAGE | INSOLUBLE RESIDUES STORAGE | TRITIUM BEARING WATER | HLW CONCENTRATE | MLW CONCENTRATE |

Fig. 7-11. Different waste forms from reprocessing (KFA Jülich)

factor of 10 by evaporation, which gives rise to some 0.5 m³ of liquid HLW solution per tonne of spent fuel (see Fig. 7-6). This concentrated HLW solution is first pumped into tanks of acid resistant steel (Fig. 7-11). These tanks stand in stainless steel pans to collect any leakages. Because of the strong radiation emitted by the HLW solutions, these tanks are installed in hot cells lined with steel. The HLW solution is recirculated in the tanks and cooled by water circulating in tube coils (< 65 °C) for removal of the decay heat generated by fission products and actinides. Standby tanks of sufficient capacity must be available. The HLW solution can be kept in these tanks for at least 20–30 years.

For solidification of the liquid HLW solution, the first process step is calcination, i.e., the expulsion of liquid from the HLW solution at high temperatures (900 °C). This produces nitrates or metal oxides. As an alternative, the HLW solution can also be denitrated by the addition of hot formic acid, which mainly produces only oxides in calcination. Four different techniques of calcination have been developed: the pot, fluidized bed, spray, and rotary tube calcination processes. Calcinates are not resistant enough to leaching. Their thermal conductivity is low and their mechanical stability is insufficient. They are intermediate products, which can be converted into vitreous or ceramic materials. Glass frit with additives of silicon, boron or phosphorus can be used for conversion into glass. After melting at high temperatures (1000–1200 °C), this results in three different types of glass: borosilicate glass, phosphate glass, and borophosphate glass. Development work so far has been concentrated chiefly on the fabrication of borosilicate glass, which permits storage temperatures up to roughly 200 °C.

The total HLW volume generated from the spent fuel discharged yearly from a 1 GW(e) LWR can be solidified in a volume of about 2.5–5 m^3 by vitrification. Consequently, 1 te$_{HM}$ of spent LWR fuel leads to 0.07–0.15 m^3 volume of glass. Borosilicate glass blocks are produced by pouring molten borosilicate glass into steel canisters. Another interesting alternative process is a method in which phosphate or borosilicate glass is imbedded in a metal matrix as small glass beads or leach-resistant ceramic granulate.

The glass blocks cast in steel canisters have volumes between 77 and 180 l. The steel canisters are welded, cooled gradually, and stored for an interim period prior to transportion to a final repository.

7.5.1.2 Solidification and Storage of Solid High Level Waste

Solid HLW is mainly composed of the fuel rod hulls and the end pieces of fuel elements as well as the insoluble residues generated in fuel dissolution. They are initially put into temporary storage silos under water. Solid HLW waste volumes amount to some 0.6 m^3/te$_{HM}$ for hulls and structural material and some 4 kg/te$_{HM}$ of insoluble residues. Conditioning for final storage is achieved by separating the solid waste and mixing it with liquid grout. This mixture is filled into canisters which, in turn, are put into shielded vessels. All spaces in the shielded vessels are filled with cement. The shielded canister is closed with a shielding lid. Other methods, such as imbedding the fuel rod hulls in lead, have been proposed and are still being developed.

7.5.1.3 Treatment of Medium Level Waste

Aqueous MLW solutions are concentrated by evaporation and treated in a denitrator. This amounts to waste volumes for further conditioning of about 0.6 m^3/te$_{HM}$ of spent fuel. Afterwards, these concentrates can be mixed with cement and filled in drums. Another technique is mixing the concentrate solution with hot bitumen. This causes any water residues to evaporate. The product is again filled in drums, which are closed with lids after cooling. Bituminization has the advantage of producing lower waste volumes than cementing.

Organic MLW solutions are treated by the phosphoric acid adduct method, which allows the kerosene to be purified and recycled. The remaining organic MLW is mixed with plastic granulate. A homogeneous solution is produced, which is filled in drums and sealed.

7.5.1.4 Treatment of Remaining Wastes

Solid *low level waste* (LLW) is concentrated by baling and burning, whereas liquid LLW is concentrated chiefly by evaporation. The concentrated waste is solidified by cementing and bituminization, filled in drums and sealed.

The separated Kr-85 is forced into pressurized steel cylinders of 50 liter volume. The krypton cylinders are dumped into storage shafts in a krypton storage facility. The shafts can accommodate, e.g., four krypton cylinders each stacked on top of each other. The krypton cylinders are cooled by air.

Tritium can be concentrated first by proper mass flow conduction in the PUREX process. For final storage, the water containing high concentrations of tritium may be either stored in tanks or pressed into isolated geological formations. This process allows tritium to be removed from the biosphere for periods long enough compared to its halflife of 12.4 a.

7.5.1.5 Waste Volumes to Be Stored from Reprocessing of Spent LWR Fuel

Tables 7-4 and 7-5 list data for conditioned waste volumes arising from a spent LWR fuel reprocessing plant as described in Section 7.2.1. The data are normalized to 1 GW(e)·a of energy produced. The following assumptions hold for Table 7-4:

Vitrified liquid HLW is stored in steel canisters with an inner diameter of 30 cm, roughly 3 m length and 1 cm wall thickness (type A canister, volume 0.18 m^3).

Fuel claddings and insoluble residues are filled into canisters with 86 cm inner diameter, 115 cm height and 2.5 cm wall thickness (type-B canister, volume 0.67 m^3). In the future, it is envisaged to use also larger containers of 2 m^3 volume.

Krypton is filled into steel cylinders of 50 l volume and 50 bar pressure.

The conditioned MLW and LLW is filled into shielded and unshielded drums of 53 cm inner diameter, 87 cm height and 0.17 cm wall thickness. Shielded drums have shielded casings of 20 cm thickness.

Table 7-4. *Waste volumes from reprocessing spent LWR fuel (INFCE)*

		Normalized to 1 GW(e)·a of energy produced
1) Vitrified high level waste[a]		
Volume	m^3	5
No. of type-A canisters		29
2) Hulls, spacers, insolubles		
Volume	m^3	22
No. of type-B canisters		33
3) Noble gases		
No. of gas cylinders		17
4) Medium level and plant maintenance waste		
Volume	m^3	27
No. of unshielded drums		54
No. of shielded drums		83
5) Low level waste		
Volume	m^3	25
No. of unshielded drums		113
No. of shielded drums		13

[a] Data in the literature on HLW volumes and type-A canister volumes may also be a factor of 2 smaller.

Table 7-5. *Total packaged waste for different reactors and fuel cycles (INFCE)*

Origin	Waste package type	LWR Once-through	LWR U/Pu cycle	FBR U/Pu cycle	HWR Once-through	U/Pu recycle	U/Th recycle	HTR U/Th recycle
U-refining, conversion, enrichment	unshielded drums	365	220	18	120	50	20	115
Fuel element fabrication	unshielded drums	200	285	318	500	750	1200	175
Nuclear power plant	unshielded drums				about 1800			
	shielded drums				about 600			
	type-B canister	3	10	7	–	–	–	8
Spent fuel conditioning (direct disposal)	shielded drums	45	–	–	80	–	–	–
	PWR type canister	53	–	–	–	–	–	–
	BWR type canister	22	–	–	–	–	–	–
	HWR type canister	–	–	–	132	–	–	–
Reprocessing plant	unshielded drums	–	320	200	–	750	1070	3470
	shielded drums	–	110	70	–	240	380	43
	type-B canister	–	22	86	–	31	49	–
	type-A canister	–	29	23	–	29	30	28
	gas cylinder	–	17	17	–	17	18	18

type-A canister: steel canister, 0.18 m³ volume, 30 cm inner diameter, 3 m length
type-B canister: 0.67 m³ volume, 86 cm inner diameter, 115 cm height
gas cylinder: 50 ltr volume
shielded and unshielded drums: 0.2 m³ volume, 53 cm inner diameter, 87 cm height
PWR type canister: contains 1 PWR spent fuel element
BWR type canister: contains 3 BWR spent fuel elements
HWR type canister: contains 72 HWR fuel bundles.

7.5.2 Radioactive Waste from Uranium-233/Thorium Fuel Reprocessing

Solid HLW in the form of cladding hulls is compacted and will have to be conditioned as described in Section 7.5.1. Liquid HLW consists mainly of the aqueous waste from the first extraction column, which contains the fission products, small amounts of actinides as well as significant amounts of dissolved aluminum and fluorides. They will have to be concentrated, calcined and converted into mixtures of stable oxides. Medium and low level wastes consist of aqueous waste, filter wastes, etc. and will have to be treated as described in Section 7.5.1.

Data on waste volumes arising from U/Th fuel reprocessing are indicated in Table 7-5 and can be compared with the corresponding values from the U/Pu fuel cycle.

7.5.3 Radioactive Waste from Reprocessing Plutonium/Uranium Fuel of LMFBR's

The bulk waste quantity in the LMFBR fuel cycle is produced in the reprocessing plant. Liquid HLW contains the fission products and actinides and has activity levels similar to those of liquid HLW of LWR fuel. Correspondingly, also waste treatment consists of the liquid HLW storage and vitrification steps described in Section 7.5.1. Conditioning of solid HLW as well as MLW and LLW is also identical.

A comparison of waste volumes arising from LMFBR's and their Pu/U fuel cycle is given in Table 7-5.

7.5.4 Wastes Arising in Other Parts of the Fuel Cycle

In addition to wastes arising in fuel reprocessing and refabrication plants, waste from other parts of the fuel cycle, i.e., uranium or thorium mining, fuel conversion, fuel enrichment, fuel fabrication and the nuclear power plant must be considered (see Table 7-5).

7.5.4.1 Uranium Ore Processing

Mill tailings are a slurry of ore residues with some of the process chemicals. Although they only contain natural uranium or thorium, they do require careful treatment. As a result of chemical treatment in ore processing, radioactive isotopes of the U-238 and Th-232 decay chains may be distributed in the biosphere (see Chapter 8.1). Of these, Ra-226 (halflife 1600 a) is most important, because it can both be ingested by way of the aqueous pathways and inhaled through its gaseous daughter, Rn-222 (halflife 3.8 d). Also Th-230 is relevant on a long term basis, being a precursor of Ra-226.

By comparison, the management of thorium mill tailings is a minor problem. The decay products of Th-232, such as Ra-228 (halflife 5.9 a) and Th-228 (halflife 1.9 a), have decay periods so short as to be of little relevance for the biosphere.

7.5.4.2 Uranium Refining, Conversion and Enrichment

Both in refining and in the conversion of U_3O_8 to UF_6 for the enrichment process small volumes of waste are produced, which contain natural uranium. In the enrichment step, depleted UF_6 is produced which, in turn, can be converted into depleted UO_2. This depleted UO_2 can be used as a fertile material in FBR's. In a pure once-through fuel cycle, this depleted UO_2 has to be treated as waste.

7.5.4.3 Fuel Element Fabrication and Nuclear Power Plants

Similar to the fabrication of MOX fuel, also plants fabricating low, medium or highly enriched uranium oxide fuels produce waste, which must be treated and stored.

Reactor power plants generate waste in the form of filters and ion exchangers. Repair and maintenance of radioactive components and the replacement of absorbed rods produces additional waste volumes.

Estimates of the total amount of waste produced by different reactors operating in different fuel cycle options are indicated in Table 7-5. This waste has to be packaged in shielded and unshielded drums, gas flasks, type-A and B canisters as described in Section 7.5.1.5. Canisters labeled PWR, BWR or HWR type would contain the spent fuel elements for direct disposal, as will be described in Section 7.6.2.

It can be seen from Table 7-5 that the number of krypton gas cylinders and the number of type-A canisters for HLW are about the same for all reactors operating in either the U-238/Pu or Th/U-233 closed fuel cycles. The number of type-B canisters from the reprocessing plant is largest (factor of 2–3) for the LMFBR fuel cycle. But the LMFBR has the smallest number of unshielded drums for uranium conversion and needs neither U-enrichment nor U-mining and ore processing. The number of shielded and unshielded drums is similar, to a factor of 2–3, for all types of reactors and fuel cycle options.

7.6 Nuclear Waste Repositories

7.6.1 Waste Disposal in Deep Geological Formations

It is commonly accepted that HLW should be stored in suitable geological formations below the continents. Such geological formations should be free from circulating groundwater, have high impermeability and good heat conductivity. Thick rock salt formations meet these requirements in an almost ideal way. In addition, they have a high plasticity so that fissures in the salt around waste canisters or drums are closed (self-sealed) and isostatic pressure conditions are maintained.

Granite, gneis and basalt formations at a depth of several hundred meters as well as argillaceous formations, like clays etc., are also attractive for nuclear waste disposal. Even tectonically very stable and extremely thick sediment packs under the Atlantic and Pacific Oceans have been proposed as disposal sites.

Table 7-6. *Time schedule of waste repository site investigation programs (H. Röthemeyer)*

Country	Host rock	Field investigation	Site definition
Belgium	clay	1975–1990	1990
Canada	crystalline igneous rock (salt, limestone, shale)	1984–1987/89	
Denmark	dome salt	1979–1981	
Federal Republic of Germany	dome salt	1979–1982	1977
The Netherlands	dome salt	1983	
Sweden	granite and gneiss	1990-ies	
Switzerland	granite, gneis, clay, anhydrite	1981–1986	
United Kingdom	granite, clay, (salt)	1976–1990	1995
USA	New Mexico, bedded salt	1981	
	Stanford, basalt	1982	
	Gulf coast, salt dome	1982	1985
	other bedded salt	1983	
	Nevada, tuff	1984	
	granite	1984	

MLW and LLW have been dumped into the Pacific Ocean by the USA and into the Atlantic Ocean by the UK. The USA stopped ocean dumping in 1971 and there are trends towards a reduction of the amounts of waste to be dumped in the oceans. Practical experience exists in Canada, France, the UK, USA and USSR with burying α-emitter free MLW and LLW in specially selected and prepared burial grounds close to the surface (shallow ground burial). At present, there is no HLW repository in deep geological formations in operation. However, there are definite plans in a number of countries to construct national nuclear waste repositories for placement of all kinds of radioactive waste in deep geological formations. Table 7-6 indicates the time schedules for repository site investigation programs and repository site definitions in different countries in the Western world. Experience with salt mines for the disposal of MLW and LLW and test results for HLW are available in the USA and the Federal Republic of Germany.

Fig. 7-12 gives an impression of an underground salt repository. It consists of shafts, access corridors and disposal rooms excavated some 600 m or more under ground, deep within the salt formation. Following the excavation of the rooms, storage holes for canistered HLW are drilled in the floors of the rooms. For HLW canisters, the necessary spacing of storage holes is about 10 to 50 m, dependent upon the storage concept applied. The reason for spreading HLW canisters over such an area is their heat generation. Since this heat is only transferred by thermal conduction, the temperature at the surface of

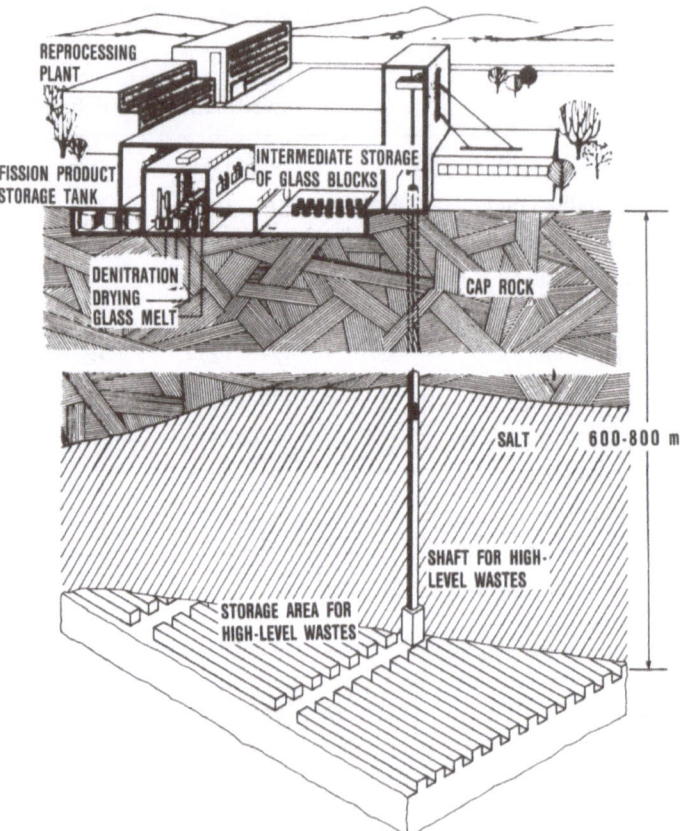

Fig. 7-12. Waste repository and supporting facilities

the canister and also of the salt in the direct vicinity of the canister must be compatible with the allowable limits and design bases in salt. Fig. 7-13 shows the different contributions to heat generation by fission products and actinides in HLW, by insoluble residues, hulls and structural material, MLW and LLW as a function of time. Over several hundred years, the heat generation by fission products decreases by several orders of magnitude. After this time period, the actinides will be mainly responsible for heat generation.

When the HLW canisters have been placed in prepared holes by remote handling techniques, the remaining void at the top of the hole is backfilled with excavated salt or other material. Canisters with negligible heat generation rates are placed five per hole. Drums with medium and low level wastes are placed in storage rooms at different locations in a salt repository. After an excavated room has been filled with the desired amount of waste, it will be backfilled with excavated salt.

7.6.2 Direct Disposal of Spent Fuel Elements

In the once-through cycle concept with direct disposal, the spent fuel elements, following interim storage for a couple of years, are conditioned in such a way

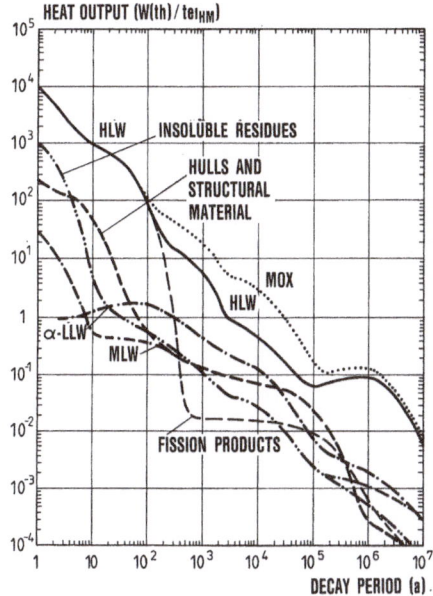

Fig. 7-13. Heat outputs of hulls and structural materials, insoluble residues, aqueous MLW, and solid LLW (alpha-waste) as compared with HLW (H.O. Haug)

that they can be stored in a repository over secular periods of time. This concept has been proposed in Canada, the USA and Sweden. Studies of the concept were carried out also in the Federal Republic of Germany. The uranium (U-235 and U-238) contained in the spent fuel elements and the plutonium in that case are no longer available for further utilization.

The spent fuel elements can either be encapsulated full size as they come out of the interim storage facility, in which case they will contain the whole fuel inventory, actinides and fission products in addition to the structural material or, in a technique proposed in Sweden, they can be disassembled and, after removal of the carriers and head pieces of the fuel elements, the fuel rods can be shortened by winding up. Only then will the shortened fuel rods be encapsulated. After encapsulation in waste containers, they will be put into a repository. The walls of the containers will have to assume long term barrier functions for the radioactive materials to be retained over sufficiently long periods of time, even in cases of water ingress in a repository.

In the Swedish concept, the fuel rods are put into a thick walled copper canister, the hollows being cast with lead. The copper canister has a wall thickness of about 20 cm. After filling, the canister is welded tight by attaching three lids. The thick walled copper sleeve is sufficiently stable against the pressures and movements of geological layers in the repository and also against corrosion due to groundwater. The copper container is to be inserted into crystalline geological strata (repository) in boreholes spaced about 6 m apart. The boreholes are surrounded by compacted bentonite rings, filled up then with a mixture of bentonite and sand and closed. The bentonite layer has a very low permeability to groundwater. Safety analyses have shown that, in case

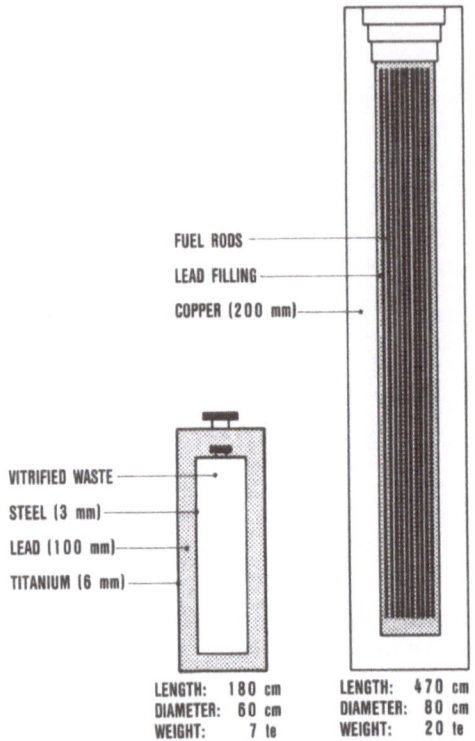

FUEL RODS

LEAD FILLING

COPPER (200 mm)

VITRIFIED WASTE

STEEL (3 mm)

LEAD (100 mm)

TITANIUM (6 mm)

LENGTH: 180 cm   LENGTH: 470 cm
DIAMETER: 60 cm   DIAMETER: 80 cm
WEIGHT: 7 te   WEIGHT: 20 te

Fig. 7-14. Direct disposal of a PWR fuel element (PWR type canister) as compared with vitrified HLW canister (K.D. Closs)

of an ingress of groundwater into the repository, the copper sleeves would be dissolved by corrosion only after 10^5–10^6 years. Dissolution of the fuel and transport of the different nuclides through the layers of rock to a well close to the repository after such long periods of time would lead to a radioactivity level in the drinking water still below the legally permitted limits (see Section 7.6.3).

Fig. 7-14 shows a schematic diagram of a PWR type spent fuel canister compared with a container for vitrified HLW. Conditioning of fuel elements produces larger volumes of high level waste than do reprocessing and vitrification. As can be compared in Fig. 7-7, the activity in Ci/te_{HM} of the spent fuel elements is an order of magnitude higher than the vitrified HLW (reprocessing case) after the fission products have decayed. Moreover, the Swedish concept of direct final storage implies very high consumption rates of lead and copper.

According to studies in the USA and the Federal Republic of Germany, conditioning of complete fuel elements is possible in single and double shell steel containers. The double shell steel container carries an outer anti-corrosive layer of corrosion resistant material, such as a titanium alloy. The inner volumes of the container may also be cast with lead. Afterwards, these containers are welded tight. When stored in salt formations, one fuel element container each is put into a borehole. The boreholes are to be spaced roughly 6 m apart.

Boreholes are filled with bentonite or salt. After final storage of all waste containers, the gallery will be backfilled with salt.

Compared with the reprocessing and vitrification concepts for HLW, the concept of direct final disposal of fuel elements may require larger storage areas in a repository. However, the storage areas required for MLW are correspondingly smaller. A comparison of the total volumes of HLW, MLW and LLW for the once-through cycle with direct disposal of fuel elements and other fuel cycles is given in Table 7-5.

7.6.3 Health and Safety Impacts of Radioactive Waste Disposal

Any discussion of the health and safety impacts on the environment arising from nuclear waste repositories involves the problem of very long time periods and the long term risk of geologic incidents or other events. Hazards to the environment can only occur if there is a release of radionuclides from the repository after a failure of all barriers around the nuclear waste. This can be initiated after a failure of the geologic confinement. A detailed assessment of potential modes of breaking the geological confinement as a consequence of tectonic and igneous activity, erosion (lowering of the land surface), meteorite impact, surface nuclear explosions and release of radioactive material from a sealed repository by sabotage or inadvertent drilling into the HLW has been summarized by the American Physical Society Study Group on Nuclear Fuel Cycles and Waste Management as well as other research groups. The result is that such potential modes would either not lead to a failure of the geological confinement within the time period of concern or that the risk can be reduced to very low levels by the selection of the site of the repository. The movement of groundwater almost universally present in the underground is the only important medium for the transport of radionuclides from the waste repository to the biosphere.

Breaking by faulting or diapirism in salt or rock formations may create a path for groundwater towards the waste. If such water reaches the biosphere, an uptake of radioactive substances is possible by ingestion. Studies using theoretical models and experimental parameters have been performed for the hypothetical case of water ingress in a salt repository with subsequent leaching of the waste and transport of radionuclides by groundwater to the environment. The first barrier against the brine would be the canister around the glass block. If multi-layer (steel-lead-titanium) canisters are used, the lifetime against corrosion would be 500–1000 years. During this period, strontium and cesium would have decayed. After destruction of the canister wall, radionuclides could enter the groundwater by leaching of the glass block. Typical leaching rates are 10^{-9} to 10^{-11} kg/m$^2 \cdot$s, which leads to time periods for full destruction on the order of 10,000 years. The transport of nuclides with the water is then mainly determined by convection and dispersion effects. Convection means radionuclide transport at groundwater velocity. Dispersion describes the diffusion and dilution of the contaminant in the groundwater. In addition, processes such as ion exchange, colloid filtration, reversible precipitation and irreversible mineralization must be accounted for. Some of the radionuclides are adsorbed very strongly in the rock or soil formations; others, like iodine and technetium, are not adsorbed at all. Model studies for a salt repository show that the mean arrival times of radionuclides of interest are on the order of a million years. During this time period, most of the nuclides will have decayed. The calculated activity of drinking water then is below the maximum permissible concentrations.

Uncertainties still inherent in such model analysis and parameter studies should certainly not be neglected. Therefore, active research is still going on. But uncertainties also have to be weighed against the occurrence probability of such hypothetical water intrusion events, which are assumed to occur perhaps once on a time scale of 10^6 years.

Selected Literature

Transport and storage of spent fuel elements

Brennelement-Zwischenlager Ahaus. Hannover: Deutsche Gesellschaft für Wiederaufarbeitung von Kernbrennstoffen (DWK). 1979.

International Nuclear Fuel Cycle Evaluation, Spent Fuel Management. Report of INFCE Working Group 6. Vienna: International Atomic Energy Agency. 1980.

The Safe Transport of Radioactive Materials. Special Issue. IAEA Bulletin *21* (6), 2–75 (1979).

Shipments of Nuclear Fuel and Waste: Are They Really Safe? Washington: US Department of Energy, DOE/EV-0004 (1977).

Spent Fuel Storage Factbook: Facts Booklet. Washington: US Department of Energy, DOE/NE-005 (1980).

Transportbehälter – Die trockene Lagerung von ausgedienten Brennelementen. Hannover: Deutsche Gesellschaft für Wiederaufarbeitung von Kernbrennstoffen (DWK). 1979.

Uranium/plutonium fuel cycle

Baumgärtner, F.: Chemie der nuklearen Entsorgung. München: Karl Thiemig. 1978.

Benedict, M., et al.: Nuclear Chemical Engineering. New York: McGraw-Hill. 1981.

Bericht über das in der Bundesrepublik Deutschland geplante Entsorgungszentrum für ausgediente Brennelemente aus Kernkraftwerken. Hannover: Deutsche Gesellschaft für Wiederaufarbeitung von Kernbrennstoffen (DWK). 1977.

Cohen, B.: High-level Radioactive Waste from Light Water Reactors. Reviews of Modern Physics *49*, 1–20 (1977).

Final Generic Environmental Statement on the Use of Recycle Plutonium in Mixed Oxide Fuel in Light Water Cooled Reactors (GESMO). Washington: US Nuclear Regulatory Commission, NUREG-002 (1976).

Fischer, U., Wiese, H.W.: KORIGEN – Ein Programm zur Bestimmung des nuklearen Inventars von Reaktorbrennstoffen im Brennstoffkreislauf. Kernforschungszentrum Karlsruhe, KfK-3014 (1982).

Haug, H.O.: Zerfallsrechnungen verschiedener mittelaktiver und actinidenhaltiger Abfälle des LWR-Brennstoffkreislaufs; Teil I: Modellmäßig abgeleitete Basisdaten, Aktivität und Wärmeleistung; Teil II: Radiotoxizitätsvergleich. Kernforschungszentrum Karlsruhe, KfK-3221 (Part I) and KfK-3222 (Part II) (1981).

Hennies, H.H., Körting, K.: Nukleare Sicherheit von Wiederaufarbeitungsanlagen. Kernforschungszentrum Karlsruhe, KfK-2399 (1976).

International Nuclear Fuel Cycle Evaluation, Reprocessing, Plutonium Handling, Recycle. Report of INFCE Working Group 4. Vienna: International Atomic Energy Agency. 1980.

Lakey, L.T.: A Safety Analysis for the Nuclear Fuel Recovery and Recycling Center. Nuclear Technology *43*, 213–221 (1979).

Leblanc, J.M., Vanden Bemden, E.: Chemical Aspects of Mixed Oxide Fuel Production. Radiochimica Acta *25*, 149–152 (1978).

Rapin, M.: Reprocessing: Experience and Outlook. In: World Nuclear Energy – Accomplishments and Perspectives, Proc. Int. Conference Washington, 17–21 November 1981. Transactions American Nuclear Society *37*, 176–183 (1981).

Thorium/U-233 fuel cycle

Feraday, M.A.: Remote Fabrication of (U-233/Th)O$_2$ Pellet-Type Fuels for CANDU Reactors. Transactions American Nuclear Society *32*, 233–234 (1979).

International Nuclear Fuel Cycle Evaluation, Advanced Fuel Cycle and Reactor Concepts. Report of INFCE Working Group 8. Vienna: International Atomic Energy Agency. 1980.

Lotts, A.L., *et al.*: HTGR Fuel and Fuel Cycle Technology. In: Nuclear Power and Its Fuel Cycle, Proc. Int. Conference, Salzburg, 2–13 May 1977, Vol. 3, pp. 433–452. Vienna: International Atomic Energy Agency. 1977.

Merz, E.: Wiederaufarbeitung thoriumhaltiger Kernbrennstoffe im Lichte proliferationssicherer Brennstoffkreisläufe. Naturwissenschaften *65*, 424–431 (1978).

Orth, D.A.: Savannah River Plant Thorium Reprocessing Experience. Nuclear Technology *43*, 63–74 (1979).

Report to the American Physical Society by the Study Group on Nuclear Fuel Cycles and Waste Management (Pines, D., ed.). Reviews of Modern Physics *50* (1), Part II (1978).

Zimmer, E., *et al.*: Aqueous Chemical Processes for the Preparation of High Temperature Reactor Fuel Kernels. Radiochimica Acta *25*, 161–169 (1978).

Uranium/plutonium fuel cycle of fast breeder reactors

Allardice, R.H. *et al.*: Fast Reactor Fuel Reprocessing in the United Kingdom. In: Nuclear Power and Its Fuel Cycle, Proc. Int. Conference, Salzburg, 2–13 May 1977, Vol. 3, pp. 615–630. Vienna: International Atomic Energy Agency. 1977.

Auchapt, P., *et al.*: The French R&D Programme for Fast Reactor Fuel Reprocessing. In: Fast Reactor Fuel Reprocessing, Proc. Symposium, Dounreay, 15–18 May 1979, pp. 51–59. London: Society of Chemical Industry. 1980.

Barret, T.R.: The Reconstruction of the Fast Reactor Reprocessing Plant, Dounreay. In: Fast Reactor Fuel Reprocessing, Proc. Symposium, Dounreay, 15–18 May 1979, pp. 17–35. London: Society of Chemical Industry. 1980.

Baumgärtner, F., Ochsenfeld, W.: Development and Status of LMFBR Fuel Reprocessing in the Federal Republic of Germany. Kernforschungszentrum Karlsruhe, KfK-2301 (1976).

Bishop, J.F., *et al.*: Fast Reactor Fuel Design and Development. In: Nuclear Power and Its Fuel Cycle, Proc. Int. Conference, Salzburg, 2–13 May 1977, Vol. 3, pp. 377–391. Vienna: International Atomic Energy Agency. 1977.

Ebert, K.H.: Die Wiederaufarbeitung von Schnellbrüter-Brennelementen. Atomkernenergie/Kerntechnik *36*, 259–263 (1980).

Funke, P., *et al.*: Weiterentwicklung des oxidischen Brennstoffes zum Schnellbrüterein-
    satz. Atomkernenergie/Kerntechnik *36*, 253–258 (1980).
International Nuclear Fuel Cycle Evaluation, Fast Breeder Reactors. Report of INFCE
    Working Group 5. Vienna: International Atomic Energy Agency. 1980.
Sauteron, J.: Technologie du retraitement des combustibles des réacteurs rapides. In:
    Nuclear Power and Its Fuel Cycle, Proc. Int. Conference, Salzburg, 2–13 May 1977,
    Vol. 3, pp. 633–645. Vienna: International Atomic Energy Agency. 1977.

Waste conditioning and waste disposal

Closs, K.D.: Vergleich der verschiedenen Entsorgungsalternativen und Beurteilung ihrer
    Realisierung. Kernforschungszentrum Karlsruhe, KfK-3000 (1980).
Environmental Aspects of Commercial Radioactive Waste Management. Washington:
    US Department of Energy, DOE/ET-0029 (1979).
Environmental Survey of the Reprocessing and Waste Management Portions of the LWR
    Fuel Cycle. Washington: US Nuclear Regulatory Commission, NUREG-0016 (1976).
International Nuclear Fuel Cycle Evaluation, Waste Management and Disposal. Report
    of INFCE Working Group 7. Vienna: International Atomic Energy Agency. 1980.
Röthemeyer, H., Closs, K.D.: High Level Waste Disposal. In: World Nuclear Energy
    – Accomplishments and Perspectives, Proc. Int. Conference, Washington, 17–21 No-
    vember 1981. Transactions American Nuclear Society *37*, 165–174 (1981).
Technology of Commercial Radioactive Waste Management. Washington: US Depart-
    ment of Energy, DOE/ET-0028 (1979).

8 Environmental Impacts and Risks of Nuclear Fission Energy

8.1 Radioactivity Releases from Nuclear Power Plants and Fuel Cycle Facilities During Normal Operation

8.1.1 Radioactivity Releases and Exposure Pathways

During normal operation of nuclear power plants and other facilities of the nuclear fuel cycle, radioactivity is released into the environment at a controlled rate. Airborne radioactivity includes the radioisotopes of the noble gases

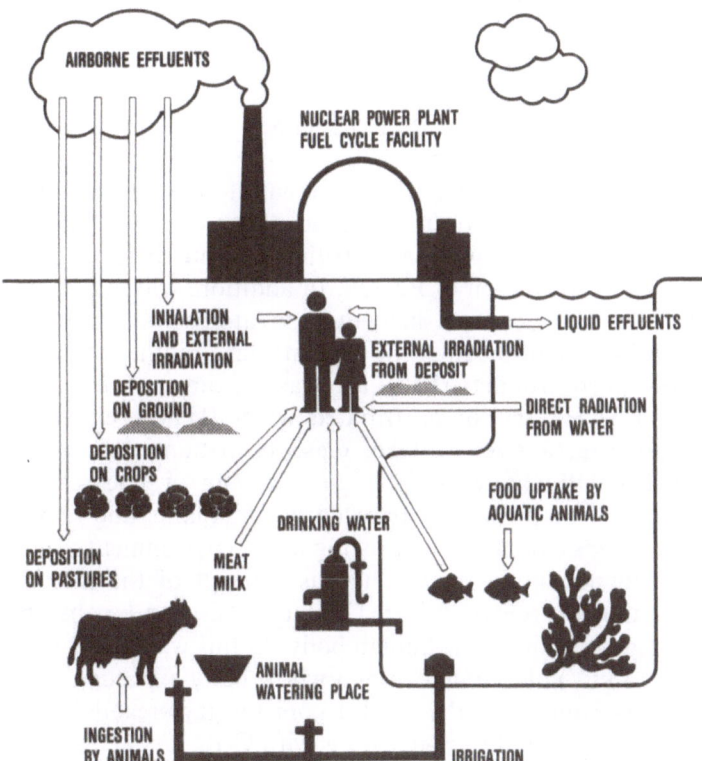

Fig. 8-1. Possible exposure pathways to man from the nuclear fuel cycle (KFA Jülich)

krypton, xenon, radon, of tritium, C-14, and also of fission product and fuel aerosols. Liquid effluents released into rivers, large lakes or the ocean contain tritium, fission products and other radioactive substances. Man may be exposed to ionizing radiation through various exposure pathways (Fig. 8-1):
– external β- and γ-radiation of the gaseous radioactive nuclides in the atmosphere (β- and γ-submersion) or by immersion in water (swimming),
– radiation from aerosol particles deposited on the ground (soil radiation),
– internal exposure following inhalation of radioactive nuclides (inhalation),
– internal exposure as a result of the intake of contaminated food or water (ingestion).

The release rate of radioactive nuclides into the environment depends on the retention mechanisms incorporated in the engineered safeguards design of a fuel cycle facility. Series connection of several containment barriers with low leak rates and other technical measures allow very high retention factors or low release factors to be attained.

Gaseous nuclides or aerosols escaping from the plant or discharged in a controlled way through a stack are diluted in the ambient atmosphere. Dilution is a function of the height of the exhaust stack, the turbulence conditions of the atmosphere, and the distance from the plant. Radionuclides are also deposited on the ground by dry and wet disposal. Aqueous radioactive discharges are diluted as a function of the quantitative relationships between the liquid effluents and the ambient water volume.

The individual radioactive nuclides can enter the human body on various pathways, where they again may have different radiological impacts.

8.1.1.1 Exposure Pathways of Significant Radionuclides

8.1.1.1.1 Tritium, Carbon-14 and Krypton. Tritium (halflife 12.4 a) is produced in the reactor core by ternary fission; on the average, about one in 10^4 fissions of U-235 is accompanied by the formation of tritium. About twice as many tritium nuclei are formed in the fission of Pu-239. In addition, tritium is generated in the coolant by neutron capture in deuterium atoms (deuterium has an abundance in natural water of 0.015 atom percent), and by the interaction of neutrons with the boron control material. It is released from nuclear reactors and reprocessing plants as HT gas or as tritiated water (HTO), either into the atmosphere or into, e.g., a river or lake. Gaseous tritium, HT, is very soon oxidized into HTO. Ultimately, any tritium escaping or released in a controlled manner thus will be present as tritiated water. Plants and animals may contain HTO/H_2O ratios close to those existing in the environment. Radioactive exposure of the human body then occurs as a result of the ingestion of food and drinking water. Moreover, tritiated water (HTO) can be absorbed by inhalation and through the skin of the human body. In this way, the ß-radiation (maximum energy 18 keV) of tritium causes a whole body exposure.

C-14 (halflife 5770 a) is built up in the reactor core by (n,p)-reactions with N-14, (n,α)-reactions with O-17, and (n,γ)-reactions with C-13. C-14 emits ß-radiation (maximum energy 16 keV). In reprocessing plants, C-14 is oxidized to $^{14}CO_2$ as the fuel is dissolved in nitric acid. In plants and animals, $^{14}CO_2/^{12}CO_2$

ratios may be established which are very close to those in the atmosphere. Radioactive exposure of the human body then occurs mainly as a result of the ingestion of food (milk, vegetables, meat). Direct inhalation and exposure from the ambient atmosphere only play minor roles.

Kr-85 (halflife 10.7 a) is a fission product. Kr-85 emissions from a fuel cycle facility are diluted in the atmosphere. Approximately 99.6% of the Kr-85 nuclei decay by emitting ß-particles with a maximum energy of 0.67 MeV. Only 0.4% of the Kr-85 nuclei decay by emitting a ß-particle (maximum energy, 0.16 MeV) and γ-radiation (0.51 MeV). There is no reduction in the airborne concentration as a result of deposition or washout. Kr-85 is only sparingly soluble in water. Its main radiological impact on the human body is due to the exposure to the skin. The inhalation of Kr-85 plays a lesser role.

8.1.1.1.2 Radioisotopes of Iodine. For releases from nuclear reactors the short-lived radioactive isotopes of iodine, I-131 (halflife 8 d) and I-133 (halflife 20 h), are important fission products. In reprocessing plants I-129 (halflife 1.7×10^7 a) remains an important radioisotope. Radioiodines in the airborne effluents from a nuclear power plant occur partly as elemental iodine and partly as an organic compound (methyl iodide). In reprocessing plants, radioiodine is mainly released with the effluent air. Airborne iodine is deposited on the surfaces of grass or vegetables. If liquid effluents containing radioiodine are discharged into rivers or lakes etc., a possible major pathway will be their accumulation in fish or plants. The human body can take up radioiodine with the inhalation of air, ingestion of vegetables or fish, and by drinking milk. The radioiodine absorbed by the human body is concentrated mainly in the thyroid. Radioiodine emits both ß- and γ-radiations.

8.1.1.1.3 Strontium and Cesium. Fission product Sr-90 (halflife 29.1 a) can be discharged into the atmosphere from the effluent air of reprocessing plants or into rivers with liquid effluents. Through the food chain (milk, vegetables, fish, meat and drinking water), Sr-90 enters the human body. Like calcium, Sr-90 is deposited preferably in bones, representing a major burden on the blood forming organs because of the long biological residence times of 18 years and the ß-radiation of 2.3 MeV maximum energy of the daughter product, Y-90 (halflife 2.7 d).

Releases of radiocesium by way of gaseous and liquid effluents from reprocessing plants also cause radiological exposures of the human body through the uptake of food, as in the case of strontium. Cs-134 (halflife 2.1 a) and Cs-137 (halflife 30 a) also emit γ-radiation in addition to ß-radiation. Cesium can largely replace potassium in living organisms and, like the latter, is distributed throughout the body in highly soluble compounds.

8.1.1.1.4 Plutonium Isotopes. Plutonium isotopes may be discharged into the atmosphere with gaseous effluents as aerosols of PuO_2 or $PuNO_3$ or into rivers together with liquid effluents. The following plutonium isotopes are of main interest: Pu-238 (halflife 87.8 a); Pu-239 (halflife 24,100 a); Pu-240 (halflife 6450 a); Pu-241 (halflife 14.4 a); Pu-242 (halflife 3.9×10^5 a). The highest burden results from inhalation, in which case plutonium is deposited in the lung. More-over, plutonium may be ingested together with vegetables, milk, meat, fish and drinking water. Plutonium taken up by way of ingestion is deposited prefera-bly in bone tissue.

8.1.1.1.5 Other Radiobiologically Significant Isotopes. In the vitrification process of high level waste, volatile oxides of the radioisotopes of ruthenium, technetium, selenium, tellurium, antimony and the higher actinides, such as americium (Am-241: halflife 433 a; Am-243: halflife 7380 a) and curium (Cm-242: halflife 163 d; Cm-244: halflife 18.1 a) may be discharged. The higher actinides can be absorbed in the body through similar exposure pathways as plutonium and can generate similar radiological burdens.

8.1.1.2 Radiation Dose

The measure of radiation absorbed is the energy dose. Its unit is called rad (*radiation absorbed dose*). It is defined as a radiation dose that deposits 100 erg of energy per gram of absorbing material:

absorbed dose
1 rad $\cong$ 100 erg/g = 10^{-5} Joule/g.
100 rad = 1 Gray (Gy)

The biological effects of the radiation dose absorbed also depend on the energy and type of radiation (γ-radiation, ß-radiation, α-particles, neutrons). To take these differences into account, an equivalent dose is defined, the unit of which is called rem (*roentgen equivalent man*) (1 rem = 1000 mrem):

equivalent dose
100 rem $\cong$ 1 Sievert (Sv)

The absorbed dose and the equivalent dose are related as follows:

equivalent dose (rem or Sv)
= absorbed dose (rad or Gy) $\times$ quality factor.

The quality factor is a measure of the relative effects of the particles in producing damage for a given energy deposition. For most of the γ- and ß-radia-tions of fission products the quality factor can be taken as unity. For thermal neutrons the quality factor is 3, for fast neutrons it is 10. For α-particle emitters entering the body and accumulating in certain tissues the quality factor is 20 (earlier value 10) according to ICRP Publication 26 (1977).

All individuals are exposed to natural background radiation, which consists of cosmic radiation, external terrestrial radiation from naturally radioactive isotopes in the soil and rocks or houses and internal radiation from naturally radioactive isotopes in the human body. The average background whole body dose in the USA is about 70–220 mrem/a, medical X-rays excluded. It varies with altitude, geographic location, etc. In the Federal Republic of Germany, the corresponding average background whole body dose is 110 mrem/a.

8.1.1.3 Permissible Radiation Exposures

The *I*nternational *C*ommission on *R*adiological *P*rotection (ICRP) has published recommendations about permissible radiation exposures of man from nuclear installations. In those recommendations, a distinction is made between occupationally exposed persons and individual members of the public. Considerable lower levels for exposure limits of the general population were set in national regulations (Table 8-1). In the Federal Republic of Germany, e.g., the limits indicated in Table 8-1 apply to all facilities of the nuclear fuel cycle; in the USA, graded limits for nuclear power plants are listed in 10 CFR 50, Appendix I while, for facilities of the uranium fuel cycle, limits are defined by the US *E*nvironmental *P*rotection *A*gency (USEPA).

Precisely defined methods of calculation exist to estimate the dose to which a person is exposed in the area of impact of a plant. In ICRP Publication 26 (1977) the earlier organ dose concept, which was based on earlier ICRP recommendations, has been replaced by a new "effective dose" concept. Under this new concept, dose contributions for different parts of the body can be summed up in accordance with the relative radiation sensitivities of organs and tissues exposed.

The USEPA has established emission limits for the uranium fuel cycle (CFR, Title 40, Part 190). These limits are to apply as of January 1983; per GW(e)·a and on a basis of a load factor of 1, they are as follows:

≤ 0.5 mCi of combined Pu-239 and other α-emitting transuranic nuclides with halflives greater than 1 year,

$\leq 5 \times 10^4$ Ci   of Kr-85,

≤ 5 mCi   of I-129.

The uranium fuel cycle for that purpose comprises only the steps of uranium milling, conversion, enrichment and UO_2 fuel fabrication, the power reactor (e.g., LWR) and the reprocessing plant. Uranium mining operations, transportation, re-use of non-uranium recovered and by-product materials as well as waste disposal operations are excluded from this definition.

Table 8-1. *Exposure dose limits for the general public from nuclear installations (exclusive of natural background and medical exposures) mrem/a)*

Organ or tissue	ICRP (1966)	Germany (BMI)[a]	USA (EPA)
Whole body (including gonads)	500	30	30
Thyroid	3000[b]	180	75
		90[c]	
Bone	3000	180	25
Skin	3000	180	25
Lung	1500	90	25

[a] These limits apply to gaseous and liquid effluents, respectively.
[b] 1500 mrem/a for children up to 16 years of age.
[c] The thyroid dose via all ingestion pathways in addition is limited to 90 mrem/a.

Above and beyond these standards defined by ICRP and national regulations, the *"as low as reasonably achievable"* (ALARA) principle must be applied to all emissions of nuclear plants. This means that practically all facilities of the fuel cycle must keep below those standards.

8.1.2 Radionuclide Effluents and Radiation Exposures from Various Parts of the Fuel Cycle

In the following sections, the radioactive effluents and the radiation exposures from important parts of the nuclear fuel cycle are discussed. The radiation exposures are analyzed on the basis of measurements and calculations for uranium mines, enrichment and fuel fabrication plants, the power reactor and the reprocessing plant with waste treatment facilities. Following a long term strategy with reprocessing of spent fuel, the use of plutonium in both LWR's and LMFBR's is discussed.

8.1.2.1 Uranium Mining and Milling

8.1.2.1.1 Radioactive Effluents from Mining and Milling. The main radioactive effluents from uranium mines and mills are radon and dust particles containing uranium and its decay products. All daughter products of uranium, except radon, are solid, emitting α- and ß-particles, mostly in combination with γ-radiation. Radon is a noble gas trapped in the crystalline structure of rock, but distributed into the atmosphere as the rock is being opened up. Radon and the other decay products of uranium enter the lung via the inhalation process. In open pit mining, radon is directly released into the atmosphere. The radon released during underground ore mining must be removed by the ventilation system of the mine. After extraction of the ore, its storage above ground prior to further processing and milling constitutes another source of radon emission.

The remaining uranium mill tailings are usually stored in a tailings impoundment located on the mill site. They and the low grade ore stockpiles can be covered with a layer of sand several meters thick. This largely prevents further releases of radon.

In EPRI studies (1979), a model mine-mill complex with an open pit mine of 500,000 m^2 of open area and a production capacity of 730,000 te of ore per annum was used as a basis. The associated uranium ore mill produces 930 te U_3O_8 per annum (average uranium ore compounds, 0.2%). The gaseous effluents and the release rates of radioactivity of this open pit mine-mill complex are given in Table 8-2. Studies conducted by the USEPA and the US Nuclear Regulatory Commission (USNRC) show radon release levels of the same order of magnitude for model open pit uranium mines of similar size. The release of radioactivity stems mainly from radon. Open pit mining and tailings are the main sources of radon emissions, while milling operations contribute only relatively little to the emissions of radioactivity.

Comparisons with the radon emissions from underground mining result in radon releases of 5400 Ci/GW(e)·a, which are considerably higher than those arising in open pit mining. This is probably due to the necessary ventilation of the tunnel system of underground mines. In the USA, uranium is extracted approximately half in open pit mining and half in underground mining.

8.1.2.1.2 Radioactive Exposure Pathways for Uranium Mines and Mills. The radioactive exposure pathways for people living in the vicinity of a uranium mine and mill complex are
– external radiation from radon gas and radioactive airborne dust,
– internal radiation after inhalation of radon gas and dust as well as ingestion of contaminated food products.

The data for the release of radioactivity in Table 8-2 and meteorological data allow the exposure doses at a certain distance from the model mining and milling complex to be calculated. Table 8-3 shows the so-called *working level month* (WLM) exposures and the lung doses for individuals from radon exposure due to model uranium mines. The exposure levels apply to individuals at a location 500 m from the model mining complex in the predominant wind direction. One WLM is equivalent to 0.5 rad or 5 rem for the lung dose (based on a radiobiological quality factor of 10 for α-particles; with the new quality factor of 20 (ICRP Publication 26), this relation changes to 10 rem). The data of Table 8-3 can be compared to the standard maximum exposure of 4 WLM/a defined by USEPA. According to this standard, no individual shall receive an exposure of more than 4 WLM in uranium mines over a period of 12 consecutive months.

Table 8-2. *Estimated radioactive effluent release rates from major sources in a model open pit mine-mill complex (EPRI) (values of Rn-222 releases may vary considerably in different mines)*

Source	Effluent	Release rate	
		(Ci/a)	(Ci/GW(e)·a)
Open pit mine	Rn-222	7000	1100
Mill stack			
Ore dust[a]	U-238 and daughter isotopes	4.4×10^{-3} (each)	6.9×10^{-4} (each)
Yellowcake	U-238 and U-234	8.8×10^{-2} (each)	1.4×10^{-2} (each)
dust[a]	Th-230	4.4×10^{-3}	6.9×10^{-4}
	Ra-226	1.8×10^{-3}	2.8×10^{-4}
	Rn-222	38	6
Tailings	Rn-222	1450[b]	230[b]
		(420)	(65)
Inactive period of mine complex after 20 years of operation	Rn-222	6000[b]	940[b]
		(1740)	(270)

[a] Dust collector efficiency more than 90%.
[b] These numbers can be reduced (values in brackets) by a factor of 0.3 by 2.5 m of tailing cover.

Table 8-3. *Working level exposures from radon-222 emissions of model uranium mines (USEPA). (Exposures apply to an individual living 500 m from a mine vent or from the edge of open pit mining area in the predominant wind direction)*

Source	Working level months per year (WLM/a)	Lung dose[a]	
		(rem/a)	(mrem/GW(e)·a)
Underground mine	7.2×10^{-2}	0.36	0.14
Open pit mine	9.6×10^{-3}	0.048	0.01

[a] One working level month (WLM) equivalent to roughly 5 rems for the lung dose.

The data in Table 8-3 apply to an open pit mine-and-mill complex providing U_3O_8 sufficient for about 4–5 GW(e)·a of PWR capacity. The underground mine also referred to has an annual production capacity for about 2–3 GW(e)·a of PWR. It is seen that underground mining, because of the higher radon emissions, represents a higher individual risk than open pit mining.

8.1.2.2 UF_6 Conversion, Enrichment, and Fuel Fabrication

Conversion of U_3O_8 into UF_6 and subsequent enrichment of the fuel do not entail any significant radiological burdens to the environment. In UF_6 conversion, gaseous effluents are passed through filters and wet sorbers. Discharges from the plant contain only traces of radioactive material. Solid waste, which contains small quantities of uranium or traces of thorium, is packaged and consigned to a licensed waste burial site.

In uranium enrichment plants U-234, U-235 and U-238 radionuclides may be emitted. U-234 occurs in natural uranium with an abundance of 0.0054 atom percent. Compared with all other emissions of the fuel cycle, emissions from enrichment plants are negligibly low.

In the fabrication of uranium fuel, the enriched UF_6 is first chemically converted into UO_2. Afterwards, the UO_2 powder is pelletized. Liquid and gaseous effluents from the fabrication plant contain certain concentrations of U-234, U-235, U-238 and Th-234 and must therefore be reprocessed and filtered. Data for radioactive effluents from UF_6 conversion, enrichment and fuel fabrication are given in Table 8-4.

8.1.2.3 Nuclear Power Plants

8.1.2.3.1 Radioactive Effluents of Nuclear Power Plants. Most of the radioactive inventory of a nuclear power plant is made up of fission products. Gaseous fission products, such as noble gases (especially krypton and xenon) and tritium, can enter the coolant through leaks in the claddings of fuel rods. They may get through the primary coolant purification system and the exhaust air system into carbon filter lines and into the exhaust air stack from where they are released into the environment. Emissions of shortlived isotopes, such as Kr-88 (halflife 2.8 h), can be minimized by adequate holdup of gaseous effluents in storage and decay tanks before release.

Table 8-4. *Radioactive emissions from U_3O_8 conversion, enrichment and uranium fuel fabrication plants (normalized to 1 GW(e)·a of energy produced (NUREG)*

	Nuclide emitted	Radioactive emissions	
		Gaseous (Ci/GW(e)·a)	Liquid (Ci/GW(e)·a)
U_3O_8 conversion plant	Uranium	2.1×10^{-3}	0.6
	Ra-226	–	4.7×10^{-2}
	Th-230	–	2×10^{-2}
Enrichment plant	U-234	9.3×10^{-4}	2.8×10^{-5}
	U-235	2.8×10^{-5}	1.1×10^{-6}
	U-238	1.1×10^{-4}	2.4×10^{-5}
Fuel element fabrication plant	Th-234	9.9×10^{-5}	2.4×10^{-2}
	U-234	8.1×10^{-4}	2×10^{-1}
	U-235	2.3×10^{-5}	5.1×10^{-3}
	U-238	9.9×10^{-5}	2.4×10^{-2}

The tritium produced in the core migrates along the grain boundaries of the fuel into the fission product gas plenum of the fuel rod. The zircaloy tubes of LWR fuel rods bind some 60% of the tritium inventory. Moreover, an oxide layer building up on the outer wall of the zircaloy cladding tube acts as a diffusion barrier to the tritium. As a consequence, more than 99.9% of the tritium formed is retained in the LWR fuel rod. Only if cladding tube failures occur, will releases of tritium into the cooling water be increased. Some of the tritium is discharged with the gaseous effluents. In water cooled reactors most of it remains in the coolant as tritiated water. Some of the tritiated water is released at a controlled rate. Improved methods under study are the concentration of tritiated water by evaporation and its prolonged storage in decay tanks. With a halflife of tritium of about twelve years, some 90% will have decayed after forty years.

Besides the radioactive noble gases and tritium, also such elements as strontium, technetium, ruthenium, silver, tellurium, antimony, iodine, cesium, barium, lanthanum and cerium are radiologically significant. Except for iodine, cesium and rubidium, these elements have only low volatilities. They may enter the primary coolant through defective fuel rod claddings and may be released into the exhaust air of the stack partly as aerosols. Non-volatile fission products can enter the liquid effluent only through the primary coolant purification system.

8.1.2.3.2 Radioactive Effluents from PWR's. Table 8-5 shows emission data of a typical PWR. This set of data was developed from an analysis of surveillance measurements conducted in German PWR plants during 1972–1979. Tritium is mainly released with liquid effluents (404 Ci/GW(e)·a), whereas only 45 Ci/GW(e)·a of tritium are emitted with gaseous effluents. Roughly 5.5 Ci/GW(e)·a

Table 8-5. *Radioactive emissions from a model PWR power plant (normalized to 1 GW(e)·a of energy produced) (KfK)*

Liquid effluents (Ci/GW(e)·a)		Gaseous effluents (Ci/GW(e)·a)	
H-3	4.0E+02	H-3	4.5E+01
		C-14	5.5E+00
		Noble Gases	
Cr-51	1.0E−01	Ar-41	9.1E+01
Mn-54	6.8E−02	Kr-85	1.3E+03
Fe-59	8.7E−04	Kr-85 m	4.8E+02
Co-57	6.2E−04	Kr-87	1.8E+00
Co-58	7.2E−01	Kr-88	2.3E+02
Co-60	6.1E−01	Kr-89	1.8E+00
Zn-65	3.8E−04	Xe-131 m	5.0E+00
Sr-89	9.7E−03	Xe-133	6.0E+03
Sr-90	4.6E−03	Xe-133 m	3.5E+02
Zr-95	5.9E−03	Xe-135	8.4E+02
Nb-95	7.2E−03	Xe-135 m	1.8E+00
Mo-99	4.4E−03	Xe-137	1.8E+00
Ru-103	7.1E−03	Xe-138	1.8E+00
Ru-106	3.4E−03		
Ag-110m	3.1E−02	Aerosols	
Sb-124	9.3E−02		
Sb-125	1.8E−03	Cr-51	1.7E−03
Te-123m	2.1E−03	Mn-54	9.5E−04
J-131	1.4E−01	Fe-59	3.2E−04
Cs-134	5.9E−01	Co-57	1.2E−05
Cs-137	9.3E−01	Co-58	7.8E−03
Ba-140	7.3E−01	Co-60	1.3E−02
La-140	1.7E−01	Sr-89	4.2E−05
Ce-141	5.7E−02	Sr-90	6.4E−06
Ce-144	2.5E−02	Nb-95	7.5E−04
Isotopes of:		Zr-95	7.0E−04
Uranium	⎫	Ru-103	8.3E−04
Neptunium	⎪	Ru-106	3.3E−05
Plutonium	2.0E−07	Ag-110m	2.2E−03
Americium	⎪	Sb-124	1.3E−03
Curium	⎭	Te-123m	7.7E−04
		J-131	4.7E−02
		Cs-134	5.6E−04
		Cs-137	1.7E−03
		La-140	1.8E−04
		Ce-141	3.4E−05
		Ce-144	4.5E−04
		Isotopes of:	
		Uranium	⎫
		Neptunium	⎪
		Plutonium	2.0E−07
		Americium	⎪
		Curium	⎭

E-01 ≙ 10^{-1} etc.

of C-14 are released from the reactor plant as carbon dioxide, methane and other hydrocarbons. If no cladding failures occur, the C-14 generated in the fuel will be released only in the reprocessing plant. Kr-85 with 1350 Ci/GW(e)·a and Xe-133 with about 6000 Ci/GW(e)·a are the main contributors as gaseous effluents. I-131 contributes up to 0.05 Ci/GW(e)·a to the radioactivity released into the atmosphere and about 0.14 Ci/GW(e)·a to the liquid effluents. Emissions with the liquid effluent are largely due to Cs-137 and Cs-134, which are released in amounts of approximately 0.9 Ci/GW(e)·a and 0.6 Ci/GW(e)·a, respectively. Aerosol releases of Cs-134 and Cs-137 through the exhaust air stack are very small by comparison.

Another group of radioactive nuclides are the radioactive corrosion products of materials in the reactor core and activated parts of the primary cooling system. Elements of radiological significance are Zr-95, Nb-95, and Mo-99. They are partly released through the exhaust air stack or with liquid effluents. Among the primary system components, the Cr-51, Mn-54, Fe-59, Co-57, Co-58, and Co-60 radioisotopes can enter the cooling system as corrosion products. Defective cladding may also cause fissile materials (uranium) and higher actinides, such as neptunium, plutonium, americium, and curium, to enter the gaseous and liquid effluents in trace concentrations. The resultant emission rates are on the order of 0.2 μCi/GW(e)·a for gaseous and liquid effluents. These contributions are negligible compared to other emissions.

8.1.2.3.3 Comparison of Radioactive Effluents from PWR's and BWR's. As BWR's do not use boric acid for burnup control, less tritium will be formed from reactions of neutrons with boron. Since they are equipped with a direct steam cycle, fission products from defective fuel rods are carried by the steam directly into the turbine. BWR's in comparison to PWR's, therefore, have a different system for the treatment of radioactive gases and liquids prior to discharge to the environment. As a consequence of these differences, they release less tritium into the environment but more activity as noble gases. The higher activity from noble gases arises mainly from short-lived radioisotopes like Kr-87 (halflife 1.3 h), Kr-88 (halflife 2.8 h) and Xe-133 (halflife 5.3 d), which decay within a short time. After this short decay time, only Kr-85 remains in the atmosphere. Technical improvements are under development to achieve longer holdup times for noble gases within BWR plants. Such possibilities are low temperature absorption on carbon or liquefaction of the noble gases.

8.1.2.3.4 Radioactive Effluents from LMFBR and Other Nuclear Power Plants (CANDU-PHWR and HTGR). The operating experience of prototype LMFBR's (PHENIX, PFR etc.) and estimates of emission data of LMFBR's show radioactive effluents to be somewhat lower than those from PWR's. Tritium, after diffusion through the stainless steel cladding of fuel rods, is chemically bound by the sodium coolant. Other non-gaseous fission products from failed fuel rods are also bound by the sodium coolant and eliminated together with radioactive corrosion products in cold traps by the sodium purification system; from there they pass into the solid waste treatment system.

Although there are inherent design differences between LWR's and CANDU-PHWR's or HTGR's, there are no major differences in their radioactive effluents during normal operation. All types of power reactors have radioactivity releases below the radiation limits described in Section 8.1.1.3. Nuclear power plants are equipped with instruments to measure continuously the amounts of gaseous and liquid radioactive effluents. These must be reported to national environmental protection agencies or nuclear regulatory commissions. Such reports are made available to the public.

8.1.2.3.5 Radiation Exposures Caused by Emissions from Nuclear Power Plants. In calculating radiation exposures it is assumed that gaseous effluents are released into the environment from a stack of 100 m height. Moreover, liquid effluents are introduced into the cooling water of a nuclear power plant and further diluted in the main canal with a water flow of 250 m^3/s. Taking into account statistical data about the weather conditions and following the different exposure pathways, it is possible to determine the radiation exposure in the specific environment of a plant. The exposure calculations below were performed on the basis of German rules and regulations. In this respect, it is assumed that a person stays in the same place throughout the year and ingests both drinking water and food from the immediate environment. Fig. 8-2 presents the exposure data from a model PWR plant normalized to 1 GW(e)·a of energy produced and indicates the organ specific doses to the bones, the thyroid, and the whole body resulting from gaseous and liquid effluents. For gaseous effluents also the skin exposure dose was determined. The different exposure doses are

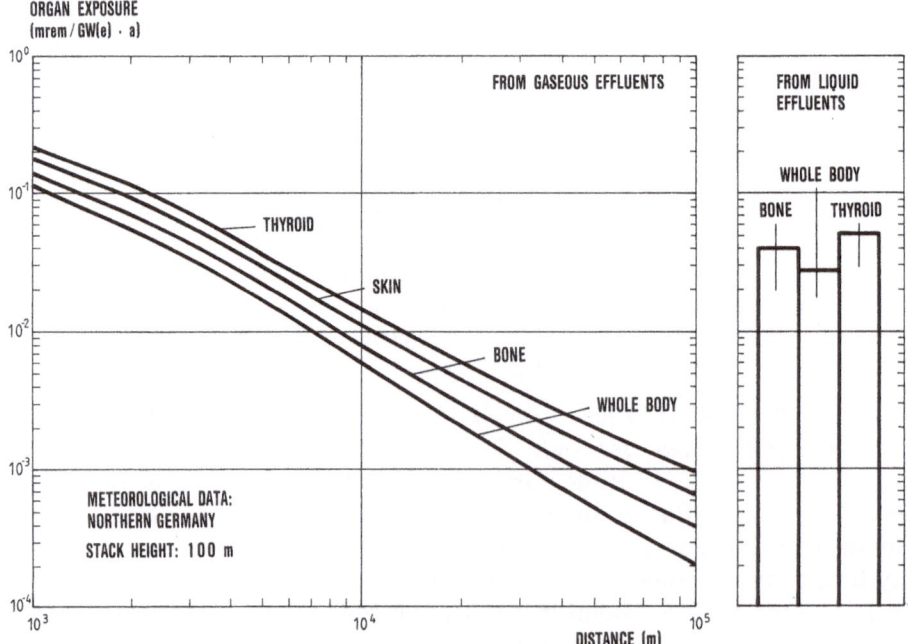

Fig. 8-2. Radiation exposure caused by emissions from a 1 GWe model PWR plant (KfK)

presented for distances up to 100 km from the model plant. For soil radiation and ingestion, nuclide enrichment in the soil over a period of fifty years is assumed. ß-submersion from emissions of Kr-85, Xe-131 and Xe-133 furnishes the main contributions to the exposure of the skin. Mainly the γ-radiation from cobalt and cesium is responsible for soil radiation. Inhalation results in minor exposures only, due to tritium and C-14. In the exposure resulting from ingestion, the radioactive C-14, tritium, strontium, and cesium nuclides play the main role. I-131 dominates in the exposure arising from the thyroid. In the immediate vicinity of the PWR facility, a maximum radiation exposure from gaseous effluents of approximately 0.22 mrem/a to the thyroid, 0.2 mrem/a to the skin, 0.15 mrem/a to the bones and 0.12 mrem/a to the whole body is found. Dose levels drop sharply with increasing distances from the reactor plant. At a distance of 10 km, the exposure is about a factor of 10 lower than in the immediate vicinity of the site. The organ exposure doses for bones, the thyroid and the whole body resulting from liquid effluents are all below 0.1 mrem/a.

The organ specific doses for an LMFBR model plant calculated with the same models and under the same assumptions as for the PWR model plant are even somewhat lower than in the case of a PWR. This is due to the differences mentioned above in the emission data, which show lower values for nearly all nuclides.

All exposure doses from PWR's or LMFBR's are far below the limits imposed by radiation protection ordinances as indicated in Section 8.1.1.2. Similar studies and survey measurements show that this holds also for other fission reactors like HWR's and gas cooled reactors.

8.1.2.4 Spent Fuel Reprocessing and Waste Treatment Centers

In the reprocessing and waste treatment plant described in Chapter 7, most of the radioactive substances released during chopping and dissolution of the fuel are retained in the facility by various engineered safeguards measures. Releases of radioactivity into the environment are controlled.

During chopping and dissolution of the fuel elements, a certain fraction of tritium enters the gaseous effluents and is released through the stack. Another part of the tritium may be retained in the zircaloy cladding hulls and will go into the solid high level wastes. The residual tritium remains in the nitric acid solution as tritiated water. After removal of the nitric acid, the remaining liquid effluent is concentrated by evaporation. Part of it can be released in a controlled way. In modern large reprocessing plants, the prolonged storage of tritium in storage tanks or its injection into deep wells may be applied.

C-14 is currently released without any retention systems. In the future, retention is possible in connection with low temperature rectification. Noble gases can largely be retained by means of low temperature rectification processes. More than 99.9% of the I-129 can be held back by silver nitrate filters.

Other radionuclides generated as aerosols in the dissolver offgas are:
– Mn-54, Fe-55, Co-60, and Ni-63, which are activated in the structural materials of the fuel elements.

Table 8-6. *Release factors of large spent fuel reprocessing centers*

Isotope	WASH-1535 (1975)	DOE/ET-0028[a] (1979)	KfK-3266[a] (1982)	KfK-3315[b] (1982)
Tritium	1	1	0.25	0.25
Carbon-14		10^{-2}	1	1
Noble gases	10^{-2}	0.1	5×10^{-2}	5×10^{-2}
Iodine	10^{-4}	10^{-3}	10^{-2}	10^{-2}
Ruthenium	10^{-9}	5×10^{-9}	5×10^{-8}	10^{-7}
Strontium	2×10^{-10}	10^{-9}	5×10^{-9}	10^{-8}
Cesium	2×10^{-10}	10^{-9}	5×10^{-9}	10^{-8}
Cerium	2×10^{-10}	10^{-9}	5×10^{-9}	10^{-8}
Actinides	5×10^{-10}	10^{-9}	5×10^{-9}	10^{-8}
Other aerosols		10^{-9}	5×10^{-9}	10^{-8}

[a] Applying to the complete facility.
[b] Applying only to the dissolver offgas system.

– fission products, such as Tc-99, Ru-103, Ru-106, Cs-137, etc.,
– aerosols of the higher actinides and the fuel, such as the isotopes of plutonium, americium and curium.

The technical processes of treating radioactive waste have been described in Section 7.5. Radioactive emissions play significant roles, especially in gaseous effluent treatment from high level waste vitrification plants. Gaseous effluents from melters can have temperatures up to 200 °C and contain up to 3% aerosols, besides water vapor and nitrogen oxides. They are passed through wet scrubbers to retain ruthenium tetroxide. Then they flow through a nitrogen oxide absorption column and *high efficiency particulate air* (HEPA) filters for ruthenium aerosol retention.

All gaseous effluent streams of reprocessing, refabrication and waste treatment plants are passed through HEPA filter lines for removal of particulates. In addition, the application of several containment barriers with low leak rates, storage of radioactive liquids or gases in decay tanks or pressurized gas cylinders allows high retention factors and, conversely, low release factors to be achieved. The retention factor denotes the ratio between radioactive inventories and radioactive releases. The release factor is the reciprocal value of the retention factor. A number of release factors applied in studies of specific spent fuel reprocessing center projects are listed in Table 8-6 for the different gaseous nuclides and aerosols. By series connecting several containment barriers and filter lines it is possible to attain retention factors of 10^8 or even 10^9 and release factors of 10^{-8} or 10^{-9}, respectively, for such aerosols as strontium, ruthenium, plutonium and higher actinides. Low temperature rectification of the radioactive noble gases, iodine absorption in silver nitrate filters as well as concentration and storage of tritiated water can also be applied to achieve the required release factors for noble gases, iodine and tritium. For waste treatment facilities the same release factors may be applied as indicated in Table 8-6.

8.1.2.4.1 Radioactive Effluents from an LWR Low Enriched UO₂ Spent Fuel Reprocessing and Waste Treatment Center. Table 8-7 shows the estimated radioactive inventories and emissions from a model reprocessing and waste treatment center for low enriched (LEU) UO_2 fuel as discharged from PWR's. The data are normalized to 1 GW(e)·a of energy produced. Gaseous and airborne radio-

Table 8-7. *Inventories of radioactivity and emissions from reprocessing and waste treatment centers for spent PWR fuel (normalized to 1 GW(e)·a of energy produced) (KfK)*

Nuclide	Release factor	Specific inventory (g/te$_{HM}$)	Emissions (Ci/GW(e)·a)
Gaseous			
H-3[a]	2.50E−01	1.49E−02	1.11E+03
C-14[a]	1.00E+00	7.15E−02	9.82E+00
Mn-54	1.00E−08	3.19E−03	7.60E−06
Fe-55	1.00E−08	3.77E+00	2.90E−03
Co-58	1.00E−08	1.28E−11	1.25E−13
Co-60	1.00E−08	1.35E+01	4.71E−03
Ni-63	1.00E−08	3.40E+01	6.46E−04
Kr-85	5.00E−02	1.70E+01	1.03E+04
Sr-89	1.00E−08	1.75E−14	1.57E−16
Sr-90	1.00E−08	5.16E+02	2.17E−02
Y-91	1.00E−08	3.34E−12	2.52E−14
Zr-95	1.00E−08	7.47E−11	4.95E−13
Nb-95	1.00E−08	9.11E−11	1.10E−12
Tc-99	1.00E−08	8.54E+02	4.46E−06
Ru-103	1.00E−07	1.31E−18	1.30E−19
Ru-106	1.00E−07	1.34E+00	1.38E−02
Ag-110m	1.00E−08	5.39E−04	7.88E−07
Sb-125	1.00E−08	1.65E+00	5.26E−04
Te-125m	1.00E−08	2.33E−02	1.29E−04
Te-127m	1.00E−08	1.25E−07	3.63E−10
Te-127	1.00E−08	4.39E−10	3.57E−10
J-129	1.00E−02	1.91E+02	1.04E−02
Cs-134	1.00E−08	1.25E+01	4.89E−03
Cs-137	1.00E−08	1.14E+03	3.06E−02
Ce-141	1.00E−08	1.26E−22	1.10E−24
Ce-144	1.00E−08	7.80E−01	7.66E−04
Pu-238	1.00E−08	1.49E+02	7.85E−04
Pu-239	1.00E−08	5.03E+03	9.63E−05
Pu-240	1.00E−08	2.38E+03	1.67E−04
Pu-241	1.00E−08	9.06E+02	2.87E−02
Np-239	1.00E−08	8.42E−05	6.00E−06
Am-241	1.00E−08	3.94E+02	4.15E−04
Am-242	1.00E−08	5.09E−06	1.27E−06
Cm-242	1.00E−08	1.29E−03	1.32E−06
Cm-243	1.00E−08	2.87E−01	4.55E−06
Cm-244	1.00E−08	1.74E+01	4.34E−04

Table 8-7 (continued)

Nuclide	Release factor	Specific inventory (g/te_{HM})	Emissions $(Ci/GW(e)\cdot a)$
Liquid			
H-3[a]	7.50E−01		3.33E+03
Sr-90	1.00E−08		2.17E−02
Ru-106	1.00E−08		1.38E−03
Cs-134	1.00E−08		4.89E−03
Cs-137	1.00E−08		3.06E−02
Pu-238	1.00E−08		7.85E−04
Pu-239	1.00E−08		9.63E−05
Pu-240	1.00E−08		1.67E−04
Pu-241	1.00E−08		2.87E−02

(Initial U-235 enrichment 3.5%, burnup 36,000 MWd(th)/te, cooling time 7a)

[a] From fuel inventory only.   E-01 $\cong 10^{-1}$ etc.

active effluents from different facilities of the reprocessing and waste treatment center are released into the atmosphere through a number of exhaust air stacks of different height. Emissions of the spent fuel storage facility at the entry of the reprocessing and waste treatment center may come from about 0.1–1% of defective fuel rods producing some tritium, Kr-85 and iodine effluents. By far most of the radioactivity, however, is emitted from the spent fuel reprocessing and waste treatment facilities. Their most important radioactivity contributions come from tritium, Kr-85, C-14, I-129, Cs-137, and, to a lesser extent, from other ß-radiation emitting fission products as well as α-particle emitting uranium, plutonium and higher-actinide aerosols. By comparison, the uranium fuel and the mixed oxide fuel fabrication facilities emit relatively little radioactivity in the form of uranium and plutonium aerosols.

The release factors shown in Table 8-7 are identical to those in the right column of Table 8-6. For the tritium release it is assumed that 25% is released to the atmosphere and 75% is diluted into river water. No credit is taken yet from tritium retention techniques, e.g., concentration and subsequent storage in tanks or pressing in deep geological formations. In case such a reprocessing and waste treatment center had to be located on a site where not enough water from a river or the ocean would be available for the dilution of tritium, the emissions into the atmosphere would have to be increased above the 25% level assumed in Table 8-7.

8.1.2.4.2 Estimated Radioactive Effluents from a Reprocessing and Waste Treatment Center for Spent PuO_2/UO_2 Fuel. Besides the reprocessing center for spent LEU fuel of LWR's discussed in the previous section, also fuel cycle alternatives should be discussed in which PuO_2/UO_2 is used as a fuel. These alternatives are the self-generating plutonium recycling (SGR) case, the Pu burner case and the LMFBR fuel cycle.

Table 8-8. *Inventories of radioactivity and emissions from reprocessing and waste treatment centers for spent PuO_2/UO_2 (self-generated) PWR fuel and spent PuO_2/UO_2 LMFBR fuel (normalized to 1 GW(e)·a of energy produced) (KfK)*

Nuclide	Release factor	Self-generated PWR fuel		LMFBR fuel	
		Specific inventory (g/te_{HM})	Emissions $(Ci/GW(e)\cdot a)$	Specific inventory (g/te_{HM})	Emissions $(Ci/GW(e)\cdot a)$
Gaseous					
H-3 [a]	2.50E−01	4.91E−02	3.97E+03	7.24E−02	5.59E+03
C-14 [a]	1.00E+00	1.72E−02	2.56E+00	6.99E−03	9.95E−01
Mn-54	1.00E−08	1.76E−01	4.56E−04	3.10E+00	7.68E−03
Fe-55	1.00E−08	7.97E+00	6.67E−03	1.69E+01	1.30E−02
Co-58	1.00E−08	9.79E−04	1.05E−05	1.24E−02	1.26E−04
Co-60	1.00E−08	1.02E+01	3.85E−03	4.58E−01	1.66E−04
Ni-63	1.00E−08	2.15E+01	4.46E−04	1.80E+00	3.26E−05
Kr-85	5.00E−02	1.88E+01	1.24E+04	1.20E+01	7.52E+03
Sr-89	1.00E−08	1.21E−03	1.18E−05	1.07E−03	9.95E−06
Sr-90	1.00E−08	4.51E+02	2.06E−02	2.35E+02	1.05E−02
Y-91	1.00E−08	7.53E−03	6.20E−05	6.73E−03	5.28E−05
Zr-95	1.00E−08	2.72E−02	2.03E−04	2.84E−02	1.95E−04
Nb-95	1.00E−08	3.43E−02	4.66E−04	3.45E−02	4.32E−04
Tc-99	1.00E−08	7.78E+02	4.42E−06	6.75E+02	3.71E−06
Ru-103	1.00E−07	1.40E−04	1.51E−05	1.82E−04	1.88E−05
Ru-106	1.00E−07	4.76E+01	5.33E−01	7.44E+01	7.97E−01
Ag-110m	1.00E−08	1.08E−01	1.72E−04	2.02E−02	3.06E−05
Sb-125	1.00E−08	6.40E+00	2.22E−03	1.19E+01	3.94E−03
Te-125m	1.00E−08	8.98E−02	5.43E−04	1.67E−01	9.60E−04
Te-127m	1.00E−08	1.50E−02	4.76E−05	2.31E−02	6.98E−05
Te-127	1.00E−08	5.27E−05	4.66E−05	8.11E−05	6.85E−05
J-129	1.00E−02	2.01E+02	1.19E−02	2.10E+02	1.19E−02
Cs-134	1.00E−08	5.83E+01	2.53E−02	8.88E+00	3.68E−03
Cs-137	1.00E−08	1.17E+03	3.42E−02	9.86E+02	2.74E−02
Ce-141	1.00E−08	1.01E−05	9.61E−08	1.17E−05	1.06E−07
Ce-144	1.00E−08	5.79E+01	5.20E−02	5.34E+01	5.44E−02
Pu-238	1.00E−08	4.37E+02	2.51E−03	1.17E+03	6.43E−03
Pu-239	1.00E−08	7.92E+03	1.65E−04	5.45E+04	1.07E−03
Pu-240	1.00E−08	5.19E+03	3.95E−04	1.96E+04	1.43E−03
Pu-241	1.00E−08	2.73E+03	9.41E−02	5.02E+03	1.60E−01
Np-239	1.00E−08	4.96E−04	3.85E−05	4.08E−04	3.02E−05
Am-241	1.00E−08	4.39E+02	5.06E−04	1.74E+03	1.91E−03
Am-242	1.00E−08	3.06E+00	9.95E−06	6.26E+01	1.95E−04
Cm-242	1.00E−08	2.07E+00	2.29E−03	5.11E+00	5.41E−03
Cm-243	1.00E−08	1.30E+00	2.26E−05	6.48E+00	1.02E−04
Cm-244	1.00E−08	1.73E+02	4.69E−03	6.81E+01	1.76E−03
Liquid					
H-3 [a]	7.50E−01		1.19E+04		1.68E+04
Sr-90	1.00E−08		2.06E−02		1.05E−02

Table 8-8 (continued)

Nuclide	Release factor	Self-generated PWR fuel		LMFBR fuel	
		Specific inventory (g/te$_{HM}$)	Emissions (Ci/GW(e)·a)	Specific inventory (g/te$_{HM}$)	Emissions (Ci/GW(e)·a)
Ru-106	1.00E−08		5.33E−02		7.97E−02
Cs-134	1.00E−08		2.53E−02		3.68E−03
Cs-137	1.00E−08		3.42E−02		2.74E−02
Pu-238	1.00E−08		2.51E−03		6.43E−03
Pu-239	1.00E−08		1.65E−04		1.07E−03
Pu-240	1.00E−08		3.95E−04		1.43E−03
Pu-241	1.00E−08		9.41E−02		1.60E−01

Self-generated PWR fuel: burnup 33,000 MWd(th)/te, cooling time 2 a.
LMFBR fuel: burnup (core + blankets) 28,500 MWd(th)/te, cooling time 2 a.

[a] From fuel inventory only.   E-01 $\cong 10^{-1}$ etc.

Table 8-8 shows the estimated radioactive inventories and the radioactivity released into the environment from a reprocessing and waste treatment center for SGR plutonium recycling fuel and for PuO_2/UO_2 LMFBR fuel on the basis of 1 GW(e)·a nuclear energy generation. The radioactive effluents for the Pu burner case are comparable to those of LMFBR fuel and will not be treated in greater detail here.

As in the case of the spent LEU fuel (see Table 8-7), tritium, Kr-85 and C-14 represent the largest fraction of the radioactivity released. In the case of LMFBR fuel, this is followed by ruthenium, plutonium, cerium, cesium, strontium, iodine and curium. A comparison between radioactive inventories of model reprocessing centers for LEU spent fuel from LWR's and for PuO_2/UO_2 fuel from LMFBR's shows that, under the assumption of equal release factors, the UO_2 fuel reprocessing plant would release a factor of two more radioactivity as Sr-90, while the LMFBR fuel reprocessing plant would release a factor of two more radioactivity as plutonium and higher actinides. Table 8-8 is based on the same release factors as presented in the right column of Table 8-6. Such release factors are considered to be feasible at the present state of the art. Future improvements in relevant technologies will certainly lead to release factors as indicated in the left column of Table 8-6.

8.1.2.4.3 Radiation Exposure Caused by Reprocessing and Waste Treatment Centers. For assessing the radiation exposure in the environment of the model spent fuel reprocessing centers mentioned above it is assumed by way of approximation that the gaseous effluents are discharged into the atmosphere from a main stack of 200 m height. Taking into account statistical data on meteorological conditions and following the exposure pathways referred to earlier, it is possible to determine organ exposure doses for every location in the environ-

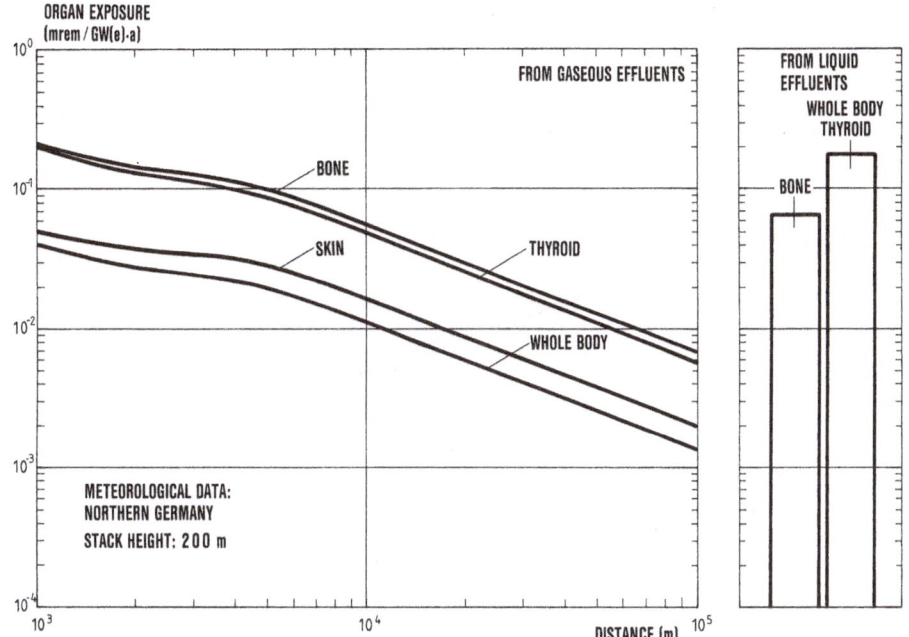

Fig. 8-3. Radiation exposure caused by emissions from reprocessing spent UO_2 LEU fuel (KfK)

ment. For this purpose, the same assumptions are applied as to the model reactor facility above. Fig. 8-3 shows the radiological organ exposures of the whole body, the thyroid and the bones from gaseous and liquid effluents of reprocessing spent UO_2 LEU fuel. For gaseous effluents also the exposure values for the skin are determined. The different exposure values are given as a function of the distance from the model reprocessing center.

Exposure dose rates from gaseous effluents are a maximum for bones and the thyroid, reaching approximately 0.2 mrem/GW(e)·a in the direct vicinity of the plant. The corresponding whole body dose rate and the dose rate for the skin are about a factor of 4–5 lower. Exposure dose rates due to liquid effluents are again in the range of 0.2 mrem/GW(e)·a for the whole body and the thyroid and roughly a factor of two lower for bones. All exposure dose rates decrease strongly as a function of distance from the reprocessing and waste treatment center.

Fig. 8-4 shows the results for reprocessing Pu-SGR type fuel and LMFBR fuel. In the direct vicinity of the model plants the exposure dose rates due to gaseous effluents in both cases, with some 0.3 to 0.4 mrem/GW(e)·a, are a maximum for the bones and the thyroid. The whole body dose rate and the dose rate for the skin are again about a factor of 6–8 lower. Compared with the UO_2 LEU fuel reprocessing and waste treatment center, the higher exposure dose rates from gaseous emissions for bones are caused by the higher airborne effluents of plutonium and the higher actinides. The exposure levels due to liquid effluents in the direct vicinity of the plants are lower for bones and roughly in the same range for the thyroid and the whole body.

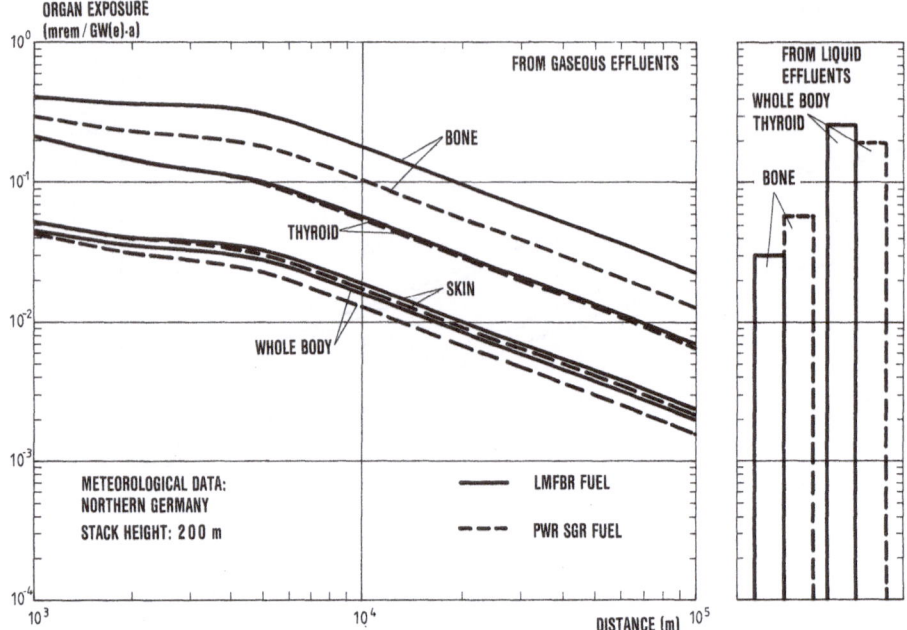

Fig. 8-4. Radiation exposure caused by emissions from reprocessing spent SGR-type PWR fuel and LMFBR fuel (KfK)

Commercial size LWR reprocessing centers with a capacity of 4 te/d of spent UO_2 or SGR-type fuel serving about 30–40 GW(e) of LWR's (see Section 7.2.1) will cause exposure levels a factor of 30–40 higher than shown in Fig. 8-3 and Fig. 8-4, respectively. Similarly, LMFBR reprocessing plants with a capacity of some 250 te/a serving about approx. 10 GW(e) LMFBR's (see Section 7.4.2) would cause exposure levels a factor of 10 higher than indicated in Fig. 8-4. These exposure doses are still below the limits imposed in radiation protection ordinances (see Section 8.1.1.2).

Improvements in iodine retention factors from a factor of 100 to a factor of 5000, as have been proved already in experiments, will reduce thyroid exposures considerably. Improved retention factors of fission products and plutonium aerosols as well as of tritium and noble gases as presented in the two columns on the left in Table 8-6 will additionally reduce the whole body exposure and the bone doses.

Fig. 8-5 is a comparison of bone exposures to gaseous and airborne effluents for the four alternatives of the U/Pu fuel cycle. The bone doses are particularly significant for this comparison, because Sr-90 as well as plutonium and the higher actinides are deposited preferably in bone tissue. There is roughly a factor of two as the difference between the lower (LWR UO_2 fuel) and the upper (Pu burner fuel) curves. The LMFBR type fuel case is very close to the upper curve, whereas the Pu SGR case is found in between. The exposure data from liquid effluents again show differences of about a factor of two. However, in this case the picture is reversed.

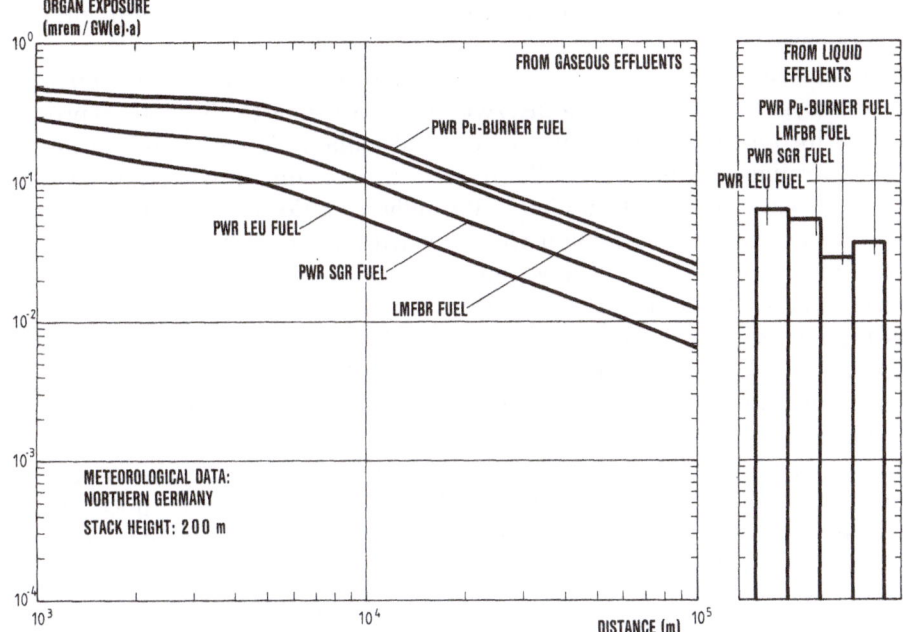

Fig. 8-5. Bone exposures caused by emissions from reprocessing in different plutonium fuel cycle options (KfK)

In the case of gaseous and airborne effluents, the differences are due to different plutonium and higher-actinide aerosol emissions, which amount to 30 mCi/GW(e)·a for the Pu burner case, 20 mCi/GW(e)·a for the LMFBR case, 10 mCi/GW(e)·a for the Pu SGR case and 2 mCi/GW(e)·a for low enriched UO_2 fuel. For the liquid effluents, Sr-90 is the radioisotope with the greatest importance. It is released in amounts of 1.3 mCi/GW(e)·a for the Pu burner case, 1 mCi/GW(e)·a for the LMFBR case, 2.1 mCi/GW(e)·a for the Pu SGR case and 2.4 mCi/GW(e)·a for the case of low enriched UO_2 fuel.

On the basis of this comparison, no significant advantage of one of the four alternatives can be concluded. Only a long term collective dose estimate could probably show differences in the opposed tendencies caused by airborne and liquid effluents.

Similar studies comparing radioactive exposures from the U/Pu fuel cycle with those of the Th/U cycle did not reveal any significant advantage of either fuel cycle alternative.

8.1.3 Long Range Accumulation of Tritium, Krypton-85, and Carbon-14

Unlike the xenon and iodine isotopes (except I-129), which have sufficiently short halflives, tritium, Kr-85 and C-14 may accumulate in the atmosphere over prolonged periods of time. Consequently, the potential severity of radioactive exposures arising from future accumulations of these radioisotopes must be studied in the light of an expanding nuclear power economy including reprocessing plants.

Tritium is produced in the atmosphere as a result of cosmic radiation interacting with oxygen and nitrogen atoms. According to estimates, this natural process provides for an overall inventory of some 50 million Curies of tritium on the earth at all times. Atmospheric thermonuclear weapons tests conducted until the early sixties generated another 1700 million Curies of tritium. From this source, at least some 300 million Curies of tritium will still be present by the year 2000. Studies of USEPA have indicated that a growing worldwide nuclear power economy of a theoretical capacity of 2000 GW(e) by the year 2000, with the necessary fuel cycle plants, would release another 450 million Curies of tritium into the environment. In similar studies of the USEPA it was further assumed that all Kr-85 and C-14 would be released into the atmosphere under the same assumptions of a growing nuclear power economy.

The tritium, Kr-85 and C-14 released was then assumed to distribute and accumulate in the atmosphere of the northern hemisphere, where most nuclear power plants will be located. The tritium, Kr-85 and C-14 would enter the human body on the pathways described in Section 8.1.1. This would lead to an additional radiological exposure per person by the year 2000 of

0.015 mrem/a	to the whole body	from tritium
0.9 mrem/a	to the skin	
0.05 mrem/a	to the lung	from Kr-85
0.025 mrem/a	to the whole body	
0.14 mrem/a	to the whole body	from C-14.

For comparison, it should be recollected that the average whole body dose from natural background radiation in the USA is about 70–220 mrem/a.

Tritium release can be restricted by concentrating tritiated water and storing it in tanks for a sufficiently long time. This reduces the above whole body exposure. A restriction of Kr-85 releases from the fuel cycle to one tenth, as demanded by USEPA (see Section 8.1.1.3), would reduce the radiation exposure by a corresponding margin. Technical processes, such as cryogenic distillation, fluorocarbon absorption and carbon adsorption, have already been developed for this purpose. The Kr-85 separated can be filled in pressurized cylinders or embedded in a metal matrix and then stored for about 150 years. It is possible that, as for Kr-85, a similar release limit will later be imposed for C-14. Such a limitation is technically feasible in connection with the low temperature rectification of Kr-85. C-14 dioxide would then be converted to calcium carbonate and could be stored over long periods of time.

8.2 Risk Assessment of Nuclear Fission Reactors

Besides providing a consistent safety design concept of nuclear fission reactor plants, as outlined for converter and breeder reactors in Chapters 4 and 5, risk assessments of nuclear power systems serve to determine the probable frequency of occurrence of events initiating failures and to assess the damage arising from such faults. Such risk assessments allow comparisons to be made with other energy systems or other technical systems in general. In addition, weak spots in the safety systems of a plant can be detected in this way.

8.2.1 Methods and Procedures

8.2.1.1 General Procedure

The first concepts of risk assessment in nuclear power plants were developed by F.R. Farmer in Great Britain in 1967. The first comprehensive study performed to determine the risk of LWR's by probabilistic methods, the US Reactor Safety Study WASH-1400, was published in 1975. Similar studies were performed later also in other countries, such as the Federal Republic of Germany.

The risk, $R_{m,i}$, resulting from a type i accident initiated by a type m event (e.g., break of a primary coolant pipe) in a reactor plant can be described in a simplified way by this relationship:

$$R_{m,i} = F_{m,i} \cdot D(C_{m,i})$$

where
– $F_{m,i}$ is the annual frequency of occurrence of a type i reactor accident initiated by a type m event,
– $C_{m,i}$ is the amount of radioactive material, expressed in Ci or Bq, released into the atmosphere from the reactor containment during a type i accident initiated by a type m event,
– D is the damage resulting from the atmospheric release of radioactivity. D depends on a number of other environmental parameters, such as atmospheric conditions, transfer factors, population distribution, etc.

The annual frequency, $F_{m,i}$, of occurrence of an accident is determined in detailed probabilistic analyses applying event tree and fault tree methods. In those studies, the failure probability of all relevant components of a safety system is taken into account (Fig. 8-6). In determining the radioactivity release, $C_{m,i}$, the sequence of accident events must be assessed as a function of time in the reactor core, the pressure vessel, and the surrounding containment. This then results in the radioactivity, $C_{m,i}$ (fission products, activation products and actinides), released into the environment from the containment (see Fig. 8-1). Subsequently, meteorological data and models of atmospheric diffusion and aerosol deposition are used to determine the radioactivity concentration and the radiation dose to which individuals in the environment of the plant are exposed, countermeasures being taken into account. Finally, health physics data are used to determine the probability of disability or death as a result of the exposure dose.

A distinction has to be made in this respect between the risk arising to an individual and the risk arising to the whole population. In calculating the individual risk, the probability of disability or death is determined for a single person at any location outside the plant. If the risk to the whole population is determined, the population distribution is used to calculate the total number of disabilities or deaths in the entire surrounding area.

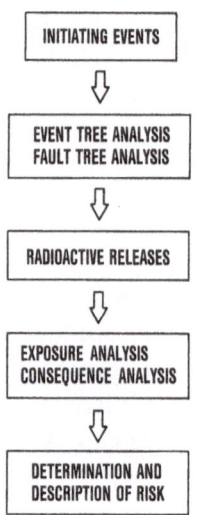

Fig. 8-6. Major steps in a risk study

8.2.1.2 Event Tree Method

An accident sequence is started by an initiating event, e.g., a break of a pipe in the primary coolant system. The safety system of the reactor reacts to this initiating event, and the consequences of the sequence of accident events are controlled, provided that the safety system functions sufficiently well. Only if components of the safety system fail on a major scale, will there be a release of radioactivity. Fig. 8-7 shows a simplified event tree for a loss-of-coolant accident in an LWR. In this case, the accident is initiated by the break of a primary pipe. This pipe rupture is assumed to occur with a frequency of f_m per annum.

The further development of this accident is then mainly determined by the availability of the electricity supply. Failure of the electric power supply to operate the *emergency core cooling system* (ECCS) is assigned the probability of p_1. Since electricity is either available or not, the probability of power being available and the ECCS functioning properly is $(1 - p_1)$. If there is no electricity, the ECCS will not work and the core, after having lost its coolant, will melt down partially or entirely for lack of cooling. In that case, there may well be major releases of radioactivity into the environment as a result of a failure of the containment.

If power is available, the next possible event will be a potential failure of the ECCS, which must be assigned the probability of p_2. The availability of the ECCS is again characterized by $(1 - p_2)$.

If fission products are released in the course of an accident, the fission product removal system mitigates the radioactivity release into the containment. The failure probability of this system is characterized by p_3, its availability by $(1 - p_3)$.

The final barrier against the release of radioactivity is the leak tightness (integrity) of the containment. The probability of this containment function

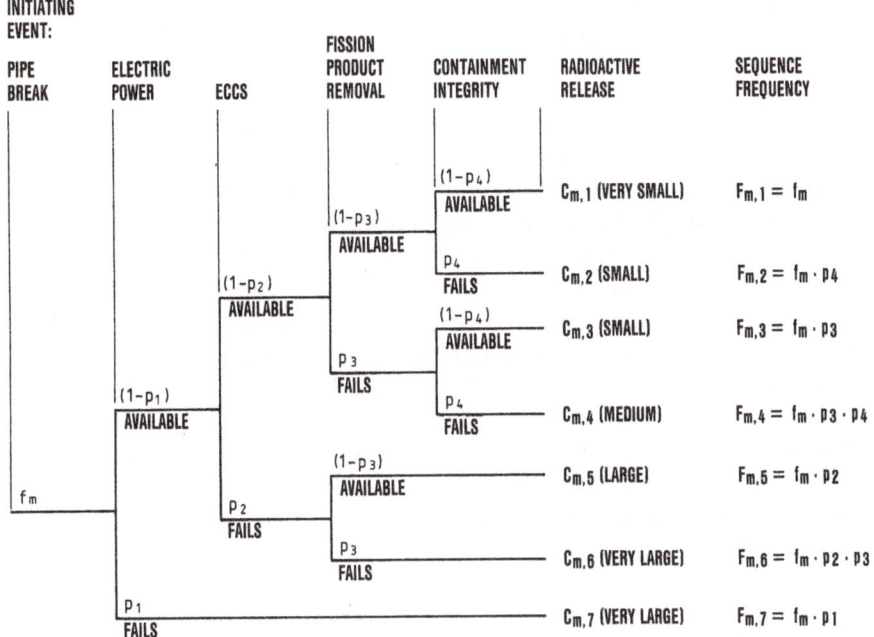

Fig. 8-7. Simplified event tree for a loss-of-coolant accident in a water cooled reactor

failing is called p_4, the availability of that function, $(1-p_4)$. If the containment integrity is preserved, releases of radioactivity can only be slight, but if the containment leaks, radioactivity can escape into the environment, depending on the size of the leak.

From the results of this simplified event tree of Fig. 8-7 it can be seen that radioactivity releases can vary between very small and very large releases, depending on the level at which the safety systems fail.

Since the individual components of the safety system are characterized by their high availabilities and, consequently, very low failure probabilities (p_1, p_2, p_3, $p_4 \ll 1$), the probability of the availability of $(1-p_1)$ etc. of these safety components can each be assumed to be approximately equal to 1. Consequently, the sequence frequency at the upper end of the branching of the event tree of Fig. 8-7 is

$$F_{m,1} = f_m(1-p_1)\,(1-p_2)\,(1-p_3)\,(1-p_4) \approx f_m.$$

The radioactive release caused in this case, $C_{m,1}$, is negligible. On the other hand, the radioactive releases, $C_{m,6}$ or $C_{m,7}$, as a consequence of a failure of the electricity supply followed by a failure of the ECCS and of the integrity of the containment would be very large, because the core would melt and a large fraction of the radioactive inventory would be released from the containment. The frequency of occurrence, however, of this maximum accident is extremely low, amounting to $F_{m,6} = f_m \cdot p_2 \cdot p_3$ and $F_{m,7} = f_m \cdot p_1$, respectively.

In a detailed event tree analysis, many more details must be considered, such as the individual functions of the ECCS, etc. Interdependencies of the different events may lead to systemic consequential failures and to the elimination of branches in an event tree.

8.2.1.3 Fault Tree Analysis

The fault tree analysis approach is used for numerical assessment of the failure probabilities of larger units of the safety system. It breaks these larger systems down into single components, concluding about the failure probability of a larger unit from the failure probabilities of such individual components by taking into account the way in which the logical functions of the single components are interrelated. Often, fault trees must be developed to such fine detail that available data on single equipment components or human error can be applied from experience. Uncertainties in reliability data are taken into account by entering not only single values, but distribution functions for the failure probabilities of single components. For other components, such as emergency power diesel systems, statistical data directly available from experience are applied. When determining the failure rate of the pressure vessel, methods of probabilistic fracture mechanics must be used in addition.

8.2.2 Releases of Fission Products from a Reactor Building Following a Core Meltdown Accident

8.2.2.1 Initiating Events

Initiating events controlled by the safety system will not contribute to risk. Accordingly, major contributions to the overall risk must be expected to arise only from those events in which the reactor core will melt down partially or completely because of an extensive failure of the safety systems. Event tree studies show that the occurrence of a major leak in a main coolant pipe followed by a failure of the respective safety systems (emergency cooling systems and afterheat removal systems) will cause the reactor core to melt down; within a few hours the molten core can even penetrate the reactor pressure vessel. In an early superheated phase of the reactor core, hydrogen will be generated in a reaction between water and the zirconium in the fuel claddings.

After having penetrated through the reactor pressure vessel, the hot core material will contact the concrete foundation slab of the reactor building and gradually melt into the concrete. This will cause water bound in the concrete to be released and react with the melt, which will generate hydrogen. Depending on the type of concrete used, also CO_2 may be released. For the further sequence of accident events it was assumed in a pessimistic estimate in the US and German risk studies that the molten core contacts and evaporates the sump water, thus increasing the pressure in the containment.

8.2.2.2 Failure of the Containment

A number of penetrations through the containment building for locks, pipes and cables may develop leaks with a certain failure rate. In this case, radioactivity could directly escape to the outside. If, on the other hand, the containment is assumed to remain tight, i.e., preserve its integrity, the core meltdown accident described above would generate vapor, H_2, CO and CO_2, and raise the pressure in the containment so that the permissible design pressure of the outer containment would be exceeded. After failure of containment integrity, radioactivity would be released into the environment.

In the US Reactor Safety Study, WASH-1400, and the German Risk Study, also the case of a potential steam explosion resulting from a contact between molten hot core material and water was discussed. This was assumed to occur with a certain probability in the bottom part of the reactor pressure vessel or after meltthrough of the concrete structures and subsequent contact with sump water. More recent considerations of the contact modes between molten core material and water and appropriate experimental results, however, show that steam explosions under the conditions of this core meltdown accident may be impossible. Estimates have shown that the containment would not fail in case of a steam explosion. In addition, it could be proved that ignition and combustion of hydrogen produced during a core meltdown would not lead to a dangerous pressure increase and the containment would preserve its integrity.

8.2.2.3 Releases of Radioactivity

Radioactivity may be released from the reactor core in the following events:
– In cladding tube failures the gaseous and highly volatile fission products are released.
– As the fuel is heated to melting temperature and melts, fission products with lower melting points will be released.
– During interaction of the melt with the concrete, aerosols are generated and, in case of a steam explosion, additional release mechanisms would be effective.

In a pipe rupture of the primary circuit, or if the reactor pressure vessel has been penetrated by the molten core, the gaseous and volatile fission products will enter the containment. They will be retained there by active removal (e.g. spray) systems and by diffusion, coagulation, condensation, sedimentation and thermophoretic processes of the aerosols. Radioactive decay makes the retention by the containment more effective, the longer its integrity can be ensured. More recent studies show that the time after which the maximum pressure would be exceeded and the containment fails would be more like 5–12 days (depending upon the concrete composition) than the 1 day assumed in the German Risk Study. Within such a time period of about 5 days or more, the concentration of airborne aerosols decreases by some four orders of magnitude.

8.2.2.4 External Events

To determine properly the frequency of occurrence of initiating events in the reactor facility, external events must be taken into account. This applies, for instance, to the frequencies of occurrence and the intensities of earthquakes and their repercussions upon important components determining the structural stability of the reactor building. For the site of a nuclear reactor plant, statistical data about past earthquake events are used for this purpose and also the geological structure of the site is taken into account. In addition, frequencies of floods, tornados, strokes of lightning, airplane crashes and explosion shock waves from nearby industrial plants or transport routes (rivers, pipelines, etc.) are determined and the potential release of radioactivity is assessed.

8.2.3 Accident Consequence Model and Human Exposure

On the basis of radioactivity releases the propagation of the air plume is simulated, taking into account the thermal energy released with the exhaust air plume. This determines the altitude of the rising radioactive cloud. This cloud is carried in the atmosphere at the velocity of the wind and is distributed normal to the transport direction. The activity concentration decreases in this process. In addition to the direction of the wind, weather conditions and precipitations play a major role. On the whole, meteorological parameters determine the area in the environment of a nuclear facility and the number of persons affected by an accident.

On the basis of location and time dependent radioactivity concentrations in the air and on the ground within the area involved, the radiation exposure of the public concerned is calculated. The following exposure pathways are taken into account:
– external exposure due to the radioactive cloud passing by,
– external exposure due to the radioactivity deposited on the ground,
– internal exposure due to the inhalation of airborne radionuclides, also after resuspension,
– internal exposure due to the activity ingested with food.

In damage assessment, emergency measures like seeking shelter in houses, evacuation and decontamination, etc., are taken into account for the environment of a plant. In the German Risk Study the following types of damage are calculated:
– somatic early damage, i.e., the number of deaths from acute radiation disease,
– somatic late damage, i.e., the number of deaths from leukemia and other cancers.

The genetic burden is expressed in terms of the collective genetically significant dose.

8.2.4 Results of Reactor Safety Studies

8.2.4.1 Results of Event Tree and Fault Tree Analysis

The methods and analyses described above were applied to both PWR's and BWR's in the US Reactor Safety Study. In the German Risk Study on Nuclear Power Plants, the PWR designed and built by Kraftwerk Union, as described in Section 4.1.1, was investigated. While the US study covers 100 reactor plants on 68 different sites, the German study applies to 25 reactor plants on 19 different sites.

Table 8-9 lists the main findings of the event tree and fault tree analyses in the German study. In analyzing all possible initiating events, mainly two groups of accidents were found to result in core damage:

Table 8-9. *Summary of the results of the event tree analysis for core meltdown (German Risk Study)*

Accident initiating event	Frequency of occurence of the initiating event per reactor year (f_m)	Failure probability of required safety functions (P_{FS})	Frequency of occurence of core meltdown per reactor year ($F_m = f_m \times P_{FS}$)
Loss of coolant accident			
– large	2.7×10^{-4}	1.7×10^{-3}	5×10^{-7}
– medium	8×10^{-4}	2.3×10^{-3}	2×10^{-6}
– small	2.7×10^{-3}	2.1×10^{-2}	5.7×10^{-5}
Loss of offsite power (emergency power case)	1×10^{-1}	1.3×10^{-4}	1.3×10^{-5}
Loss of main feedwater supply	8×10^{-1}	4×10^{-6}	3×10^{-6}
Emergency power case with small leak at pressurizer	2.7×10^{-4}	2.6×10^{-2}	7×10^{-6}
Other transients with small leak at pressurizer	1×10^{-3}	2×10^{-3}	2×10^{-6}
Anticipated transient event without scram	3×10^{-5}	3×10^{-2}	1×10^{-6}

$$\Sigma\ 8.55 \times 10^{-5}$$

– loss-of-coolant accidents initiated by a leak or break in the reactor coolant system,
– transients leading to an imbalance between the heat generated in the core and the heat removed from the core.

The data in Table 8-9 result from a detailed study of some seventy different accident sequences. A loss of main coolant flow through a small leak in a coolant pipe dominates all other contributions. This reflects the fact that small leaks can occur more frequently than large ones. The second largest contribution comes from transient accident sequences, where the loss of the offsite power supply plays an important role. The contribution from a large leak in a coolant pipe is relatively small, because the safety systems have been optimized already to cope with this accident. The sum total of all contributions adds up to an overall core meltdown frequency of about 9×10^{-5} per reactor year.

The results in Table 8-9 can be combined with results obtained on accident consequences from the core meltdown and containment impact analyses to assess the probability of occurrence of a specific type of accident and the respective release of radioactivity from the reactor containment. The analysis can be simplified by grouping into release categories accident sequences in the same containment failure mode. This procedure is presented in Table 8-10. In addition to the frequencies of occurrence, also the schedule of radioactivity release from

Table 8-10. *Characteristic quantities of release categories (German Risk Study)*

Release category	Description	Time of release (h)	Duration of release (h)	Release height (m)	Energy released (10^6 KJ/h)	Frequency of release (per year)
1	Core meltdown with steam explosion	1	1	30	540	2×10^{-6}
2	Core meltdown, large leak in containment (300 mm diameter)	1	3	10	15	6×10^{-7}
3	Core meltdown, moderate leak in containment (80 mm diameter)	2	3	10	1	6×10^{-7}
4	Core meltdown, small leak in containment (25 mm diameter)	2	3	10	–	3×10^{-6}
5[a]	Core meltdown, overpressurization failure, failure of the stack filters	0 1 25	1 1 1	10 10 10	– – 200	2×10^{-5}
6[a]	Core meltdown, overpressurization failure	0 1 25	1 1 1	100 100 10	– – 200	7×10^{-5}
7	Loss-of-coolant accident, large leak in containment	0	1	10	9	1×10^{-4}
8	Controlled loss-of-coolant accident	0	6	100	–	1×10^{-3}

[a] Since the release takes place over a longer time period, the release fractions are given separately for three time intervals.

the containment after the onset of an accident and the fractions of fission products released are indicated. These fission product fractions refer to the total radioactive inventory of the PWR core as shown, e.g., in Table 2-4.

Release category 1 indicates the frequencies of occurrence and activity releases for a core meltdown accident followed by a steam explosion. This release category includes the highest activity releases which would occur one hour after the initiating event and the subsequent core meltdown. As mentioned above, more

Released fraction of core inventory

Xe-Kr	I(org)	I_2-Br	Cs-Pb	Te-Sb	Ba-Sr	Ru[b]	La[c]
1.0	7.0×10^{-3}	7.9×10^{-1}	5.0×10^{-1}	3.5×10^{-1}	6.7×10^{-2}	3.8×10^{-1}	2.6×10^{-3}
1.0	7.0×10^{-3}	4.0×10^{-1}	2.9×10^{-1}	1.9×10^{-1}	3.2×10^{-2}	1.7×10^{-2}	2.6×10^{-3}
1.0	7.0×10^{-3}	6.3×10^{-2}	4.4×10^{-2}	4.0×10^{-2}	4.9×10^{-3}	3.3×10^{-3}	5.2×10^{-4}
1.0	7.0×10^{-3}	1.5×10^{-2}	5.1×10^{-3}	5.0×10^{-3}	5.7×10^{-4}	4.0×10^{-4}	6.5×10^{-5}
2.0×10^{-5}	1.8×10^{-7}	1.8×10^{-5}	4.7×10^{-5}	3.6×10^{-7}	5.5×10^{-9}	–	–
2.3×10^{-2}	1.6×10^{-4}	9.6×10^{-4}	6.7×10^{-4}	6.7×10^{-4}	8.0×10^{-5}	5.5×10^{-5}	8.8×10^{-6}
9.8×10^{-1}	6.8×10^{-3}	9.6×10^{-3}	4.5×10^{-4}	7.7×10^{-4}	4.7×10^{-5}	5.3×10^{-5}	9.5×10^{-6}
2.0×10^{-5}	1.8×10^{-9}	1.8×10^{-8}	4.7×10^{-8}	3.6×10^{-10}	5.5×10^{-12}	–	–
2.3×10^{-2}	1.6×10^{-6}	9.6×10^{-7}	6.7×10^{-7}	6.7×10^{-7}	8.0×10^{-8}	5.5×10^{-8}	8.8×10^{-9}
9.8×10^{-1}	6.8×10^{-3}	9.6×10^{-3}	4.5×10^{-4}	7.7×10^{-4}	4.7×10^{-5}	5.3×10^{-5}	9.5×10^{-6}
1.7×10^{-2}	3.7×10^{-5}	5.3×10^{-3}	1.3×10^{-2}	2.5×10^{-5}	2.5×10^{-7}	0.	0.
4.6×10^{-4}	1.0×10^{-8}	1.2×10^{-8}	2.1×10^{-8}	4.1×10^{-11}	4.1×10^{-13}	0.	0.

[b] Includes Ru, Rh, Co, Mo, Tc.
[c] Includes Y, La, Zr, Nb, Ce, Pr, Nd, Np, Pu, Am, Cm.

recent studies have shown such steam explosions to be impossible. However, they have been included within the US and German risk studies as pessimistic assumptions with a mean probability of occurrence of $2 - 10^{-2}$.

Release categories 2 through 4 comprise core meltdown accidents with failure of the integrity of the containment. In these cases, openings in the outer containment of 25–300 mm equivalent diameter are assumed. The radioactivity release in these cases is assumed to be considerably higher than in release categories

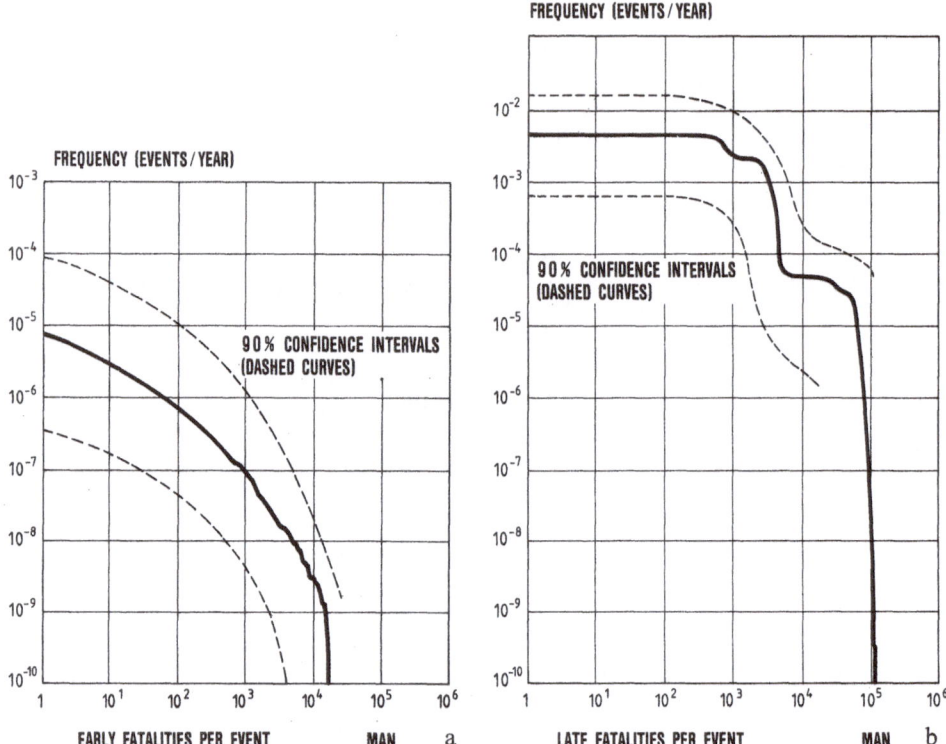

Fig. 8-8a and b. Complementary cumulative distribution function of early and late fatalities for 25 reactor units (German Risk Study)

5 and 6 representing core meltdown with late containment overpressure failure. Release categories 7 and 8 cover loss-of-coolant accidents controlled by the emergency cooling systems. Since, in these cases, the reactor core is cooled sufficiently by the emergency core cooling systems, the fuel element claddings will be damaged only partially and the reactor core will not melt. In release category 7, a large leak (300 mm equivalent diameter) was assumed to exist in the reactor containment. In release category 8, only a minor leakage was assumed, which corresponds to ten times the design value for the reactor containment. In this case, only minimum amounts of radioactivity are released.

8.2.4.2 Results of Accident Consequence Models

8.2.4.2.1 The German Risk Study. Fig. 8-8 shows the complementary, cumulative frequency distribution of early and late fatalities, which could be caused by radiation exposure after accidental radioactivity release. These results apply to the German Risk Study, which covers 25 PWR's of 1300 MW(e), as described in Section 4.1.1.

The results show that early fatalities caused by a core meltdown accident in 25 reactor plants would occur only with a frequency of occurrence of approx. 7×10^{-6} per year. Referred to one year of reactor operation, this frequency of occurrence is 1/25, i.e., approx. 3×10^{-7} per reactor year. If this value is compared with the frequency of occurrence of about 9×10^{-5} per year of core meltdown accidents (sum total of probabilities of occurrence of release categories 1 to 6 in Table 8-10), it is seen that early fatalities will occur in less than 1% of the core meltdown accidents. On the one hand, this can be explained by the fact that only rapid releases after core meltdown (release categories 1 through 4 of Table 8-10) will release sufficiently high radioactivities. On the other hand, only under certain weather conditions (e.g., precipitations) will the concentrations building up be high enough to cause early fatalities.

A large number of early fatalities will occur if major releases are encountered on sites with high population densities, the wind blows in the direction of the sector with the highest population density, and rain falls in the immediate vicinity, thus creating high concentration levels on the ground. The study shows that the largest number of approximately 16,600 early deaths (release category 1) could only occur with a frequency of occurrence of 5×10^{-10} per year. If release category 1 (steam explosion) were not taken into account, the maximum damage would amount to approx. 5700 early deaths with a frequency of occurrence of 1.4×10^{-10} per year.

In the German Risk Study, an attempt was made to quantify the reliability of these results as shown in Fig. 8-8. The dotted lines indicate the calculated 90% confidence intervals of the results.

Late fatalities can occur at all dose levels since a linear dose rate independent dose response function without threshold value is assumed according to ICRP (1977). The occurrence of late fatalities is therefore not restricted to only the immediate vicinity of the reactor plant, as in case of early fatalities. Fig. 8-8b indicates the number of late fatalities and the corresponding frequencies of such events. The maximum number of late fatalities amounts to about 107,000 deaths for release category 1 (core meltdown followed by a steam explosion). If the steam explosion were excluded, the maximum number amounted to some 35,000 deaths. However, it must be mentioned that a considerable fraction (between 16 and 90%, depending on the release category) of the late fatalities is due to low radiation exposures of <5 rem, which would correspond to the exposure resulting from 50 years of natural background radiation (see Section 8.1.1). The late fatalities shown in Fig. 8-8b would occur over a period of several decades.

In Table 8-11, the expectation values of collective damage are listed for the different release categories 1 through 8. The expectation values of collective damage are defined as the average annualized number of early or late fatalities. The total expectation value or overall risk for the 25 reactor plants assumed in Germany is 5.1×10^{-4} per year for early fatalities and 9 per year for late fatalities. Release categories 5 through 8 do not contribute to the expectation value of early fatalities. On the other side release categories 1 and 7 are the main contributors to the expectation value for late fatalities.

If, for comparison, the dose-response function applied in the German Risk Study as a basis for assessing late fatalities is also applied to natural background radiation, the corresponding expectation value would be 760 fatalities per year in Germany. The collective risk calculated for late fatalities due to reactor

Table 8-11. *Expectation values of collective damage (early and late fatalities) and of collective genetically significant doses corresponding to 25 reactor units in Germany (German Risk Study)*

Release category	Expectation value of collective damage		Expectation value of collective genetically significant dose, (man-rem/a)
	Early fatalities (per year)	Late fatalities (per year)	
1	4.0×10^{-4}	2.2	1.1×10^4
2	1.0×10^{-4}	2.8×10^{-1}	1.6×10^3
3	2.5×10^{-6}	6.4×10^{-2}	3.9×10^2
4	5.7×10^{-6}	8.9×10^{-2}	5.3×10^2
5	0	2.9×10^{-1}	1.2×10^3
6	0	7.3×10^{-1}	1.9×10^3
7	0	5.4	4.0×10^4
8	0	$\ll 1$	3.0
Total	5.1×10^{-4}	9.0	5.6×10^4

Table 8-12. *Consequences of reactor accidents as a function of the median frequency of their occurrence (WASH-1400)*

Frenquency (events per reactor year)[a]	Early fatalities	Cases of early illness	Latent cancers	Thyroid nodules	Genetic effects
5×10^{-5} [b]	–	1	–	–	–
10^{-6}	–	300	170	1,400	25
10^{-2}	110	3,000	460	3,500	60
10^{-8}	900	14,000	860	6,000	110
10^{-9}	3,300	45,000	1,500	8,000	170

[a] These are median probabilities, with uncertainties of a factor of 5 in either direction.
[b] 5×10^{-5} per year is the estimated median frequency of meltdown, per reactor.

accidents thus is about two orders of magnitude lower than the corresponding value expected from exposure to natural radiation.

The collective genetically significant doses as defined by ICRP are listed in the last column of Table 8-11. They total 5.6×10^4 man-rem/a. For comparison, this number must be seen against the exposure to natural radiation, which amounts to 6.1×10^6 man-rem/a. This again is two orders of magnitude lower.

8.2.4.2.2 The US Reactor Safety Study. The US Reactor Safety Study, WASH-1400, was performed for 100 reactor plants (PWR's and BWR's) in the United States. It was published in 1975, thus preceding the German Risk Study. Table 8-12 summarizes its most important findings. All frequencies of occurrence have already been referred to one reactor year. Compared with the results of the German Risk Study, and aside from slightly different safety designs of German LWR's, it was mainly the meteorological data, the population density, the purely linear dose-risk relationship as well as protective measures and countermeasures, which differed in the two studies.

The basic statements included in the findings in these two studies are approximately identical up to probabilities of occurrence of 10^{-9} per reactor year. The slightly higher figures for early fatalities in the German Risk Study must be seen in the light of the much lower probabilities of occurrence of approximately 10^{-11}. In addition, the values in the US study listed in Table 8-12 are median values, whereas the results of the German study discussed above are mean values (mean values are about a factor of 2 higher than median values).

The frequency of core meltdown accidents was determined to be 5×10^{-5} per reactor year in the US Reactor Safety Study. The largest number of early fatalities shown is approx. 3300, with a probability of occurrence of 10^{-9} per reactor year. The cases of early illness vary between 300, with a frequency of occurrence of 10^{-6} per reactor year, and 45,000, with a frequency of occurrence of 10^{-9} per reactor year. Corresponding values are shown in Table 8-12 also for cases of latent cancer and thyroid nodules.

The US Reactor Safety Study was reviewed by a number of organizations. Criticism was leveled mainly against the treatment of common mode failures. Protective systems are usually arranged such that a failure of one system is backed up by the operation of another (principle of redundancy and diversity). If the event that caused one protective system to fail also caused the other to fail, that would be a common mode failure. Data of common mode failures are very difficult to validate. Other points of criticism referred to the possibility that the uncertainties and error bands in accident frequencies and consequences could be larger than reported. Risks contributed by effects of war or sabotage also were not taken into account.

Risk studies of this kind offer valuable indications of those parts of reactor safety concepts which can still be improved. Moreover, valuable findings can be drawn from the results on accident consequences with respect to planning for emergency measures.

8.2.4.2.3 More Recent Improvements in Risk Studies. As mentioned in the previous sections, rather simplified and pessimistic assumptions were made in the German Risk Study and the US Reactor Safety Study about the sequence of accident events in which the molten core penetrates the concrete and finally contacts the sump water, causing it to evaporate. These assumptions lead to a relatively early loss of containment integrity and, consequently, to overestimating the consequences and risks. More recent theoretical studies and preliminary experimental results indicate that
– large steam explosions leading to early containment failure seem to be impossible,
– the penetration of the molten core into the concrete does not necessarily lead to contacts with the sump water. A loss of containment integrity occurs only after time periods of approximately 5 to 12 days,

– for reasons of aerosol physics, a considerable percentage of radioactive aerosols will have settled within the containment under fog or rainlike conditions and shortlived radioisotopes will have decayed away by the time containment failure occurs after a few days. The release of radioactivity into the environment then decreases by several orders of magnitude.

If these preliminary results can be confirmed by larger scale experiments, the consequences and risks of core meltdown accidents as described in the previous sections (see Fig. 8-8) can probably be decreased by about two orders of magnitude.

8.2.5 Risk Studies of Other Types of Reactors

Risk studies on other types of reactors have not yet been carried out to the same level of detail as on LWR's. Only on LMFBR's and HTR's, the same methods were employed in similar studies.

The results of the LMFBR risk studies are not too different from those of the LWR risk studies, aside from the fact that the LMFBR risk studies had been conducted only on single prototype LMFBR's (CRBR, USA, and SNR 300, Germany). Compared with LWR's, LMFBR's have emergency core cooling systems that inherently dissipate the afterheat by natural convection. However, greater attention must be devoted to potential failures of the shutdown systems following, e.g., pump failure, because such accidents would lead to core meltdown. Moreover, the leaktightness of the containment after a core meltdown accident is of great importance.

Fig. 8-9a shows the most important results of the Risk Study on the German SNR 300 prototype LMFBR. Unlike the German LWR Risk Study, this one contains no early fatalities, since no lethal dose levels were found to exist in the inhabited area around the SNR 300. Consequently, Fig. 8-9a only indicates late fatalities and the corresponding frequencies of such events. As in the LWR risk studies, a linear dose rate independent dose response function without threshold values was assumed according to ICRP (1977). The largest contribution to the late fatalities is not caused by a core disruptive accident, but by a cooling failure in the sodium cooled spent fuel store.

The overall frequency of core meltdown (disruption) events is as low as approximately 2×10^{-6}/a. In seventy percent of all core meltdown events the reactor vessel is not destroyed. Only in less than one percent of the cases will there be rapid destruction of the vessel and damage to the inner containment structure. The relatively low frequency of core meltdown cases in the SNR 300 is mainly due to the high reliability of the two diverse shutdown systems.

However, when comparing the results of the SNR 300 Risk Study with those of the LWR risk studies (see Section 8.2.4.2) it should be taken into account that the SNR 300 findings include greater uncertainties because of the prototype character of the plant.

Risk studies on HTR's were performed in the Federal Republic of Germany on an HTR-1160 conceptual reactor design with block type fuel elements. For this type of reactor the main risk is dominated by accidents which are initiated by a failure of electrical power supply followed by temperature transients with

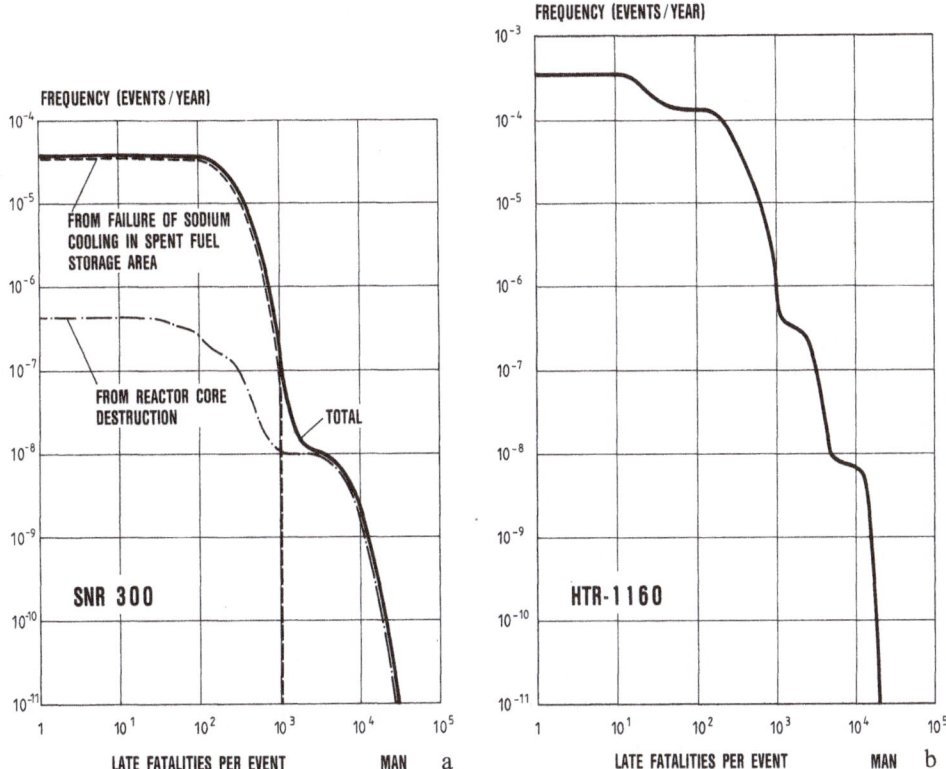

Fig. 8-9a and b. Complementary cumulative distribution function of late fatalities for the SNR 300 LMFBR demonstration plant and an HTR-1160 conceptual reactor design (GRS, KFA Jülich, KfK)

failure of the integrity of the containment after about four days. The release of radioactivity in these cases is somewhat smaller than the results described in Section 8.2.4 for the LWR risk study. As in the LMFBR case no early fatalities were found in the inhabited area around the plant which was assumed to be located at Schmehausen, the site of the German prototype THTR 300. Fig. 8–9b indicates late fatalities and the corresponding frequencies of accidents for the HTR-1160 risk study. However, as in the LMFBR case it should be taken into account that these findings include greater uncertainties than the LWR risk studies because of the prototype character of HTR plants.

8.2.6 Risk Studies of Fuel Cycle Plants

Besides risk studies of nuclear power plants, studies must also be performed on the whole fuel cycle for a comprehensive assessment to be made of the risk associated with nuclear energy. Until now, only the LWR U/Pu fuel cycle has been under development on an industrial scale. Accordingly, preliminary risk studies have so far been carried out on this LWR fuel cycle.

The only study to cover the entire LWR fuel cycle by the same methodology as the US Reactor Safety Study was published by EPRI in the USA in 1979.

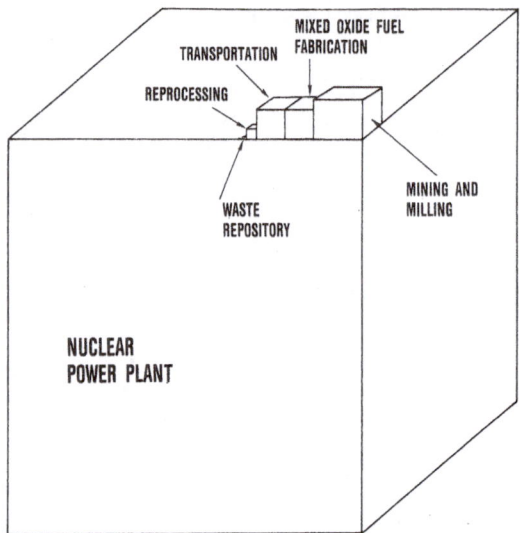

Fig. 8-10. Radiological risk of different parts of the fuel cycle (EPRI)

The preliminary result of that study is that the risk attaching to the fuel cycle is only approximately 1% of that associated with the reactor plant itself.

Fig. 8-10 shows a comparison of risks associated with the reactor plant and other parts of the fuel cycle, the volumes of the cubes representing a measure of comparable risks.

The technical reasons underlying the lower nuclear risk of the fuel cycle compared with the LWR plant are mainly these:
– In certain parts of the fuel cycle (mining, milling, enrichment, fuel fabrication), there are no radioactive fission products.
– In other parts of the fuel cycle (reprocessing, waste disposal, transport), the radioactivity of fission products has already decayed by a sizable margin.
– Nowhere in the fuel cycle, except in the reactor core, has the nuclear fuel a power density high enough to cause potential meltdown.

In the steps of mining, milling and purification of uranium ores, no particular radiological accidents can occur. However, the release of gaseous radon, dust of ores, and ore tailings must be taken into account. This aspect is dealt with in Section 8.1.2. In the process of U-235 enrichment, no radiological accidents must be expected either.

In the reprocessing and mixed oxide fuel refabrication plants, the EPRI Study applied the methods of event tree and fault tree analyses to recognize the dominant accident categories listed in Table 8-13, the consequences of which were investigated in particular.

For shipments of irradiated fuel elements, plutonium oxide or radioactive waste to a waste repository, the EPRI Study mainly took into account accidents occurring in the categories of loss of neutron shield, and loss of confinement with fuel damage and fire.

The risk analysis for the waste repository also included transport accidents, elevator drop, final filter failure, and airplane crash into the waste storage area.

Table 8-13. *Categories and consequences of accidents in the nuclear fuel cycle (EPRI)*

Reprocessing plant	MOX fuel refabrication plant
• Loss of water in fuel storage pool	• Hydrogen explosion in sintering furnace
• Explosion and fire in a solvent treatment ion exchange bed	• Ion exchange resin fire
	• Dissolver explosion in scrap recovery
• Criticality accident in processing cell	• Loaded final filter failure
	• Criticality accident
• Hydrogen explosion in a HLW tank	• Plutonium shipping container damage
• Fire in LLW storage	• Earthquake and tornado in excess of design basis
• Drop of fuel assembly	
• Explosion in HLW calciner	• Aircraft crash into head end area
• Failure of krypton storage cylinder	

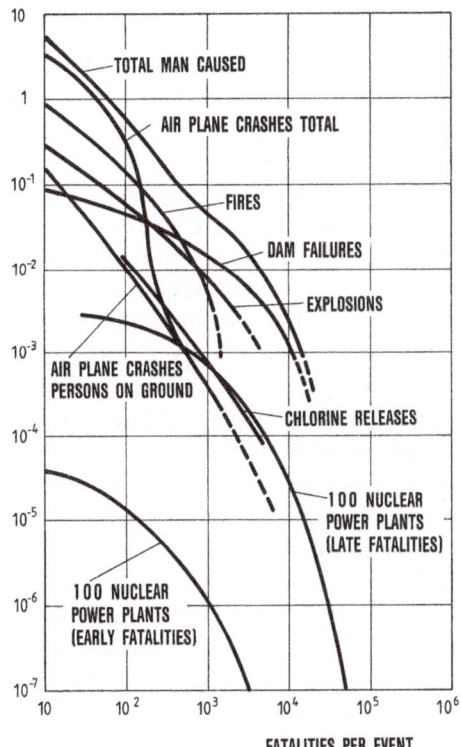

FREQUENCY (EVENTS/YEAR)

TOTAL MAN CAUSED

AIR PLANE CRASHES TOTAL

FIRES

DAM FAILURES

EXPLOSIONS

AIR PLANE CRASHES PERSONS ON GROUND

CHLORINE RELEASES

100 NUCLEAR POWER PLANTS (LATE FATALITIES)

100 NUCLEAR POWER PLANTS (EARLY FATALITIES)

FATALITIES PER EVENT

Fig. 8-11. Complementary cumulative distribution functions of man caused event fatalities compared with early and late fatalities for 100 reactor units (WASH-1400)

8.2.7 Comparison with Risks of Other Technical Systems

Besides determining the actual risk resulting from the presence of a nuclear power plant, the question arises for the level of risk that is acceptable to the public. Compared with other technical systems in an industrial society, risk studies for nuclear reactors indicate relatively large damage with extremely low probabilities of occurrence. This is similar to the risk arising from large dam failures or accidents with chlorine chemical plants and leads to the principal problem that single large accidents appear to be less tolerable to society than many smaller accidents. One example of this attitude is the attention given to single train failures or large aircraft crashes as compared to the larger number of deaths from automobile accidents.

Fig. 8-11 compares the risk of early and late fatalities resulting from 100 LWR's in the USA (WASH-1400) with the risks resulting from man-made events in other technical systems. The figure shows that the risk of early fatalities caused by large, but extremely improbable, accidents in nuclear reactors is very much lower than any other risk caused by other industrial plants.

8.3 The Risk of Nuclear Proliferation and Possibilities of Its Mitigation

8.3.1 History

With the peaceful uses of nuclear energy, nuclear materials, know-how and facilities are built up and disseminated all over the world which, in principle, makes it possible for individual states to develop nuclear weapons. A growing spread of nuclear weapons would add to the hazard of acts of nuclear war. This is a risk inherent in the exploitation of nuclear energy. However, nuclear weapons can be developed and manufactured without the peaceful exploitation of nuclear energy. This has been borne out by historical developments so far in all nuclear weapon countries: the USA, USSR, UK, France and China. Accordingly, the proliferation of nuclear weapons cannot simply be prevented by restrictions on the peaceful uses of nuclear energy.

The inherent proliferation risk in the use of nuclear energy was recognized at the very beginning of the development of nuclear power, and a number of proposals have been made and measures taken in the course of time to prevent proliferation. The period up until 1953 can be regarded as a phase of complete classification of any kind of utilization of nuclear power. As early as in 1945/46, the idea was being ventilated in the USA to make the peaceful utilization of nuclear power accessible to other states while, at the same time, preventing the proliferation of nuclear weapons. The proposal contained in the so-called Acheson-Lilienthal report provided for the establishment of an "international atomic development authority," which was to manage or possess all nuclear activities, i.e., enjoy an international monopoly in the field of nuclear power utilization. In 1946, the USA submitted a proposal to the Atomic Energy

Commission of the United Nations. That proposal, which became known as the "Baruch Plan," failed because it called for too far-reaching a surrender of national sovereignty, and so classification was maintained.

The worldwide utilization of nuclear power began with the "Atoms for Peace" program announced by US President Eisenhower before the General Assembly of the United Nations in December 1953, under which a promotion of nuclear power utilization was planned in conjunction with control measures. One major constituent part of the plan was the establishment of an "*I*nternational *A*tomic *E*nergy *A*gency (IAEA)," which was to promote and, at the same time, monitor all international cooperation in the field of nuclear technology. After a series of negotiations, the IAEA Statute was submitted for signature in October 1956. Article II of that Statute reads, inter alia: "The Agency ensures, so far as it is able, that assistance provided by it or at its request or under its supervision or control is not used in such a way as to further any military purpose" (IAEA Statute).

Controls were defined in agreements between IAEA and the countries it supported. The guideline used was the INFCIRC/66 document, "The Agency's Safeguards System." It provided for controls of plants and materials. A system of records, reports and inspections was created, IAEA only verifying compliance with agreements by the signatory states, i.e., to not abuse nuclear power for military purposes.

That system had been designed for the surveillance of single small reactor plants below 100 MW(e) power output. It was found to be inadequate for a quickly expanding commercial nuclear technology with reactors of several 100 MW(e) power and facilities of the fuel cycle. It had been drafted in terms too general and too vague and left a wide scope for inspection activities with almost complete arbitrariness in assessing whether "compliance" existed or not. On the other hand, it was felt that safeguards against diversions of fissile material for non-peaceful uses were insufficient. For it remained at the discretion of countries to build and operate nuclear facilities without IAEA controls. As a consequence, the Treaty of Tlatelolco was concluded in 1967 for the Latin American countries, and the *N*on-*P*roliferation *T*reaty (NPT) was negotiated and signed in 1968. It entered into force in May 1970. This is a summary of its contents:

Each non-nuclear weapon state that becomes party to the NPT binds itself not to acquire nuclear weapons or other nuclear explosives (Article II). It also binds itself to conclude an agreement with IAEA for the application of safeguards to all its peaceful nuclear activities with a view to verifying the fulfillment of its obligations under the treaty (Article III). In return, the treaty recognizes the right of all parties to participate in the fullest possible exchange of equipment, materials, and scientific and technological information for the peaceful uses of nuclear energy; in other words, all parties are guaranteed full access to peaceful nuclear technology (Article IV). The parties also undertake to pursue negotiations in good faith towards nuclear disarmament (Article VI) and reaffirm their determination to achieve the discontinuance of all tests of nuclear weapons (Preamble); these latter commitments apply principally to the nuclear weapon states themselves.

The safeguards required under the NPT shall be applied to all source and special fissionable material in all peaceful nuclear activities within the territory of such state, under its jurisdiction, or carried out under its control anywhere.

The structure and the contents of a verification agreement are contained in a recommendation by IAEA, which was to constitute the basis of negotiations, but in fact represents the contents of all agreements. This was documented in INFCIRC/153.

8.3.2 The IAEA Safeguards System

While the older regulations provided for the surveillance of plants and nuclear materials, the system proposed in INFCIRC/153 is concentrated on the surveillance of nuclear material: "The objective of safeguards is the timely detection of diversion of significant quantities of nuclear material from peaceful nuclear activities to the manufacture of nuclear weapons or of other nuclear explosive devices or for purposes unknown, and deterrence of such diversion by the risk of early detection." (INFCIRC/153, Article 28).

For practical purposes, a factor much more important than deterrence to the countries involved is the finding by IAEA that no diversion has taken place, the so-called assurance of non-diversion. Often this is the condition for being supplied with nuclear material, representing a general confidence building measure.

In order to document that no nuclear material was diverted, states bind themselves to build up national material accountancy systems, in which nuclear materials inventories are continuously recorded by accountancy and measurements. Inventories and inventory changes are reported to IAEA in a system of records and reports.

For this purpose, the signatory country to the NPT first of all reports to IAEA design information about all nuclear facilities on its territory. Next, the material balance areas and the types of balancing and reporting are defined in facility attachments for each plant. IAEA verifies the information provided by plant operators by inspections and independent measurements in a defined maximum scope. If there are discrepancies, attempts are made to settle them. The system is to prevent and discover, respectively, diversions of nuclear materials by a country. Every state is obliged to establish a physical protection system so as to prevent diversion of nuclear material at a sub-national level. Recommendations for the physical protection of nuclear material are given by IAEA in document INFCIRC/225. IAEA has invited individual states to join "the convention on the physical protection of nuclear material" opened for signature on March 3, 1980.

8.3.2.1 Material Balance Measurements

For information to be generated about the nuclear material inventory in a plant, the initial inventory and the quantities coming in and going out must be known. This requires stocktaking of the real inventory and continuous measurement of all incoming and outgoing nuclear materials streams. Since measurements contain errors, statements about inventories will become less and less precise as time goes on, and new inventories will have to be determined. If a new inventory value is within the error limits of material balance measurements, it is stated that all material is still present. However, such statement implies an uncertainty, which will be explained in more detail below. The material balance is described by

$$I_2 = \Sigma M_i - \Sigma M_o + I_1 + MUF$$

with
I_1 = initial inventory
I_2 = inventory measured after time t
ΣM_i = sum of material inputs over time t
ΣM_o = sum of material outputs in time t
MUF= *material unaccounted for.*

MUF includes measuring errors, process losses and, if applicable, also diverted material. If MUF exceeds the measuring error, the IAEA inspector can state a loss of material. However, such diversion may have been due only to a particularly large statistical variation and no actual losses may have occurred. In that case, the IAEA inspector's statement would be a so-called "false alarm." On the other hand, despite a loss of material, statistical variations may have made MUF so small as to leave the loss undiscovered.

The relations connecting detection probability, rate of false alarms, detectable quantity and measuring accuracy are explained in Fig. 8-12. If a quantity M_0 is present, the probability, of obtaining the quantity M in a measurement is given by

$$P(M) = \frac{1}{\sqrt{2\pi} \cdot \sigma} \cdot e^{-\frac{(M-M_0)^2}{2\sigma^2}}$$

with
σ = standard deviation,

if a Gaussian distribution is assumed for the errors.

With 90% probability, M will be in the range of $M_0 \pm 1.65\,\sigma$. If $M < M'$, with $M' = M_0 - 1.65\,\sigma$, an alarm is initiated in the example given, i.e., a discrepancy between the operator's data and the measurement is reported. In five percent of the cases, this may be a false alarm due to statistical variations. The rate of false alarms, α, is around five percent.

If there is a loss of material such that actually only a quantity M_1 is available, with $M_0 - M_1 = 3.3\,\sigma$, a loss is discovered with 95% probability. The detection probability is $1-\beta = 95\%$, $\beta = 5\%$. Losses of $M_0 - M \geq 3.3\,\sigma$ are discovered with $p \geq 95\%$ probability at a false alarm rate of 5%. If the false alarm rate is reduced to 0.14% with a detection probability of 95%, detectable losses are $M_0 - M \geq 4.65\,\sigma$ (Fig. 8-12). This likewise applies to an improvement in detection probability.

The criteria to be met by safeguards systems thus are defined by

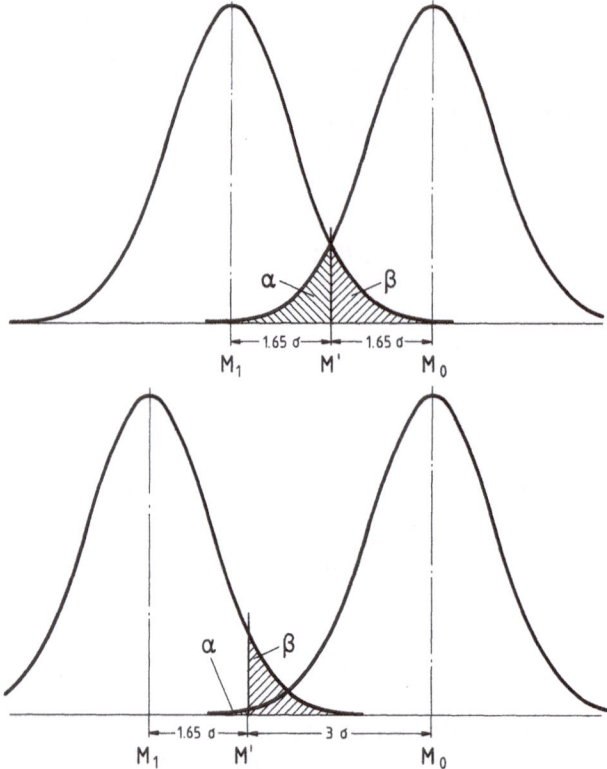

Fig. 8-12. Relations between detection probability, rate of false alarms, detectable quantity and measuring accuracy (KfK)

(1) the missing amount to be discovered (significant quantity),
(2) the detection time (timely detection),
(3) the detection probability for a significant amount,
(4) the permissible rate of false alarms.

The "significant quantity" and "timely detection" concepts have been quantified in the course of the implementation of IAEA safeguards agreements. The *Standing Advisory Group on Safeguards Implementation* (SAGSI) to IAEA has confirmed, on a preliminary basis, values for "significant quantities," which are indicated in Table 8-14. They are derived from so-called threshold amounts of special fissile material, which are defined as approximate quantities needed for nuclear explosive devices.

In order to concentrate safeguards activities upon materials particularly important with respect to proliferation, the "effective kilogram" concept was introduced in addition.

"Effective kilogram" means a special unit used in safeguarding nuclear material. The quantity in "effective kilograms" is obtained by taking for plutonium its weight in kilograms; for uranium, the ratio of "effective kilograms" to the weight in kilograms is shown in Fig. 8-13 as a function of enrichment. The maximum inspection expenditure for a plant is a function of the quantity of nuclear material, expressed in effective kilograms, of that plant.

Table 8-14. *Quantities of nuclear material of safeguards significance (IAEA)*

Material	Quantity of safe-guards significance	Threshold amounts	
"Direct use" material			
Pu	8 kg	Pu ($>$95% Pu-239)	8 kg
U-233	8 kg	U-233	8 kg
Uran ($\geqq$20% U-235)	25 kg	U ($>$90–95% U-235)	25 kg
(Plus rules for mixtures where appropriate)			
"Indirect use" material			
Uranium ($<$20% U-235)[a]	75 kg		
Thorium	20 t		
(Plus rules for mixtures where appropriate)			

[a] Including natural and depleted uranium.

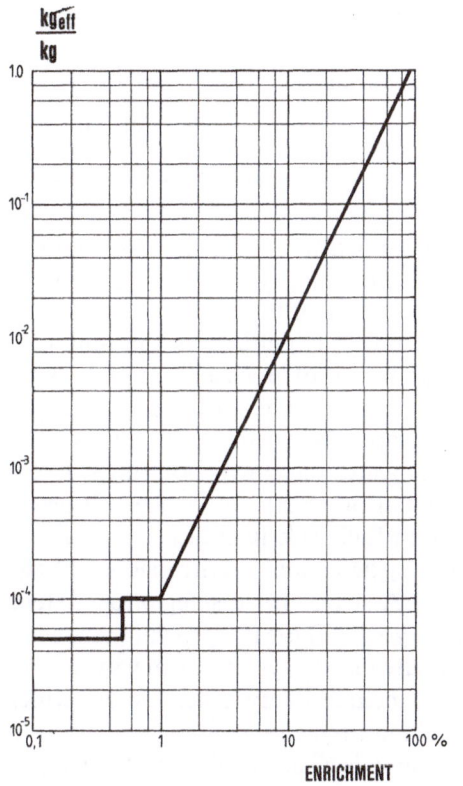

Fig. 8-13. Ratio of "effective kg" and "actual kg" as a function of enrichment (KfK)

Table 8-15. *Estimated material conversion times (IAEA)*

Material classi-fication	Beginning material form	End process form	Estimated conversion time
1	Pu, HEU, or U-233 metal	Finished plutonium or uranium metal components	Order of days (7–10)
2	PuO_2, $Pu(NO_3)_4$ or other compounds. HEU or U-233 oxide or other pure compounds. MOX or other non-irradiated pure mixtures of Pu or U. (U-233 + U-235) > 20%. PU, HEU and/or U-233 in scrap or other miscellaneous impure compounds.	Finished plutonium or uranium metal components	Order of weeks (1–3)
3	Pu, HEU or U-233 in irradiated fuels	Finished plutonium or uranium metal components	Order of months (1–3)
4	U containing < 20% U-235 and U-233; thorium		Order of one year

Criteria for "timely detection" were also defined by SAGSI as a preliminary recommendation to IAEA. According to that preliminary recommendation, the "detection time" should correspond, in orders of magnitude, to the "conversion time" for fissile material. The "conversion time" is defined as the minimum time required to convert different chemical forms of nuclear material to the metallic components of a nuclear explosive device. The "detection time" is defined as the maximum time which may elapse between a diversion and its detection by IAEA safeguards. Table 8-15 indicates the conversion times. IAEA uses these values as guidelines while additional practical experience will be acquired in implementing safeguards in different fuel cycle facilities.

With these preliminary guidelines on "timely detection" of diversions of "significant quantities" of nuclear material, IAEA strives for a safeguards system, which has a high probability of meeting these goals. There is a growing tendency, however, not to base the credibility of a safeguards system primarily on the degree to which these goals are met but to include many other factors which confirm compliance of a state with the NPT requirements.

The a priori probability of detection sought is usually 90% or higher and most often 95%. Since a statement of diversion represents a grave incident, the rate of false alarms must be very low. No precise value has as yet been defined.

In addition to the "abrupt diversion" case, to which the "detection time" applies, a case must also be considered of somebody constantly trying to divert small quantities so that, e.g., after one year, a "significant amount" will have accumulated (protracted diversion). Discovering such action imposes the most stringent requirements on the accuracy of the measurements.

In order to ensure that really all material has been considered in a material balance, containment and surveillance measures are adopted. That is to say, containers and fuel elements are sealed and transport processes are surveyed optically. Seals can also be used to reduce the measuring expenditure in inventory taking. Since sealed containers with undamaged seals still contain their material, no new quantitative assay is necessary.

8.3.3 Safeguards Techniques

Implementing the safeguards concept on the basis of measurements and material balances assumes the existence of suitable measuring techniques. As a consequence, a worldwide effort was initiated around 1968 to develop methods of measurement, measuring equipment, seals and cameras. It was also tried, in a parallel effort, to cast into more precise terms the basic theoretical principles of the system, verify its applicability to specific plants, improve the mathematical methods of data analysis, and develop systems for handling the large amounts of data.

All material balancing is based on sampling where the material is present in an open and homogeneous form, and on chemical analysis to determine the concentration of nuclear material which, combined with weight or volume measurements, indicates the mass of nuclear material. These techniques have long been applied by plant operators for operational reasons and have been developed to high levels of precision. Measuring accuracies are between 0.1% and 1%. Verification by IAEA consists in taking and analyzing samples independently in order to verify the data furnished by an operator. Often this is done by random sampling. Weight measurements (e.g., in fuel rods) can be checked in principle by having the inspector bring along a standard only known to him and having it weighed. Volumetric assays can be conducted by adding to the solution a "spike" of the inspector and subsequently assaying for the concentration. Since not all parts of a chemical plant handling radioactive material, such as a reprocessing plant or plutonium fuel fabrication plant, will be accessible, there are certain limits to the verifiability of samples.

Another problem is raised by the transition from the area in which the material is present in a open form to the identifiable "items." In this respect it must be ensured, by investing a limited amount of inspection effort, that precisely only the material analyzed was used in the final product. For this purpose, non-destructive assaying techniques have been developed, which allow inspectors to determine the contents of nuclear material in a final product.

Another application of non-destructive assaying techniques is in containers with scrap and waste, which must be included in a balance. In general, operators have no interest in precise measurements of those items, especially if such material is only stored temporarily. For this reason, particularly non-destructive assays required more development work for safeguards processes.

Table 8-16. *Major gamma ray signatures for the fissionable isotopes (T.D. Reilly, J.L. Parker)*

Isotope	Energy (keV)	Intensity (l/g·s)
U-235	185.72	4.3×10^4
U-238	1001.10	1.0×10^2
	766.40	3.9×10^1
Pu-238	766.40	1.5×10^5
	152.77	6.5×10^6
Pu-239	413.69	3.4×10^4
	129.28	1.4×10^5
Pu-240	–	–
Pu-241	207.98	2.0×10^7
	164.59	1.8×10^6
	148.60	7.5×10^6
Am-241	59.54	4.6×10^{10}
Pu-242	–	–

Table 8-17. *Gamma ray attenuation in various materials to e^{-1}*

Gamma Energy		186 keV	414 keV
Low density waste	cm	7.7	≈ 105
H_2O	cm	7.1	9.6
Al	cm	3.0	4.07
Fe	cm	0.84	1.4
UO_2	cm	0.065	0.40

The radiation emitted by the nuclear material is used for measurement; only neutral particles, i.e., gamma rays and neutrons, have the necessary penetrating capability. The easiest way is to use the characteristic radiation of the material investigated. Table 8-16 shows the major gamma ray signatures for uranium and plutonium isotopes. The most important ones are the U-235 185.7 keV and the Pu-239 413.7 keV gamma lines. After calibration, gamma ray counting allows the investigator to determine immediately the amount of a certain isotope in the sample. A problem is encountered when strong gamma ray attenuation is present. Table 8-17 shows material thicknesses in cm in which gamma rays mentioned above are attenuated to e^{-1}.

A number of sophisticated techniques have been developed to flatten the response or determine the γ-ray attenuation of homogeneous or nearly homogeneous materials. Nevertheless, the main areas of application of passive gamma counting are low density waste and liquid samples. The 185.7 keV γ-emission from a thick uranium sample ($d \geqslant 0.065$ cm) allows the uranium enrichment of the sample to be determined, since the γ-emission is proportional to the enrichment.

Table 8-18. *Neutron emission rates*

Isotope	Halflife $T_{1/2}(a)$ spont. fiss.	$\bar{\nu}$ spont. fiss.	Neutron emission rates (n/g·s) spont. fiss.	(n/g·s) (α,n)reaction[g]
U-233	$>3\cdot10^{17}$	1.90	$<1.9\cdot10^{-4}$[h]	11
U-235	$1.8\cdot10^{17}$[b]	1.88[a]	$5.9\cdot10^{-4}$	$2.5\cdot10^{-3}$
U-238	$9.86\cdot10^{15}$[c]	2.00[a]	$1.1\cdot10^{-2}$	$3.9\cdot10^{-4}$
Pu-238	$5.0\cdot10^{10}$[d]	2.21[a]	$2.45\cdot10^3$	$2.0\cdot10^4$
Pu-239	$5.5\cdot10^{15}$[b]	2.30[a]	$2.31\cdot10^{-2}$	72
Pu-240	$1.32\cdot10^{11}$[d]	2.15[a]	$8.97\cdot10^2$	265
Pu-242	$7.0\cdot10^{10}$[d]	2.14[a]	$1.67\cdot10^3$	4.4
Am-241	$1.15\cdot10^{14}$[e]	2.45	1.2	$4\cdot10^3$
Cm-242	$6.6\cdot10^{6}$[d]	2.51[a]	$2.08\cdot10^7$	$3.9\cdot10^6$
Cm-244	$1.27\cdot10^{7}$[f]	2.68[a]	$1.15\cdot10^7$	$9.5\cdot10^4$

Sources:

[a] Manero, F., Konshim, V.A., INDC(NDS)-34/G (1972).
[b] Segré, E., Phys, Rev. *86*, 21 (1952).
[c] Roberts, J.H., Phys. Rev. *174*, 1482 (1968).
[d] Nuclear Data Sheets, Vol. 4, No. 6 (1970).
[e] Gold, R., Phys. Rev. C1, 738 (1970).
[f] Barton, D.M., Jaffey, A.H., J. Inorg. Nucl. Chem. *32*, 769 (1970).
[g] with $3.15 \times 15\cdot10^{-8}$ neutrons per alpha.
[h] NUREG 002 (1976).

Fast neutrons have a much higher penetration capability than γ-rays and are emitted in spontaneous fission of many heavy nuclei. Table 8-18 lists neutron emission rates by spontaneous fission and (α,n)-processes in oxygen of the most important isotopes. If the isotopic composition of a material is known, the amount of that material in a sample can be determined by neutron counting, provided the system has been calibrated. The neutron yield of (α,n)-processes depends on many details and therefore is not well suited as a signature for plutonium or uranium. To eliminate the (α,n)-contribution, use is made of the fact that fission neutrons are mostly emitted in pairs, whereas only one neutron is emitted in the (α,n)-reaction. Thus, a coincidence logic, which only counts neutron pairs, suppresses the (α,n)-neutrons.

Neutron coincidence counting is widely used for assaying of plutonium contaminated waste. Its main drawback is the fact that primarily Pu-240 is measured and the uncertainty in isotopic composition limits the achievable accuracy of plutonium assays. For uranium, the neutron emission rate is too low. In addition, even small amounts of curium render the method inapplicable, since the neutron emission rate of curium is four orders of magnitude higher than that of plutonium.

If no "passive" assay technique is available, a so-called "active interrogation" has to be applied. In this method, fissions are induced in the material of interest by an external neutron source and the neutrons or gamma rays released in the fission process are detected. If neutrons are detected, it is necessary to discriminate between the induced fission neutrons and the source neutrons. Several methods are available:

(1) Discrimination by energy. When the energy of the source neutrons is below the threshold of the detector, only the induced fission neutrons are counted.

(2) Discrimination by time. Delayed neutrons are counted in the time interval when the source is switched off.

(3) Discrimination by multiplicity. The high multiplicity of gamma and neutron emissions in the fission process is used in a manifold coincidence measurement.

A large number of active interrogation techniques and instruments have been proposed and developed, but their implementation in existing safeguards systems is limited chiefly to an oscillating Cf source with delayed neutron counting.

In the areas of containment and surveillance, use can be made of developments in other fields, such as physical protection. This relates, for instance, to cameras and portal monitors. Fuel element seals had to be specially developed. The most promising technique is the identification of randomly distributed discontinuities in a matrix by ultrasonic wave propagation and echo registration. The characteristics of these echos vary with each variation of the discontinuity position in the matrix. Such a matrix forms part of a cap seal for LWR fuel bundles and has to be destroyed when the seal is removed. The seal may contain a fuel element identification known only to the inspector.

8.3.4 Safeguards Implementation

The Treaty on the Non-Proliferation of Nuclear Weapons was opened for signature on July 1, 1968. By the end of 1981, 114 states including three nuclear weapon states (USA, UK, USSR) had signed the treaty (Fig. 8-14). In March 1971, the basic principles of safeguards agreements with IAEA were settled and documented in INFCIRC/153. By 1981, there were safeguards agreements in force in 69 countries with significant nuclear activities. IAEA also applied safeguards in ten non-nuclear weapon states not parties to the NPT, namely Argentina, Brazil, Chile, Columbia, the Democratic People's Republic of Korea, India, Israel, Pakistan, South Africa and Spain. Only in four of these ten countries (India, Israel, Pakistan and South Africa), both unsafeguarded as well as safeguarded nuclear facilities were in operation or under construction. Unsafeguarded facilities in these four countries were or will be capable of producing weapons grade nuclear material. With these four exceptions noted, almost the entire nuclear industry known outside the nuclear weapon states is thus under the safeguards control of IAEA.

By late 1981, some 130 power reactors, 171 research reactors and critical assemblies, 4 conversion plants, 38 fuel fabrication plants, 6 reprocessing plants, 4 enrichment plants, 19 separate storage facilities, and 40 others (mostly research and development facilities) were under safeguards.

In spite of many difficulties, ranging from manpower availability, late reporting, incomplete measurements to inconsistencies in the data formats of state accountancy systems, IAEA in all cases was able to state that no nuclear material

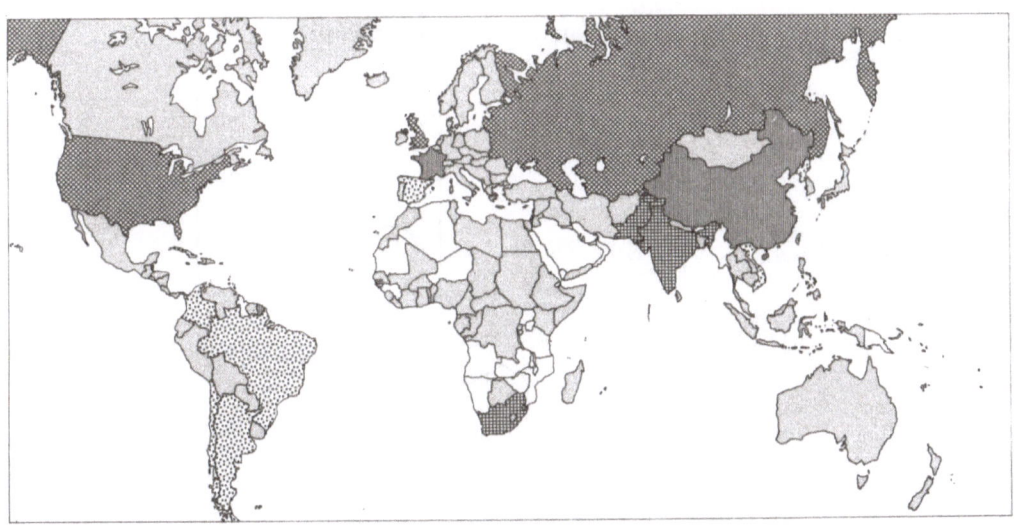

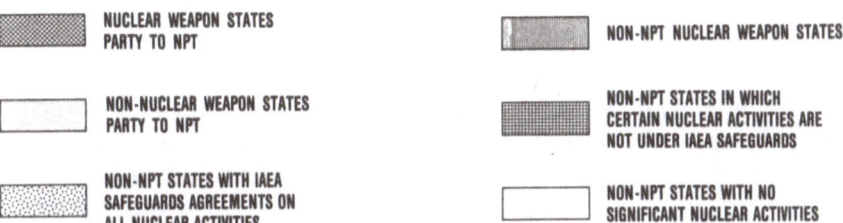

Fig. 8-14. Status of IAEA international safeguards in different countries (as of October 1981) (IAEA)

had been diverted from facilities dedicated to the peaceful application of nuclear energy.

In the majority of facilities, the nuclear material is in the form of separate items only, thus making item counting, serial number identification and surveillance by visual inspection or cameras adequate safeguards measures. This has greatly facilitated the implementation of safeguards. Attaching seals to the fuel elements will improve tamper resistance in this area.

Some reactors require additional safeguards measures. For instance, pebble bed reactors do not permit fuel element identification, so active neutron interrogation is used for nuclear materials assays. Similarly, in fast breeders, fuel elements may be under an inert atmosphere or under sodium and thus not accessible for direct identification. Passive neutron counting during fuel element transportation in the reactor area and sealing of the core after loading is proposed here.

Safeguarding such facilities as enrichment plants, fuel fabrication plants and reprocessing plants, where the nuclear material is present in an unsealed form, presents greater difficulties. Reprocessing facilities (Fig. 8-15) are typically divided into three material balance areas, which include the spent fuel storage

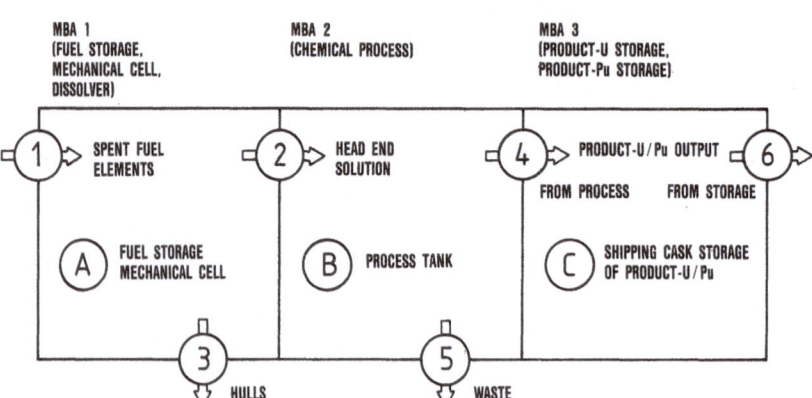

Fig. 8-15. Material balance areas and key measuring points in a reprocessing facility (KfK)

area, the process area, and the product storage area. In the spent fuel storage area (MBA1), the spent fuel elements arriving from the reactor plant can be verified. However, their fissile material content can be determined by non-destructive techniques only to an accuracy of a few percent. A better quantitative analysis is possible in the second material balance area (MBA2), which includes the whole chemical process area. Finally, MBA3 contains the storage area for the purified end product.

For MBA's 1 and 3, the statement that no material has been diverted can be based on containment and surveillance measures, such as seals and cameras. For MBA2, the difference between output and input has to be measured and compared with the inventory change according to Section 8.3.2.1. This requires inventory taking in the whole process area, which includes a washout of the process equipment; this is time consuming and expensive. Thus, only one or two inventory takings per annum are acceptable from the plant operator's point of view. This may conflict with the timeliness of detection.

No problem exists in small pilot size plants, where inventory changes stay below one significant amount of nuclear material. Typically, also the MUF accumulated over one year is less than a significant amount.

However, problems may arise in commercial size reprocessing plants. Let us assume, for illustration, a plant with an annual throughput of 1000 te of heavy metal, of which roughly 1% is plutonium. The plutonium throughput then would be 10,000 kg per annum. The largest error in the material flow measurement is attributable to the input accountability tank, where the best values achievable at present are $2\,\sigma = \pm 1\%$, including sampling, chemical analysis, and volumetric assay. This leads to an accumulated $3\,\sigma$-error over one year of ± 150 kg of plutonium, disregarding other sources of error. Getting the accuracy down to below 8 kg of plutonium would require taking inventories twenty times a year, which is economically unacceptable. A significant improvement in input accountability measurement does not seem to be feasible in the near term and, moreover, would not solve the problem with respect to the many other sources of error.

Similar problems may be encountered in large fuel fabrication plants with plutonium bearing fuel, or in enrichment plants. Advanced safeguards concepts are therefore being discussed.

8.3.5 Advanced Approaches

8.3.5.1 Near-Real Time Accountancy and Extended Containment/Surveillance Systems

Two different approaches have been proposed to solve the problems outlined above: near-real time accountancy and containment/surveillance (c/s) systems. In near-real time accountancy, inventory taking at the expense of a complete halt of the process is replaced by an essentially continuous determination of the plant inventory from readings of the process instruments, such as level indicators of vessels, densitometers or flow meters, combined with periodic analyses of samples from the vessels. Although those measurements may not be as accurate as those normally used for material balance purposes, they do solve the problem of abrupt diversion and they greatly facilitate the detection of protracted diversion. One reason for this result are the relatively small inventories of nuclear materials in large commercial plants, compared with throughput. For instance, for the reprocessing plant mentioned above, the daily throughput would be about 50 kg of plutonium, a typical inventory being around 500 kg of plutonium, of which only 26 kg are in the process equipment and thus difficult to verify, the balance being in storage vessels. In view of the cancelation of systematic errors in the differences between measurements, 1% accuracy seems feasible, which would allow the material balance of ten days to be close to ± 10 kg of plutonium. This appears to be reasonable with respect to the safeguards objectives. In addition, the large number of measurements conducted throughout a year allows trends to be discovered, which are well below the measuring error, provided that sophisticated statistical methods are applied. This helps in detecting protracted diversion. Problems are posed by the fairly large inspection effort required, the limited ability of the inspector to verify the origins of the samples received for chemical analysis, and the detailed insight the inspector obtains into plant operation (problem of intrusion).

A totally different approach is the so-called "extended containment and surveillance concept," where the inspection effort is concentrated on measurements and surveillance at the periphery of a facility, which has the character of a containment. In the pure form of this concept, the statement by the inspector that no nuclear material has been diverted is no longer based on a material balance and on verification of the presence of the material, but on the fact that no diversion of nuclear material across the periphery of the facility has been detected.

This system requires no overly accurate input accountability measurements, since the material in the main process stream can be sealed and traced as it leaves the plant. Only the waste streams, which contain about 1% of the nuclear material, have to be assayed. This greatly relaxes the accuracy requirements for the measurements. On the other hand, all inputs and outputs to the plant, including the inactive ones, must be controlled. Extensive uses of cameras and doorway monitors are characteristic of this approach. If the monitor system of a passage fails, even for a short time, the inspector is unable to assess whether nuclear material has been diverted, and how much. In addition, the inspector must ensure that there are no clandestine exits that could have been used to divert nuclear material.

The system has two advantages: It can be used in parallel, as a national physical protection system, to detect theft on a subnational basis, and it fulfills the requirement of non-intrusiveness. However, some combination with materials accountancy is said to be needed, where the c/s-system covers the abrupt diversion aspect. Extensive research and development work is still required before the system can be implemented.

Even if advanced safeguards systems succeed in meeting the requirement of timely detection of significant amounts of nuclear material, there still remains political concern about large scale nuclear fuel cycle plants in non-nuclear weapon states. There is the possibility of a state suddenly quitting the NPT and immediately possessing large amounts of nuclear material for weapons use.

Various measures have been proposed to reduce this risk, including international plutonium storage and diversion resistant fuel cycles.

8.3.5.2 International Plutonium Storage

"*International Plutonium Storage* (IPS)" is a possibility already provided for in the IAEA Statute (Article XII) to report to IAEA any surplus plutonium, i.e., plutonium at present not required in the nuclear fuel cycle, for storage by the Agency under international control. Various legal and administrative models are presently being discussed. The purpose, to prevent stockpiling, can only be achieved if the whole pathway of plutonium use is monitored for non-stockpiling. This leads to five major activities of an IPS regime:
(1) Registering the accumulating quantities of separated plutonium.
(2) Delivery of excess amounts of plutonium to the IPS.
(3) Storage of the excess plutonium in the IPS.
(4) Return of plutonium from the IPS.
(5) Verification of use.

Complete registration of the plutonium is provided for under the IAEA safeguards systems anyway. This could be used for IPS, if a few changes were made in the reporting system. Such changes, for instance, would include data about the owner of the material, and the distinction of "separated plutonium."

Storage sites envisaged are the output stores of large reprocessing plants or the storage facilities attached to fuel fabrication plants. Physical protection is the responsibility of the country on the territory of which the storage facility is located. However, only a multinational management of that storage facility will have access to the stores.

The most difficult points, and the severest restrictions of sovereign rights, relate to the return and verification of use of the plutonium. A problem is also posed by plutonium used for research purposes and the influence that could be exerted on research programs.

8.3.6 Proliferation Aspects of Different Fuel Cycles

8.3.6.1 Quantities of Fissile Nuclear Material

The types and quantities of fissile nuclear material vary in the different types of nuclear reactors and fuel cycle options (see Chapter 6). Low, medium or highly enriched uranium as well as U-233 or plutonium can be used as fissile fuels for reactor cores. The quantity of fuel coming under the NPT safeguards system thus depends on the type of reactor and the fuel cycle, but mainly also on the total installed capacity.

Two quantities are of particular interest:
– the total inventory of fissile nuclear material to be safeguarded within the fuel cycle,
– the total flow of fissile nuclear material to be handled and transported during reprocessing and refabrication.

LWR's operating in the once-through fuel cycle with direct repository storage of spent fuel have the lowest mass flow of fissile nuclear material. However, in direct final storage of spent LWR fuel elements (see Section 7.6.2) all the plutonium generated during reactor operation is accumulated in the waste repository. As the radioactivity of the fission products within the spent LWR fuel elements stored decays as a function of time, increased safeguards and physical protection measures will become necessary.

Plutonium recycle LWR's have slightly higher mass flows of fissile nuclear material. As the plutonium generated will be burned in the LWR core, only about 1% losses of the plutonium during reprocessing and refabrication will go as losses into the vitrified HLW to be stored in the final waste repository.

The highest mass flow of fissile nuclear material is found in PuO_2/UO_2 fueled FBR's working in symbiosis with LWR's. The relatively high fissile material mass flow and the high fissile material inventory are due to the relatively high fissile material inventories of FBR cores and the fissile materials simultaneously tied up in reprocessing and refabrication plants.

All thermal reactors and breeders working in symbiosis together in the Th/U-233 cycle must be classified in an interim class between the plutonium fueled FBR and the LWR recycle.

8.3.6.2 Technical Measures to Improve Diversion Resistance

Since the beginning of the use of nuclear energy, technical means have been sought to render the use of uranium and plutonium fuels impossible for direct military application. Dilution of U-235 or U-233 with U-238 (isotopic denaturing) to some extent can serve as such a means of protection for enriched uranium

Table 8-19. *Critical masses of metallic U-233, U-235 and plutonium. (Spherical geometry with natural uranium reflector) (T.B. Taylor)*

	Enrichment (%)	Reflector thickness (cm)	Critical mass of core (kg)
U-235	100	15	15
	80	15	21
	60	15	37
	40	15	75
	20	15	250
	10	15	1300
U-233	100	10	5.7
Pu-239	100	10	4.5
Pu (rg1)[a]	100	10	6
Pu (rg2)[a]	100	10	13

[a] Reactor grade (rg) plutonium,
rg1 = 70% Pu-239/Pu-241
        29% Pu-240, 1% Pu-242
rg2 = 50% Pu-239/Pu-241
        40% Pu-240, 10% Pu-242

fuel. Table 8-19 shows the critical mass of U-235/U-238 metal spheres (surrounded by a reflector) as a function of U-235 enrichment. Below 20% enrichment of U-235, the critical mass of enriched uranium spheres rises prohibitively. Consequently, uranium with <20% U-235 enrichment cannot be used directly in nuclear weapons. For U-233, this threshold is around 12% enrichment.

A proposal on this basis envisaged that uranium fuel in fresh fuel elements should always have <20% U-235 enrichment and <12% U-233 enrichment, respectively. Reactors with highly enriched fuel (HEU), as described in Chapter 6, would then be banned. Fresh fuel elements with <20% U-235 or <12% U-233 enrichment would have the advantage that, after diversion, this medium enriched uranium fuel would be useful for military applications only after additional isotopic enrichment.

Similarly, denaturing of Pu-239 through the addition of Pu-240 has been discussed. Although Pu-240 cannot be split by thermal neutrons, it has a fairly large fission cross section at higher energies. Therefore, it cannot be used for dilution in the same sense as U-238 (Table 8-19). However, Pu-240 is a spontaneous neutron emitter at a rate of about 900 neutrons per second per gram (see Table 8-18), which limits the application of plutonium with a higher Pu-240 content for military purposes. For this reason, a distinction is often made between "weapons grade" and "reactor grade" plutonium. While it was hoped that these physical properties would help establish technical means to improve the "proliferation resistance" of reactor grade plutonium, it is often asserted that Pu-240 is no technical fix to prohibit the application of reactor grade plutonium in military sectors.

It had been suggested that γ-radiation could offer some advantages with respect to containment and surveillance methods. Fresh mixed oxide fuel elements could be pre-irradiated shortly prior to their shipment to the reactor plant. They could also be spiked by adding highly radioactive nuclides to the fuel during fabrication (spiking). Radiation emitted by the fuel element would be increased, which would enhance its protection.

Partial reprocessing was proposed as a measure to provide a radiation barrier for the fuel elements. This partial reprocessing method was called the CIVEX process, in contrast to the existing PUREX process (see Section 7.2.1). Applying a CIVEX process was based on the idea that plutonium in its pure form should exist nowhere in the fuel cycle. The radiation would come from fission products, which would not be fully separated from the fuel during chemical reprocessing.

Co-processing is a modification of the PUREX process such that the plutonium is not separated from the uranium. Plutonium nitrate is not produced in such a reprocessing plant. Instead, plutonium nitrate and uranyl nitrate remain together.

All these technical measures are ultimately found to be incapable of improving the resistance to proliferation very much. Denaturing would hold only as long as advanced efficient small scale enrichment techniques remain unavailable to a diverter. Moreover, dilution with U-238 will always lead to the generation of plutonium in the reactor core during operation, which could be separated by reprocessing. Pre-irradiation, spiking and partial reprocessing, although adding to the protection against possible diversion, would render fuel element fabrication and materials accounting for safeguards purposes more difficult because of the enhanced radiation levels. Co-processing could have advantages in protecting the nuclear fuel against diversion by subnational groups or national governments. However, even small modifications to the reprocessing facility could result in the production of plutonium nitrate.

8.3.7 International Agreements and Institutional Arrangements

At present, a number of international agreements provide a framework for international cooperation in the application of nuclear energy and its fuel cycle. As mentioned before, the principal agreements are:
– the Statute of IAEA (entry into force 1956),
– the Euratom Atomic Energy Community established by the treaty of Rome, March 1957,
– the Treaty of Tlatelolco, February 1967,
– the Non-Proliferation Treaty (NPT), July 1968.

In addition to these international agreements, other future institutional arrangements are possible:

Multinational enrichment projects, international reactor projects and multinational fuel cycle centers. Examples of multinational enrichment and reactor projects already exist in Europe. Studies of so-called *regional nuclear fuel cycle centers* (RNFCC) were performed by IAEA and the Commission of the European Community. The RNFCC concept offers the advantage of co-location of large scale reprocessing and refabrication facilities. In principle, an RNFCC could also be co-located with the waste disposal site. Smaller states could participate and cooperate at all levels in a large RNFCC. Control and transfer of nuclear material would be improved because the RNFCC would be multinational.

8.3.8 Remaining Proliferation Risk

In view of the manifold possibilities open in handling nuclear materials and, also, of the limited capabilities of IAEA, given its legal, manpower and financial constraints, it could be thought that a state, as a matter of principle, would always be in a position to bypass the IAEA safeguards system and divert undetected a quantity of plutonium or highly enriched uranium significant from the safeguards point of view. The practical risk of such diversion, however, is relatively low at present.

Diversions of nuclear materials would require modifications in process flow-sheets, tampering with records, and manipulating measured samples or measuring systems. As a consequence, careful control and inspection of nuclear material flows will always imply a high risk of discovery for such a state, especially if many persons were involved. In view of the grave political consequences of discovery it is, therefore, highly unlikely that any government would go this way.

However, it should also be mentioned at this point that growing political instability in certain areas of the world could change that picture in the future. In addition, countries that are parties to the NPT could still withdraw on three months' notice, if they wished.

As the Second Review Conference on the Non-Proliferation Treaty in 1980 showed, it is essential that the nuclear weapons states make further progress in nuclear arms control and reach agreement on a treaty banning all nuclear tests. In addition, most non-nuclear weapons states were found to be dissatisfied with the manner and extent of implementation of Article IV of the NPT, which confirms their "right to participate in," as well as their undertaking to facilitate, "the fullest possible exchange of equipment, materials and scientific and technological information for the peaceful use of nuclear energy," and which places a corresponding obligation on the parties to assist each other, particularly the developing countries.

In a future growing nuclear economy it will therefore be essential to promote international cooperation in the whole nuclear fuel cycle. Open and trustful cooperation is imperative
– between countries exporting nuclear technology and less developed countries importing nuclear know-how and nuclear facilities,
– between uranium supplier and uranium consumer countries,
– between weapon states and non-weapon states.

All countries with nuclear development programs therefore should become parties to the NPT.

The IAEA safeguards system as described above must first be fully implemented in the near future and can then be further improved. By voluntarily accepting and implementing the IAEA safeguards system, the non-nuclear weapon states fully meet the essential obligations of the NPT, thereby contributing greatly to political stabilization in the world.

Selected Literature

Radioactivity releases during normal operation

Allgemeine Berechnungsgrundlagen für die Strahlenexposition bei radioaktiven Ableitungen mit der Abluft oder in Oberflächengewässern. Bonn: Bundesministerium des Innern, Gemeinsames Ministerialblatt *21*, 369–436 (1979).

Baumgärtner, F., *et al.*: Stand der Technik bei der Behandlung des Auflöserabgases einer Wiederaufarbeitungsanlage. Kernforschungszentrum Karlsruhe, KfK-2615 (1978).

Bericht über das in der Bundesrepublik Deutschland geplante Entsorgungszentrum für ausgediente Brennelemente aus Kernkraftwerken. Hannover: Deutsche Gesellschaft für Wiederaufarbeitung von Kernbrennstoffen (DWK), 1977.

Final Environmental Statement Related to Construction and Operation of Clinch River Breeder Reactor Plant. Washington: US Nuclear Regulatory Commission, NUREG-0139 (1977).

Final Generic Environmental Statement on the Use of Recycle Plutonium in Mixed Oxide Fuel in Light Water Cooled Reactors (GESMO). Washington: US Nuclear Regulatory Commission, NUREG-002 (1976).

Glasstone, S., Jordan, H.W.: Nuclear Power and Its Environmental Effects. LaGrange Park, Ill.: American Nuclear Society. 1980.

Halbritter, G., Leßmann, E.: Vergleich der Strahlenexposition aus Emissionen von Modell-Brennstoffkreisläufen für den Druckwasserreaktor und den Schnellen Brutreaktor. Kernforschungszentrum Karlsruhe, KfK-3315 (1982).

Halbritter, G., *et al.*: Beitrag zu einer vergleichenden Umweltbelastungsanalyse am Beispiel der Strahlenexposition beim Einsatz von Kohle und Kernenergie zur Stromerzeugung. Kernforschungszentrum Karlsruhe, KfK-3266 (1982).

Halbritter, G., *et al.*: Contribution to a Comparative Environmental Impact Assessment of the Use of Coal and Nuclear Energy for Electricity Generation. In: Health Impacts of Different Sources of Energy, Proc. Int. Symposium, Nashville, Tenn., 22–26 June 1981, pp. 229–247. Vienna: International Atomic Energy Agency. 1982.

International Commission on Radiological Protection. Recommendations of the International Commission on Radiological Protection. ICRP Publication 26. Oxford: Pergamon Press. 1977.

Jackson, P., *et al.*: Radon-222 Emissions in Ventilation Air Exhausted from Underground Mines. Washington: US Nuclear Regulatory Commission, NUREG-CR-0627 (1979).

Neilson, K.: Prediction of Net Radon-222 Emission from a Model Open Pit Uranium Mine. Washington: US Nuclear Regulatory Commission, NUREG-CR-0628 (1979).

Preliminary Safety and Environmental Information Document. Washington: US Department of Energy, DOE/NE-0003/7 (1980).

Proposed Final Environmental Impact Statement on LMFBR's. Washington: US Energy Research and Development Agency, WASH-1535 (1975).

Radiological Impact Caused by Emissions of Radionuclides into Air in the USA.. Washington: US Environmental Protection Agency, US-EPA 520/7-79-006, Preliminary Report (1979).

Regulating Guide 1.109, Revision 1, Calculation of Annual Doses to Man from Routine Releases of Reactor Effluents for the Purpose of Evaluating Compliance with 10 CFR, Part 50, Appendix I. Washington: US Nuclear Regulatory Commission. 1977.

Report to the American Physical Society by the Study Group on Nuclear Fuel Cycles and Waste Management (Pines, D., ed.). Reviews of Modern Physics *50* (1), Part II (1978).

Schüttelkopf, H.: Radioökologische Aspekte der Entsorgung. In: Chemie der nuklearen Entsorgung (Baumgärtner, F., ed.), Vol. 1, pp. 155–181. München: Karl Thiemig. 1978.

Status Report on EPRI Fuel Cycle Accident Risk Assessment. Palo Alto, Cal.: Electric Power Research Institute, EPRI-NP-1128 (1979).

Technology for Commercial Radioactive Waste Management. Washington: US Department of Energy, DOE/ET-0028 (1979).

Uranium Fuel Cycle (CFR, Title 40, Part 190). Washington: US Environmental Protection Agency. 1977.

Wilhelm, I.: Spaltjodabtrennung in Kernkraftwerken und Wiederaufarbeitungsanlagen. Kernforschungszentrum Karlsruhe, KfK-2244 (1975).

Risk assessment

Bayer, A., Heuser, F.W.: Basic Aspects and Results of the German Risk Study. Nuclear Safety *22*, 695–709 (1981).

Der Bundesminister für Forschung und Technologie: Deutsche Risikostudie Kernkraftwerke, Eine Untersuchung zu dem durch Störfälle in Kernkraftwerken verursachten Risiko. Köln: TÜV Rheinland, 1979.

Bunz, H., *et al.*: The Role of Aerosol Behavior in Light Water Reactor Core Melt Accidents. Nuclear Technology *53*, 141–146 (1981).

Canvey Island Report: Summary of an Investigation of Potential Hazards from Operations in the Canvey Islands/Thurrock Area. London: UK Health and Safety Executive. 1978.

Dunster, H.J.: The Approach of a Regulatory Authority to the Concept of Risk. IAEA Bulletin *22* (5/6), 123–128 (1980).

Ehrhardt, J., *et al.*: Unfallfolgen- und Risikoabschätzungen im Anschluß an die probabilistische Sicherheitsstudie für ein Kernkraftwerk mit einem HTR großer Leistung (HTR-1160). Kernforschungszentrum Karlsruhe, KfK-3382 (1982).

Farmer, F.R.: Reactor Safety and Siting: A Proposed Risk Criterion. Nuclear Safety *8*, 539–548 (1967).

Hamilton, L.D.: Comparative Risks from Different Energy Systems: Evolution of the Methods of Studies. IAEA Bulletin *22* (5/6), 35–71 (1980).

Hennies, H.H., *et al.*: Ablauf und Konsequenzen eines Kernschmelzenunfalls. Atomwirtschaft/Atomtechnik *26*, 168–176 (1981).

HTGR Accident Initiation and Progression Analysis Phase II. Report for the USA Department of Energy. San Diego, Cal.: General Atomic, GA-A 15000 (1978).

Levenson, M., Rahm, F.: Realistic Estimates of the Consequences of Nuclear Accidents. Nuclear Technology *53*, 99–110 (1981).

Lewins, E.E.: Nuclear Power Reactor Safety. New York: John Wiley. 1977.

Piper, H., *et al.*: Clinch River Breeder Reactor Plant Safety Study. Nuclear Safety *19*, 316–329 (1978).

Reactor Safety Study: An Assessment of Accidents Risks in US Commercial Nuclear Power Plants (Rasmussen, N.C., ed.). Washington: US Nuclear Regulatory Commission, WASH-1400 (NUREG-75/014) (1975).

Risikoorientierte Analyse zum SNR-300 – Bericht der GRS. München: Gesellschaft für Reaktorsicherheit (GRS), GRS-S1 (1982).

The Risk Assessment Review Group to the USNRC. Washington: US Nuclear Regulatory Commission, NUREG-CR-04000 (1978).

Status Report on EPRI Fuel Cycle Accident Risk Assessment. Palo Alto, Cal.: Electric Power Research Institute, EPRI-NP-1128 (1979).

Safeguards

The Agency's Safeguards System (1965, as provisionally extended in 1966 and 1968). Vienna: International Atomic Energy Agency. INFCIRC/66/ Rev. 2 (1968).

The Agency's Statute (as amended up to 1 June 1973). Vienna: International Atomic Energy Agency. 1980.

Avenhaus, R.: Material Accountability: Theory, Verification, Applications. New York: John Wiley. 1978.

Böhnel, K.: Determination of Plutonium in Nuclear Fuels Using the Neutron Coincidence Method. Kernforschungszentrum Karlsruhe, KfK-2203 (1975). Translated as AWRE TRANS 70 (54/4252) (1978).

Crane, T.W.: Measurement of Uranium and Plutonium in Solid Waste by Passive Photon or Neutron Counting and Isotopic Neutron Source Interrogation. Los Alamos National Laboratory, LA-8294-MS (1980).

Crutzen, S.J.: Ultrasonic Techniques Suitable for Fuel Storage Surveillance and Containers Unique Identification and Integrity Checks. In: Proc. ESARDA 1st Annual Symposium, Brussels 25–27 April 1979, pp. 89–94. Ispra: Joint Research Center, Commission of European Communities, ESARDA 10 (1979).

de Volpi, A.: Proliferation Plutonium and Policy. Oxford: Pergamon Press. 1979.

European Safeguards Research and Development Association. Proc. ESARDA Annual Symposium on Safeguards and Nuclear Material Management. 1) Brussels, 25–27 April 1979 (ESARDA 10); 2) Edinburgh, 26–28 March 1980 (ESARDA 11); 3) Karlsruhe, 6–8 May 1981 (ESARDA 13). Ispra: Joint Research Centre, Commission of the European Communities.

IAEA Safeguards – An Introduction. Vienna: International Atomic Energy Agency, IAEA/SG/INF 3 (1981).

Lovett, J.E.: Nuclear Materials: Accountability, Management and Safeguards. LaGrange Park, Ill.: American Nuclear Society. 1974.

Nuclear Safeguards Technology 1978, Proc. Symposium, Vienna, 2–6 October 1978. Vienna: International Atomic Energy Agency. 1979.

The Physical Protection of Nuclear Materials. Vienna: International Atomic Energy Agency, INFCIRC/225/Rev. 1 (1977).

Reilly, T.D., Parker, J.L.: A Guide to Gamma-ray Assay for Nuclear Material Accountability. Los Alamos National Laboratory, LA-5794-M (1975).

Reilly, T.D., et al.: The Enrichment Meter, a Simple Method for Measuring Isotopic Enrichment. Los Alamos National Laboratory, LA-4605-M (1970).

Safeguards in the Seventies – A Bibliography of LANL Safeguards R and D Publications 1970–1979. Los Alamos National Laboratory, LA-8663-MS (1980).

Safeguards Techniques, Proc. Symposium, Karlsruhe, 6–10 July 1970. Vienna: International Atomic Energy Agency. 1970.

Sher, R., Untermyer, S.: The Detection of Fissionable Materials by Nondestructive Means. LaGrange Park, Ill.: American Nuclear Society. 1980.

Special Issue for the Second Non-Proliferation Treaty. Review Conference, August 1980. IAEA Bulletin 22 (3/4), 1–101 (1980).

The Structure and Content of Agreements between the Agency and States Required in Connection with the Treaty on the Non-Proliferation of Nuclear Weapons. Vienna: International Atomic Energy Agency, INFCIRC/153 (1971).

Taylor, T.B.: Nuclear Safeguards. In: Annual Review of Nuclear Science, Vol. 25, pp. 406–421. Palo Alto, Cal.: Annual Reviews Inc. 1975.

Treaty for the Prohibition of Nuclear Weapons in Latin America, 14 February 1967. New York: United Nations, UN Treaty Series No. 9068 (1967).

The Treaty on the Non-Proliferation of Nuclear Weapons; London, Moscow, Washington, 1 July 1968. Vienna: International Atomic Energy Agency, INFCIRC/140 (1970).

Subject Index

Satz und Umbruch: Universitätsdruckerei H. Stürtz AG, D-8700 Würzburg

Already published in the series "Topics in Energy"

Energy Demand: Facts and Trends
A Comparative Analysis of Industrialized Countries

By **B. Chateau** and **B. Lapillonne,**
Institut Economique et Juridique de l'Energie, Grenoble, France

1982. 26 figures. XV, 280 pages.
ISBN 3-211-81675-5

The relationship between economic development and energy demand is investigated in this book. It gives a detailed analysis of the energy demand dynamics in industrialized countries and compares the past evolution of the driving factors behind energy demand by sector and by end-uses for the main OECD countries: residential sector (space heating, water heating, cooking ...), tertiary sector, passenger and goods transport by mode, and industry (with particular emphasis on the steel and cement industry).
This analysis leads to a more precise understanding of the long-term trends of energy demand; highlighting the influence on these trends of energy prices, especially after the oil price shocks, and of the type of economic development pattern.
This book summarizes the authors' eight years' study of energy demand analysis and forecasting and energy conservation. It gathers together the rich supply of information—both qualitative and quantitative—they had access to.

Springer-Verlag Wien New York